Encyclopedia of Soybean: Food, Nutrition and Health

Volume V

Encyclopedia of Soybean: Food, Nutrition and Health Volume V

Edited by **Albert Marinelli and Kiara Woods**

New York

Published by Callisto Reference,
106 Park Avenue, Suite 200,
New York, NY 10016, USA
www.callistoreference.com

Encyclopedia of Soybean: Food, Nutrition and Health
Volume V
Edited by Albert Marinelli and Kiara Woods

International Standard Book Number: 978-1-63239-300-5 (Hardback)

Printed in the United States of America.

Contents

Preface

Every book is a source of knowledge and this one is no exception. The idea that led to the conceptualization of this book was the fact that the world is advancing rapidly; which makes it crucial to document the progress in every field. I am aware that a lot of data is already available, yet, there is a lot more to learn. Hence, I accepted the responsibility of editing this book and contributing my knowledge to the community.

In this book, the technologies and novel features for the applicability of soybean and its products are elucidated with the help of comprehensive information. Soybean can be employed for various purposes like biodiesel generation, medicinal purposes, textile, human food, etc. The book encompasses a variety of topics and consists of contributions which cover food, nutrition and health. The contributors of this book have diverse backgrounds and they possess years of experience in this field. This book would be of interest to agro-scientists, cultivators, students and researchers.

While editing this book, I had multiple visions for it. Then I finally narrowed down to make every chapter a sole standing text explaining a particular topic, so that they can be used independently. However, the umbrella subject sinews them into a common theme. This makes the book a unique platform of knowledge.

I would like to give the major credit of this book to the experts from every corner of the world, who took the time to share their expertise with us. Also, I owe the completion of this book to the never-ending support of my family, who supported me throughout the project.

Editor

Food, Nutrition and Health

Storage of Soybeans and Its Effects on Quality of Soybean Sub-Products

Ernandes Rodrigues de Alencar[1] and Lêda Rita D'Antonino Faroni[2]
[1]Faculdade de Agronomia e Medicina Veterinária,
Universidade de Brasília, Brasília, Distrito Federal, Brasil
[2]Departamento de Engenharia Agrícola,
Universidade Federal de Viçosa, Viçosa, Minas Gerais,
Brasil

1. Introduction

World soybean production in the 2009/2010 harvest was roughly 260 million tons, and the major producers were the United States, Brazil and Argentina, producing 91.4, 69.0 and 57.0 million tons, respectively (USDA, 2011). Given the significant world production of soybeans, quality is essential for the sectors involved in production and/or processing of this commodity. Quality is an important parameter for commercialization and processing of the grains and can affect the value of the product and its derivatives.

During post-harvest stages, soybeans are subjected to qualitative and quantitative losses due to several external factors. These factors may be physical, such as temperature and humidity; chemical, such as oxygen supply; and biological, such as bacteria, fungi, insects and rodents (BROOKER et al., 1992). According to BAILEY (1974), secure storage retains the qualitative and quantitative aspects of the grains, creating conditions unfavorable to the development of insects, rodents and microorganisms. Grain storage in the natural environment of tropical regions, according to ABBA & LOVATO (1999), presents major problems as a result of the temperature and relative humidity when compared with temperate or cold climates.

Soybeans are composed of roughly 20% lipids and are susceptible to qualitative deterioration processes via degradation of these compounds when stored improperly and can result in serious damage to the food industry. According to NARAYAN et al. (1988a), physical, chemical and biochemical alterations may occur in soybeans, depending on conditions and storage time. The qualitative changes of soybeans during storage contribute to the loss of oil and meal quality (ORTHOEFER, 1978), as well as other derivatives such as tofu and soymilk (NARAYAN et al. 1988b; LIU, 1997; HOU & CHANG, 1998; KONG et al., 2008).

2. Soybean storage

The objective of storage is to preserve the characteristics that grains present after harvest, therefore it is possible to obtain and market sub-products with satisfactory quality. Vitality of the grains can be preserved and the grinding quality and nutritive properties of the food can be maintained (BROOKER et al., 1992).

2.1 Soybean quality during storage

The pursuit of quality grain and sub-products should be a priority for producers, processors and for distributors of these products. According to BROOKER et al. (1992), the main characteristics that determine soybean quality are: low and uniform moisture content; low percentage of foreign material, discoloration, susceptibility to breakage, damage by heat (internal cracks), insect and fungal damage, elevated values of density, oils and protein concentration, and seed viability. Some factors can affect these characteristics such as the environmental conditions during grain formation on plants, season and harvesting system, drying system, techniques of storage, transport and characteristics of the species, and the variety.

The grain mass is an ecological system in which deterioration is the result of interaction between physical, chemical and biological variables (internal and external). The rate of deterioration during storage depends on the rates of change of these variables, which are directly affected by temperature and water content, and also by their inter-relationship with the grain and the storage structure (SINHA & MUIR, 1973). Insects, mites, rodents and fungi are the main biological factors responsible for qualitative and quantitative losses in stored grains, where development of these organisms is influenced by environmental factors such as temperature and relative humidity (PADIN et al., 2002).

2.2 Principal variations which affect quality of stored grains

Among the many variables that affect the storability of grains and their sub-products, moisture content and temperature are highlighted, associated with the storage time. Moisture content can be considered the most important factor on the quality of stored grain. ACASIO (1997) suggested that grains with moisture contents greater than 13% w.b. must be dried to reduce risk of deterioration in the form of dry matter loss by respiration, fungi attack, spontaneous heat production and reduction of germination percentage. Table 1 shows periods of safe storage for soybean with different moisture contents (BARRE, 1976 as cited in ACASIO, 1997).

Moisture content (%) w.b.	Safe storage period
10.0 – 11.0	4 years
10.0 –12.5	1-3 years
12.5 – 14.0	6-9 months
14.0 – 15.0	6 months

Source: BARRE (1976) as cited in ACASIO (1997)

Table 1. Safe storage period for soybeans.

Another determinant variable in the quality of stored products is temperature. When it comes to storage of soybeans, temperature not only affects the development of fungi but can promote chemical changes such as hydrolytic and oxidative rancidity. This physical variable also affects the development and reproduction of insect pests, where the optimum temperature for most species is between 27 and 35 °C. Soybeans with water contents between 14 and 14.3% w.b. and maintained at a temperature of between 5 and 8 °C can be stored for two years without development of fungi, while grain stored at 30 °C can be infected by fungi within a few weeks and severely damaged after six months of storage (ACASIO, 1997). It is emphasized that decision making must take into consideration the ideal conditions for grain storage, analyzing the combination of the variables of moisture and temperature, and not each one separately.

2.3 Qualitative parameters of soybeans and alterations during storage
2.3.1 Bulk density
Bulk density of grains is defined as the ratio between mass and volume (kg m^{-3}). This parameter generally increases with the decrease in moisture content of the product, except for coffee, paddy rice and barley (SILVA et al. 2009). It is emphasized that this trend depends on the percentage of damaged grains, initial moisture content, temperature reached during drying, final moisture content and the grain variety (BROOKER et al., 1992). Bulk density can be used as a qualitative indicator and the decrease in its value during storage may be associated with quality losses.

ALENCAR et al. (2009) studied the effect of different combinations of moisture contents and temperatures on the quality of stored soybeans (Fig. 1). The authors observed that the bulk density remained almost constant at different combinations of moisture content and temperature, except for grains stored with 12.8 and 14.8% moisture content and temperature of 30 and 40 °C, respectively. According to the authors, the decrease in bulk density of stored grain with 12.8% moisture content at the temperature of 30 °C (Fig. 1A), was due to infestation by *Plodia interpunctela* and *Sitotroga cerealella* whose optimum conditions of temperature and relative humidity are 30 °C and 75%, respectively (MBATA & OSUJI, 1983, MASON, 2006, HANSEN et al., 2004). On the other hand, the decrease in grains stored at 14.8% moisture content and temperature of 40 °C (Fig. 1B), was attributed to the development of fungi, where a high incidence (87%) of *Aspergillus glaucus* was verified.

2.3.2 Germination
Germination can be defined as the appearance of the first visible signs of growth or root protrusion, and is affected by several factors, including attack by insects, fungal infection, temperature, moisture and damage to the grains or seeds (BLACK, 1970, as cited in AL-YAHYA, 2001). The germination percentage has been used as an indicator of deterioration in different types of grains during storage.

HUMMIL et al. (1954) studied the qualitative deterioration of wheat grain stored with different moisture contents and inoculated or not with fungi at different temperatures. These authors observed a rapid deterioration of the wheat grains stored at 18% w.b. They verified that the process was more pronounced at a temperature of 35 °C. KARUNAKARAN et al. (2001) stored wheat with moisture contents between 15 and 19% w.b. at different temperatures in order to verify the time of safe storage, using the germination percentage as a quality standard. Results obtained for the water content of 17% w.b. at temperatures of 25, 30 and 35 °C were equal to 15, 7 and 5 days, respectively. Qualitative deterioration of soybeans stored with initial moisture contents between 9.8 and 13.8% w.b. in tropical conditions (30 °C and 82% RH) was simulated by Locher & Bucheli (1998). These authors confirmed a marked decrease in the germination percentage between 5 and 9 months of storage, where this behavior was more pronounced in seeds with greater initial moisture contents. Bhattacharya & Raha (2002) studied alterations in soybeans stored with moisture content of 14.0%, in the presence of different fungi species. The germination percentage of soybeans after 10 months of storage was zero. GUNGADURDOSS (2003), when studying the viability of soybean seeds under different storage conditions concluded that temperature

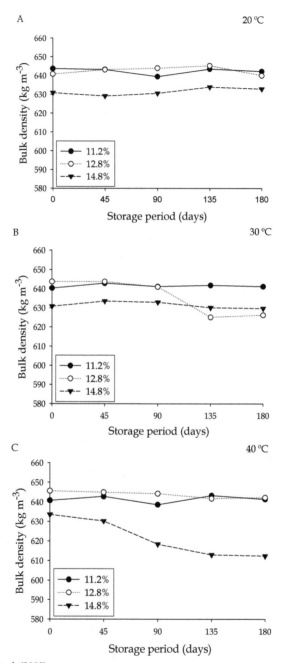

Source: ALENCAR et al. (2009)

Fig. 1. Average measurements of bulk density for soybeans stored with moisture contents of 11.2, 12.8 and 14.8% (w.b.) at the temperatures of 20, 30 and 40 °C.

was the predominant factor in maintaining the viability of soybean seeds. The effect of different combinations of moisture content and temperature on germination percentage of soybeans was evaluated by ALENCAR et al. (2008), during 180 days of storage (Fig. 2). It was verified that there was a decrease in the percentage of germinated grains, where this trend was less pronounced in grains stored with 11.2 and 12.8% moisture contents and temperatures of 20 °C (Fig. 2B).

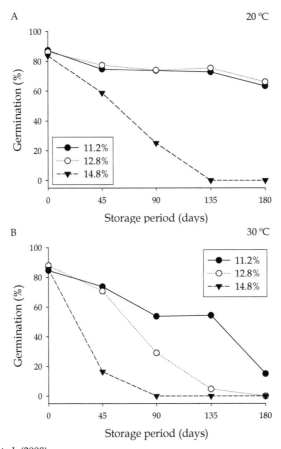

Source: ALENCAR et al. (2008)

Fig. 2. Average values of germination percentage of the soybeans with moisture contents of 11.2, 12.8 and 14.8% (w.b.) at the temperature of 20 and 30 °C, along the period of storage.

2.3.3 Electrical conductivity

According to SANTOS et al. (2004), the deterioration of grain is considered all and any degenerative change after the grain has reached its maximum quality, as evidenced by genetic damage, loss of integrity of system membranes, selective reduction of capacity, lipid peroxidation, leaching of solutes, changes in respiratory activity of the grains and seeds, changes in enzyme activity and protein synthesis, the inability to maintain the electrochemical gradient and loss of cellular compartmentalization and accumulation of toxic substances. Membrane damage is the initial event of degenerative changes in grains and seeds (DELOUCHE, 2006). According to HESLEHURST (1988), determination of electrical conductivity can be used to evaluate vigor, since the value of the conductivity is related to the amount of ions leached into solution, which is directly associated with cell membrane integrity. Poorly structured membranes and damaged cells are usually associated with the deterioration process of grains and seeds. Losses in germination and vigor in aged grains and seeds, according to LIN (1990), are correlated with increased electrolyte leaching, which increases with the decrease of membrane phospholipids. The lowest values corresponding to the lower release of exudates, indicate a high physiological potential (greater vigor), revealing a lower intensity of disorganization of cell membrane systems (VIEIRA et al., 2002).

ALENCAR et al. (2008) used electrical conductivity as a qualitative parameter for soybeans stored with moisture contents of 11.2, 12.8 and 14.8%, at temperatures of 20, 30 and 40 °C for 180 days (Table 2 and Fig. 3). In general, the authors observed a tendency for increased electrical conductivity during storage, where this trend was more pronounced as the moisture content and temperature were increased. It is reinforced that for the soybeans stored with 11.2 and 12.8% moisture content and temperature of 20 °C, electrical conductivity remained almost constant.

Temperature (°C)	Moisture content (%)	Regression equation adjusted	R^2
20	11.2	$\hat{y} = 153.4$	
	12.8	$\hat{y} = 147.1$	
	14.8	$\hat{y} = 149.4 - 0.2029X + 0.0087X^2$	0.89
30	11.2	$\hat{y} = 145.4 - 0.0403X$	0.91
	12.8	$\hat{y} = 145.5 - 0.1477 + 0.0045X^2$	0.87
	14.8	$\hat{y} = 143.7 + 2.8360X - 0.0056X^2$	0.94
40	11.2	$\hat{y} = 159.6 - 0.4581X + 0.0053X^2$	0.97
	12.8	$\hat{y} = 175.5 + 3.768X - 1.002X^2$	0.88
	14.8	$\hat{y} = 194.2 + 4.403X - 1.530X^2$	0.84

Source: ALENCAR et al. (2008)

Table 2. Regression equations adjusted for electrical conductivity of the solution containing the soybeans with moisture contents of 11.2, 12.8 and 14.8% (w.b.) at temperatures of 20, 30 and 40 °C along the storage period (X), and respective coefficients of determination.

2.3.4 Color of the grains

Appearance of the grains is considered a critical and decisive factor in the commercialization process. The color of soybeans, according to SINCLAIR (1992), has been used as an indicator of quality, and discoloration is indicative of physical or chemical alterations, presence of metabolites or other unfavorable characteristics. According to this author, changes in color of the soybeans are caused mainly by microorganisms, although changes in climatic conditions can enhance or affect color of the grain, but is not the main cause of the problem. In the United States upper limits are established for the classification of soybeans with distinct colors of yellow that is predominant, but may be green, black,

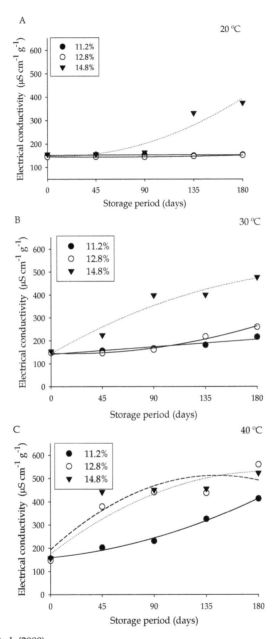

Source: ALENCAR et al. (2008)

Fig. 3. Regression curves of electrical conductivity (µS cm⁻¹ g⁻¹) of the solution which contains soybeans stored with moisture contents of 11.2, 12.8 and 14.8% w.b. at the temperatures of 20, 30 and 40 °C.

brown or bicolored (USDA, 2006). The percentage limits of grain characterized as other colors for soybeans in types 1, 2, 3 and 4 are 1.0, 2.0, 5.0 and 10.0%, respectively. These values indicate that product is of very poor quality.

The darkening of soybeans, according to SAIO et al. (1980), is an important qualitative indicator of deterioration during storage, and for LIU (1997), the variation in color characterizes the aging of the grains and is associated with qualitative changes such as reducing the germination percentage. HOU & CHANG (2004) evaluated alterations in the color of soybeans stored with 5.4% moisture content in different conditions of temperature and relative humidity. These authors observed a significant darkening, according to variation of the Hunter coordinates, for the soybeans only when stored at 30 °C and 84% relative humidity (Table 3). ALENCAR et al. (2009) studied the change in color of the soybeans stored with moisture contents of 11.2, 12.8 and 14.8% at temperatures of 20, 30 and 40 °C for 180 days. The authors evaluated the color difference (ΔE), from values of Hunter L, a and b coordinates, and found a significant increase, where this increase is more pronounced when grains are stored with elevated moisture content and under increased temperature (Fig. 4 and Table 4). This trend of increasing color difference is directly related to the increase in the percentage of damaged grains, which are considered by Brazilian law as serious defects. The damaged grains are defined as grains or pieces of grain that present visible damage and have an accentuated dark brown color, affecting the cotyledon (MAPA, 2007). Alterations in the color of soybeans can also be viewed from the aspect of soybean flour obtained from these grains (Fig. 5).

Period (month)	Hunter L value	Hunter a value	Hunter b value
0	51.04±0.51[a]	4.05±0.04[g]	15.58±0.09[a]
1	49.56±0.55[c]	4.22±0.15[f]	15.33±0.11[a]
2	50.34±0.37[b]	4.58±0.06[e]	15.38±0.11[a]
3	48.67±0.40[d]	4.83±0.16[d]	14.59±0.25[b]
4	48.38±0.09[d]	5.19±0.06[b]	14.76±0.03[b]
5	45.14±0.44[e]	5.53±0.07[a]	12.87±0.13[c]
6	43.37±0.18[f]	4.81±0.07[d]	11.19±0.13[e]
7	45.37±0.59[e]	5.05±0.15[c]	12.53±0.22[d]
8	41.99±0.55[g]	4.28±0.09[f]	10.57±0.28[f]
9	38.97±0.94[h]	3.87±0.18[h]	8.75±0.58[g]

Values followed by the same letter in the column are not statistically different at 5% probability
Source: HOU & CHANG (2004)

Table 3. Color of soybeans stored at 30 °C and 84% relative humidity.

2.4 Soybean oil

Soybean oil has emerged as one of the products obtained from processing of the grain, being one of the major products of this nature in the world market. It is most utilized to prepare food for humans and pets. Because of its properties it is suitable for a wide range of applications including use in margarines, salad oil, mayonnaise, and other food products (MORETTO & FETT, 1998). Virtually all soybean oil is extracted by solvent and commercial extraction techniques have remained unaltered since the early nineteenth century (ERICKSON & WIEDERMANN, 2006). Table 5 shows the main components of crude and refined soy oil, according to ERICKSON & WIEDERMANN (2006).

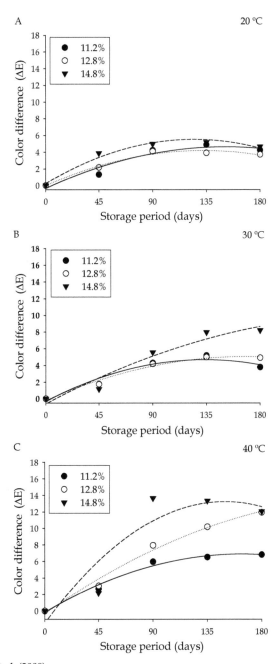

Source: ALENCAR et al. (2009)

Fig. 4. Regression curves of the color difference of the soybeans stored with moisture contents of 11.2. 12.8 and 14.8% (w.b.) at the temperatures of 20, 30 and 40 °C.

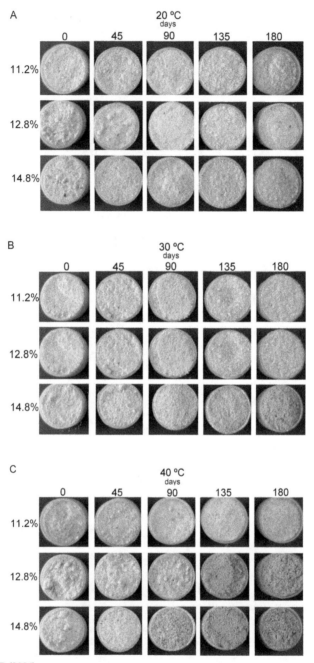

Source: ALENCAR (2006)

Fig. 5. Visual aspect of whole soybean flour obtained from soybeans with moisture contents of 11.2, 12.8 and 14.8% (w.b.) at the temperatures of 20, 30 and 40 °C, during storage.

Temperature (°C)	Moisture content (%)	Regression equation adjusted	R^2
	11.2	$\hat{y} = -0.358 + 0.0663X - 0.0002X^2$	0.91
20	12.8	$\hat{y} = -0.031 + 0.0639X - 0.0002X^2$	0.78
	14.8	$\hat{y} = 0.227 + 0.0852X - 0.0003X^2$	0.78
	11.2	$\hat{y} = -0.329 + 0.0753X - 0.0003X^2$	0.86
30	12.8	$\hat{y} = -0.202 + 0.0627X - 0.0002X^2$	0.91
	14.8	$\hat{y} = -0.619 + 0.0759X - 0.0001X^2$	0.92
	11.2	$\hat{y} = -0.169 + 0.0864X - 0.0003X^2$	0.95
40	12.8	$\hat{y} = -0.330 + 0.1020X - 0.0002X^2$	0.96
	14.8	$\hat{y} = -1.488 + 0.1972X - 0.0007X^2$	0.82

Source: ALENCAR et al. (2009)

Table 4. Regression equations adjusted for the color difference of the soybeans with moisture contents of 11.2, 12.8 and 14.8% (w.b.) at 20, 30 and 40 °C during the storage period (X), and respective coefficients of determination.

Component	Crude oil	Refined oil
Triglycerides (%)	95.0 – 97.0	99.0
Phosphatides (%)	1.5 – 2.5	0.003 – 0.045
Unsaponifiable matter (%)	1.6	0.3
Sterols (%)	0.33	0.13
Tocopherols (%)	0.15 – 0.21	0.11 – 0.18
Hydrocarbons (squalene) (%)	0.014	0.01
Free fatty acids (%)	0.3 – 0.7	< 0.05
Trace metals		
Iron (ppm)	1.0 – 3.0	0.1 – 0.3
Copper (ppm)	0.03 – 0.05	0.02 – 0.06

Source: ERICKSON & WIEDERMANN (1989)

Table 5. Principal components of crude and refined soy oil.

2.4.1 Qualitative parameters of oils and alterations resulting from storage conditions of the grains

In all stages of oil and fat processing various analysis are needed for quality control. In the refining process, for example, determining the percentage of free fatty acids is necessary in the neutralization step, or as a qualitative indicator (O'BRIEN, 2004). Other widely used analyses as quantitative indices of oils and fats are: peroxide value, iodine value, color, saponification number, water content and others.

2.4.1.1 Free fatty acids

During the storage of grains, the lipid fraction is slowly hydrolyzed by water at high temperature (physical process) or by natural lipolytic enzymes or those produced by bacteria and/or fungi, contributing to the hydrolytic rancidity of the product (ARAÚJO, 2004). Increase in the content of free fatty acids from lipids occurs by the action of lipase and phospholipase enzymes present in the soybeans or produced by the associated microflora, which contribute to the breaking of ester linkages of triglycerides (ZADERNOWSKI et al.,

1999). Thus, the percentage of free fatty acids is an important indicator of quality throughout the processing of oils and fats. O'BRIEN (2004) stated that hydrolytic rancidity can affect taste, odor and other characteristics of oil. This author stresses that vegetable oils may present relatively high contents of free fatty acids if the grains or seeds present damages due to procedures in the field or incorrect storage practices, being that high values of free fatty acids can cause excessive losses in refining. WILSON et al. (1995) claimed the refining losses between 1 and 1.5% are considered normal; however, such losses may reach 4% or more for greater levels of free fatty acids.

Many authors have related the increase in free fatty acid percentages to storage conditions. The variation in percent free fatty acids in crude oil extracted from soybeans stored with different moisture contents was observed by FRANKEL et al. (1987). Soybeans stored with 13% (w.b.) resulted in lower increases in the percentage of free fatty acids when compared with the values obtained by grains stored with 16 and 20% (w.b.) moisture content. With regards to the crude oil extracted from the soybeans stored at 13% (w.b.), it was verified that the increase in the free fatty acid percentage was from 0.2 to 1.25% after 49 days of storage; in the crude oil obtained from the grains stored with moisture contents of 16 and 20% w.b., increase was from 0.5 to 2.0% after 27 days and from 0.6 to 2.3% after 28 days, respectively. NARAYAN et al. (1988a) verified the increase in free fatty acid percentage in soybeans stored at different temperature conditions (between 16 and 40 °C) and relative humidity (between 50 and 90%), obtaining average values equal to 0.69, 4.32, 5.37 and 9.85% after 12, 24, 36 and 108 months of storage. ALENCAR et al. (2010) evaluated the effect of different combinations of temperature and moisture content on the percentage of free fatty acids of crude oil extracted from soybeans stored for 180 days. The authors adopted the grain moisture contents of 11.2, 12.8 and 14.8% and temperatures of 20, 30 and 40 °C (Fig. 6 and Table 6), and generally observed a significant increase in free fatty acid content of crude oil, except for the grains with moisture content of 11.2% at 20 °C. The increasing trend in the percentage of free fatty acids was more pronounced as water content and temperature increased, and the greatest percentage of free fatty acids from crude oil was 12.5% for grain stored at 14 8% moisture content after 180 days.

Temperature (°C)	Moisture content (%)	Regression equation adjusted	R^2
20	11.2	$\hat{y} = 0.41$	
	12.8	$\hat{y} = 0.370 + 0.0012X$	0.85
	14.8	$\hat{y} = 0.438 + 0.0069X$	0.80
30	11.2	$\hat{y} = 0.352 + 0.025X$	0.82
	12.8	$\hat{y} = 0.332 + 0.035X$	0.72
	14.8	$\hat{y} = 0.307 + 0.0251X$	0.86
40	11.2	$\hat{y} = 0.277 + 0.0066X$	0.96
	12.8	$\hat{y} = 0.440 + 0.0121X$	0.87
	14.8	$\hat{y} = -0.294 + 0.0692X$	0.84

Source: ALENCAR et al. (2010)

Table 6. Regression equations adjusted for free fatty acids of oil extracted from soybeans with moisture contents of 11.2, 12.8 and 14.8% (w.b.) at the temperatures of 20, 30 and 40 °C along the storage period (X), and respective coefficients of determination.

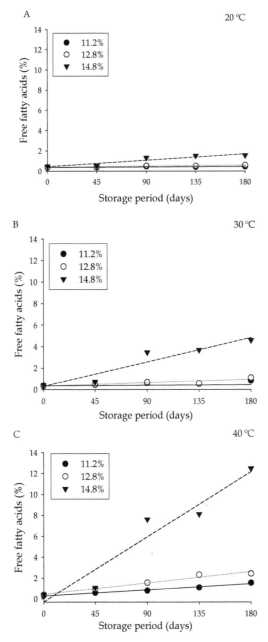

Source: ALENCAR et al. (2010)

Fig. 6. Regression curves of the percentage of free fatty acids (%) of crude oil extracted from soybeans stored with moisture contents of 11.2, 12.8 and 14.8% (w.b.) at temperatures of 20, 30 and 40 °C.

2.4.1.2 Peroxide index

Lipid oxidation is a spontaneous and inevitable phenomenon, according to SILVA et al. (1999), with direct implications on the market value of either the fatty bodies, or of all the products formulated from them, and peroxidation is the main cause of deterioration of fatty bodies (lipid materials and greases). It is the main cause of deterioration of oils and fats, and the hydroperoxides formed from the reaction between oxygen and unsaturated fatty acids are the primary products. Although these compounds do not exhibit taste or odor, they are rapidly decomposed even at room temperature into aldehydes, ketones, alcohols, hydrocarbons, esters, lactones and furans, causing unpleasant taste and odor in oils and fats (O'BRIEN, 2004; EYS et al., 2006). Other consequences of lipid oxidation in foods are changes in nutritional value, functionality, and also in the integrity and safety of the product via the formation of potentially toxic polymer compounds (SILVA et al., 1999; ARAÚJO, 2004; NAZ et al., 2004; RAMALHO & JORGE, 2006). According to HOU & CHANG (2004), the appearance of off-flavors (unpleasant aroma and taste) in soybean products can be partially attributed to lipid peroxidation.

One of the methods used to determine the degree of oxidation in fats and oils is the peroxide index. The peroxide index (PI) is a measure of oxidation or rancidity in its initial phase, as shown in Table 7 and measures the concentration of substances (in terms of milliequivalents of peroxide per thousand grams of sample) which oxidize potassium iodide to iodine is widely used in determining the quality of oils and fats, showing good correlation with taste (O'BRIEN, 2004).

Range	Degree of oxidation
<1	Freshness
1< PI<5	Low oxidation
5<PI<10	Moderate oxidation
10<PI<20	High oxidation
>20	Poor flavor

Source: O'BRIEN (2004)

Table 7. Classification of the degree of oxidation of soybean oil in accordance with the peroxide index (PI, meq kg^{-1}).

Works are encountered in literature that report the effect of different soybean storage conditions on the peroxide index of crude oil. NARAYAN et al. (1988a) studied the evolution of the peroxide index of crude oil extracted from soybeans stored at different temperatures and humidities. Average values observed for the peroxide index were 18, 40, 65 and 98 meq kg^{-1} after 12, 24, 36 and 108 months. ALENCAR et al. (2010) evaluated the peroxide value in crude oil obtained from soybeans stored with moisture contents of 11.2, 12.8 and 14.8%, at the temperatures of 20, 30 and 40 °C for 180 days. The authors verified an increase for all combinations of water content and temperature, where the highest values were obtained as the water content and temperature were increased (Table 8).

2.4.1.3 Color of the oil

The color and appearance of oils and fats, according to O'BRIEN (2004), are not monitored only due to the visual character, but also because they are related to the cost of processing and quality of the final product. Most oils present a reddish-yellow color as the result of the

Temperature (°C)	Moisture content (%)	Storage period (days)				
		0	45	90	135	180
	11.2	1.51	2.17	2.87	2.81	2.68
20	12.8	1.45	2.09	3.29	3.57	2.76
	14.8	1.52	3.58	4.33	3.62	7.84
	11.2	1.34	2.57	2.80	3.14	2.79
30	12.8	1.52	2.66	2.75	3.52	2.96
	14.8	1.48	4.25	3.51	5.89	8.09
	11.2	1.87	2.73	3.64	6.60	7.64
40	12.8	1.48	2.58	8.37	9.85	14.54
	14.8	1.30	4.47	11.77	13.88	14.76

Source: ALENCAR (2006)

Table 8. Average values of the peroxide index of crude oil extracted from soybeans stored at 20, 30 and 40 °C and moisture contents of 11.2, 12.8 and 14.8% (w.b.) during storage.

presence of carotenoids and chlorophyll. However, some crude oils may present a relatively high pigmentation due to damage of the raw material in the field, storage or processing failures; alterations in color indicate qualitative deterioration of the oil.

Alterations in color of the crude oil obtained from soybeans stored under different conditions were evaluated by ALENCAR et al. (2010). In this study different combinations of water content (11.2, 12.8 and 14.8%) and temperature (20, 30 and 40 °C) were obtained for the grains stored for 180 days, and the qualitative photometric index of the oil was analyzed. The authors observed a significant increase in the photometric color index for all combinations of water content and temperature, as for the temperature of 30 °C (Fig. 7 and Table 9). WILSON et al. (1995) associated an increase in the photometric color index to the percentage of grains damaged by fungi. It is emphasized that the degumming of crude oil extracted from seriously damaged grains is hampered and the refined oil is darker than that obtained from healthy kernels, as well as greater losses in refining (LIST et al., 1977).

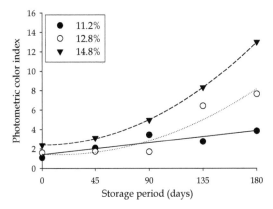

Source: ALENCAR et al. (2010)

Fig. 7. Regression curve of the photometric color index of crude oil extracted from soybeans stored with moisture contents of 11.2, 12.8 and 14.8% (w.b.) at the temperature of 30 °C.

Temperature (°C)	Moisture content (%)	Regression equation adjusted	R^2
	11.2	$\hat{y} = 1.41 + 0.0138X$	0.46
30	12.8	$\hat{y} = 1.46 - 0.0067X + 0.00024X^2$	0.74
	14.8	$\hat{y} = 2.40 - 0.0007X + 0.00033X^2$	0.85

Source: ALENCAR et al. (2010)

Table 9. Regression equations adjusted for the photometric color index of the oil extracted from the soybeans with moisture contents of 11.2, 12.8 and 14.8% (w.b.) at the temperature of 30 °C during the storage period (X), and respective coefficients of determination.

2.5 Effects of different storage conditions on the quality of other sub-products derived from soybeans

The quality of soybeans can also influence the qualitative parameters of other sub-products, including soymilk and tofu. SAIO et al. (1980) evaluated qualitative parameters of soymilk and tofu made from soybeans stored with a moisture content of 10.61%, and different combinations of temperature (15, 25 and 35 °C) and relative humidity (60 and 90%) during 180 days. The authors verified, resulting from the storage conditions adopted, significant changes in color and pH of the soymilk and hardness of tofu. The physicochemical quality of tofu, obtained from soybeans stored at temperatures of 3-4, 20 and 30 °C and relative humidities of 86.0, 57.0 and 84.0%, respectively, was assessed by HOU & CHANG (2004). For grains stored at 30 °C and relative humidity of 84.0%, the authors observed a reduction in yield (512g/100g of grains at time zero to 71g/100 g of grains after 7 months of storage) and alterations in texture with increasing hardness and color (Table 10). ACHOURI et al. (2008) evaluated the quality of the soymilk obtained from soybeans stored for 10 months at 18 °C and 50% relative humidity. Under these storage conditions the authors observed no significant change in the water uptake factor and pH of the soymilk, but there was significant variation in color and total volatiles. KONG et al. (2008) evaluated the physicochemical quality of soymilk and tofu made from soybeans stored with moisture contents between 6 and 14% in different combinations of temperature (40 to 50 °C) and humidity (55 to 80%). For the soymilk a decrease in pH and protein content was verified, this tendency being more accentuated as temperature and relative humidity increased. Reduction in the pH of the soymilk was observed for the soybeans stored under temperatures between 22 and 50 °C and relative humidity between 55.0 and 80.0%. It is highlighted that the protein content of the soymilk reduced by 24.0% in grains stored for 10 months at 40 °C. With regards to tofu, KONG et al. (2008) observed a significant reduction in yield for grains stored at 30 and 40 °C, as well as alterations in texture and color of the product. Also according to these authors, there is a strong relationship between the color of the grains and tofu, with respect to the Hunter (L, a and b) coordinates, as shown in Table 11.

3. Conclusion

The combination of high grain moisture and temperature during soybean storage accelerates the deterioration process of the sub-products of soybean. Proposed preventive measures of post-harvest handling to reduce risks of quality loss in soybean grains and sub-products are: store soybean grains with moisture content up to 15% (w.b.) at 20 °C without

Period (month)	Hunter L value	Hunter a value	Hunter b value
0	87.13±0.37[a]	-0.50±0.17[e]	13.57±0.26[b]
1	87.30±0.05[a]	-0.31±0.03[e]	13.57±0.09[b]
2	85.20±0.18[b]	0.50±0.25[d]	13.36±0.09[b]
3	85.13±0.12[b]	0.71±0.10[d]	13.95±0.13[b]
4	84.09±0.41[c]	1.06±0.17[c]	13.11±0.23[b]
5	83.60±0.41[d]	1.04±0.04c	12.63±0.25[c]
6	82.30±0.22[e]	1.71±0.06[b]	12.97±0.23[bc]
7	70.14±0.49[f]	4.56±0.12[a]	15.90±0.04[a]

Values followed by the same letter in the column do not differ statistically at 5% probability
Source: HOU & CHANG (2004)

Table 10. Tofu color obtained from soybeans stored at 30 °C and 84% relative humidity.

Coordinate	Adjusted equations	R^2
L	$\hat{y} = 19.275 + 1.1084X$	0.623
a	$\hat{y} = 9.0894 + 1.3217X$	0.546
b	$\hat{y} = 9.929 + 0.4453X$	0.125

Source: KONG et al. (2008)

Table 11. Regression equations which relate the Hunter L, a and b coordinates of tofu (y) and soybean grains (X).

risk of deterioration up to 180 days; in regions with temperatures around 30 °C, store soybean with moisture content up to 13% (w.b.); do not store soybean with moisture content above 11% (w.b.) in regions where the grain mass temperature can reach 40 °C with the risk of accelerating deterioration of grains and sub-products.

4. Acknowledgements

Part of the data presented in this chapter was obtained during Master's research of the first author, as part of the graduate program in Agricultural Engineering of the Universidade Federal de Viçosa (UFV).

5. References

Abba, E.J. & Lovato, A. (1999). Effect of seed storage temperature and relative humidity on maize (*Zea mays* L.) seed viability and vigour. *Seed Science and Technology*, Vol.27, No.1, (April 1999), pp. 101-114, ISSN 0251-0952

Acasio, A. (September 2010). *Handling and storage of soybeans and soybean meal*, 21.09.2010, Available from http://www.feedmachinery.com

Achouri, A.; Boye, J.I. & Zamani, Y. (2008). Soybean variety and storage effects on soymilk flavour and quality. *International Journal of Food Science and Technology*, Vol.43, No.1, (January 2008), pp. 82-90, ISSN 1365-2621

Alencar, E.R. (2006). *Efeitos das condições de armazenagem sobre a qualidade da soja (Glycine max (L.) Merrill) e do óleo bruto*. Dissertação, Mestrado em Engenharia Agrícola, Universidade Federal de Viçosa, Viçosa, Brasil

Alencar, E.R.; Faroni, L.R.D; Lacerda Filho, A.F.; Garcia, L.F.& Meneghitti, M.R. (2008). Qualidade fisiológica dos grãos de soja em função das condições de armazenamento, *Engenharia na Agricultura*, Vol 16, No.3, (April 2008), pp. 155-166, ISSN 1414-3984

Alencar, E.R.; Faroni, L.R.D; Lacerda Filho, A.F.; Peternelli, L.A. & Costa, A.R. (2009). Qualidade dos grãos de soja armazenados em diferentes condições, *Revista Brasileira de Engenharia Agrícola e Ambiental*, Vol.13, No.5, (September 2009), pp. 606-613, 2009, ISSN 1807-1929

Alencar, E.R.; Faroni, L.R.D.; Peternelli, L.A.; Silva, M.T.C. & Costa, A.R. (2010). Influence of soybean storage conditions on crude oil quality, *Revista Brasileira de Engenharia Agrícola e Ambiental*, Vol.14, No.3, (March 2010), pp. 303-308, ISSN 1807-1929

Al-Yahya, S.A. (2001). Effect of storage conditions on germination in wheat. *Journal Agronomy & Crop Science*, Vol.186, No.4, (June 2001), pp. 273-279, ISSN 0931-2250

Araújo, J.M.A. (2004). *Química de Alimentos: Teoria e Prática*, Editora UFV, ISBN 978-85-7269-351-6, Viçosa, Brasil

Bailey, J.E. (1974). Whole grain storage, In: *Storage of cereal grains and their products*, C.M. Christensen, (Ed.), 333-360, AACC, ISBN 0-913250-05-8, St. Paul, United States

Bhattacharya, K. & Raha, S. (2002). Deteriorative changes of maize, groundnut and soybean seeds by fungi in storage. *Mycopathologia*, Vol.155, No.3, (November 2002), pp. 135-141, ISSN 0301-486X

Brooker, D.B.; Bakker-Arkema, F.W. & Hall, C.W. (1992). *Drying and storage of grains and oilseeds*, Springer, ISBN 0442205155, New York, United States

Delouche, J. (June 2006). *Germinação, deterioração e vigor da semente*, 26.06.2006, Available from http://www.seednews.inf.br/portugues/seed66/artigocapa66.shtml

Erickson, D.R. & Wiedermann, L.H. (April 2006). *Soybean oil: modern processing and utilization*, 03.04.2006, Available from http://www.asaim-europe.org/Backup/pdf/sboprocess.pdf

Eys, J.E.; Offner, A. & Bach, A. (June 2006). *Manual of quality analyses for soybean products in the feed industry*, 12.06.2006, Available from http://www.asaim-europe.org/Backup/Library/library_e.htm

Frankel, E.N.; Nash, A.M. & Snyder, J.M. (1987). A methodology study to evaluate quality of soybeans stored at different moisture levels. *Journal of the American Oil Chemists' Society*, Vol.64, No.7, (July 1987), pp. 987-992, ISSN 0003-021X

Gungadurdoss, M. (2003). Improvement of seed viability of vegetable soybean (*Glycine max* (L) Merrill). In: Food and Agricultural Research Council, 2003, Réduit, Mauritius. Proceedings...: Réduit, Mauritius: Lalouette, J.A., BACHRAZ (Eds.), 117-123.

Hansen, L.S.; Skovgard, H. & Hell, K. (2004). Life table study of *Sitotroga cerealella* (Lepidoptera: Gelichiidae), a strain from West Africa. *Journal of Economic Entomology*, Vol.97, No.4, (August 2004), pp. 1484-1490, ISSN 1938-291X

Heslehurst, M.R. (1988). Quantifying initial quality and vigour of wheat seeds using regression analysis of conductivity and germination data from aged seeds. *Seed Science and Technology*, Vol.16, No.1, (April 1999), pp. 75-85, ISSN 0251-0952

Hou, H.J. & Chang, K.C. (1998). Yield and quality of soft tofu as affected by soybean physical damage and storage. *Journal of Agricultural and Food Chemistry*, Vol.46, No.12, (November 1998), pp. 4798-4805, ISSN 0003-021X

Hou, H.J. & Chang, K.C. (2004). Storage conditions affect soybean color, chemical composition and tofu qualities. *Journal of Food Processing and Preservation,* Vol.28, No.6, pp. 473-488, (December 2004), ISSN 0145-8892

Hummil, B. C.W.; Cuendit, L. S.; Christensen, C.M. & Geddes, W.F. (1954). Grain storage studies XIII; comparative changes in respiration, viability, and chemical composition of mold-free and mold-contaminated wheat upon storage. *Cereal Chemistry,* Vol.31, No.2, (March 1954), pp. 143-150, ISSN 0009-0352

Karunakaran, C.; Muir, W.E.; Jayas, D.S.; White, N.D.G. & Abramson, D. (2001). Safe storage time of high moisture wheat. *Journal of Stored Products Research,* Vol.37, No.3, (July 2001), pp. 303-312, ISSN 0022-474X

Kong, F.; Chang, S.K.C.; Liu, Z. & Wilson, L.A. Changes of soybean quality during storage as related to soymilk and tofu making. *Journal of Food Science,* Vol. 73, No.3, (April 2008), pp. S134-S144, ISSN 1750-3841

Lin, S.S. (1990). Alterações na lixiviação eletrolítica, germinação e vigor da semente de feijão envelhecida sob alta umidade relativa do ar e alta temperatura. *Revista Brasileira Fisiologia Vegetal,* Vol.2, No.2, (May 1990), pp. 1-6, ISSN 1806-9355

List, G.R.; Evans, C.D.; Warner, K.; Beal, R.E.; Kwolek, W.F.; Black, L.T. & Moulton, K.J. (1977). Quality of oil from damaged soybeans. *Journal of the American Oil Chemists' Society,* Vol.54, No.1, (January 1977), pp. 8-14, ISSN 0003-021X

Liu, K. (1997). *Soybeans: chemistry, technology and utilization,* Chapman & Hall, ISBN 0834212994, New York, United States

Locher, R. & Bucheli, P. (1998). Comparison of soluble sugar degradation in soybean seed under simulated tropical storage conditions. *Crop Science,* Vol.38, No.5, (September 1998), pp. 1229-1235, 0011-183X

MAPA. Ministério da Agricultura, Pecuária e Abastecimento. (2007). *Instrução Normativa N° 11, de 15 de maio de 2007,* Brasília, Brasil

Mason, L.J. (May 2006). *Indianmeal moth Plodia interpunctella (Hubner),* 30.05.2006, Available from
http://www.entm.purdue.edu/entomology/ext/targets/e-series/EseriesPDF/E-223.htm

Mbata, G.N. & Osuji, F.N.C. (1983). Some aspects of the biology of *Plodia interpunctella* (Hübner) (Lepidoptera: Pyralidae), a pest of stored groundnuts in Nigeria. *Journal of Stored Products Research,* Vol.19, No.3, (July 1983), pp. 141-151, ISSN 0022-474X

Moretto, E. & Fett, R. (1998). *Tecnologia de oleos e gorduras vegetais,* Varela, ISBN 85-85519-41-X, São Paulo, Brasil

Narayan, R.; Chauhan, G.S. & Verma, N.S. (1988a). Changes in the quality of soybean during storage. Part 1 – Effect of storage on some physico-chemical properties of soybean. *Food Chemistry,* Vol.27, No.1, (January 1988), pp. 12-23, ISSN 0308-8146

Narayan, R.; Chauhan, G.S. & Verma, N.S. (1988b). Changes in the quality of soybean during storage. Part 2 – Effect of soybean storage on the sensory qualities of the products made therefrom. *Food Chemistry,* Vol.30, No.3, (December 1988), pp. 181-190, ISSN 0308-8146

Naz, S.; Sheikh, H.; Siddiqi, R. & Sayeed, S.A. (2004). Oxidative stability of olive, corn and soybean oil under different conditions. *Food Chemistry,* Vol.88, No.2, , (November 2004), pp. 253-259, ISSN 0308-8146

O'Brien, R.D. (2004). *Fats and Oils Formulating and Processing for Applications*, CRC Press, ISBN 0849315999, Boca Raton, United States

Orthoefer, F.T. (1978). Processing and utilization. In: *Soybean: physiology, agronomy and utilization*, A.G. Norman, (Ed.)., 219-246, Academic Press, ISBN 0-12-521160-0, New York, United States

Padín, S.; Bello, G.D. & Fabrizio, M. (2002). Grain loss caused by *Tribolium castaneum*, *Sitophilus oryzae* and *Acanthoscelides obtectus* in stored durum wheat and beans treated with *Beauveria bassiana*. *Journal of Stored Products Reseach*, Vol.38, No.1, (January 2002), pp. 69-74, ISSN 0022-474X

Ramalho, V.C. & Jorge, N. (2006). Antioxidantes utilizados em óleos, gorduras e alimentos gordurosos. *Química Nova*, Vol.29, No.4, (July 2006), pp. 755-760, ISSN 0100-4042

Saio, K.; Nikkuni, I.; Ando, Y.; Osturu, M.; Terauchi, Y. & Kito, M. (1980). Soybean quality changes during model storage studies. *Cereal Chemistry*, Vol.57, No.2, (March 1980), pp. 77-82, ISSN 0009-0352

Santos, C.M.R.; Menezes, N.L. & Villela, F.A. (2004). Alterações fisiológicas e bioquímicas em sementes de feijão envelhecidas artificialmente. *Revista Brasileira de Sementes*, Vol.26, No.1, (July 2004), pp. 110-119, ISSN 0101-3122

Silva, F.A.M.; Borges, M.F.M. & Ferreira, M.A. (1999). Métodos para avaliação do grau de oxidação lipídica e da capacidade antioxidante. *Química Nova*, Vol.22, No.1, (January 2006), pp. 94-103, ISSN 0100-4042

Silva, J.S.; Berbert, P.A.; Rufato, S. & Afonso, A.D.L. (2008). Indicadores da qualidade dos grãos. In: *Secagem e Armazenagem de Produtos Agrícolas*, J.S. Silva. (Org.), 63-108, ISBN 9788562032004, Aprenda Fácil, Viçosa, Brasil

Sinclair, J.B. (1995). Reevaluation of grading standards and discounts for fungus-damaged soybean seeds. *Journal of the American Oil Chemists' Society*, Vol.72, No.12, (December 1995), pp. 1415-1419, ISSN 0003-021X

Sinha, R. N. & Muir, W.E. (1973). *Grain storage: part of a system*, Avi Pub. Co., ISBN 087055123X, Westport, United States

USDA. (May 2006). *U.S. soybean inspection*, 31.05.2006, Available from http://www.usda.gov/gipsa/reference-library/brochures/soyinspection.pdf

USDA. (March 2011). *U.S. soybean inspection*, 14.03.2011, Available from http://www.usda.gov/oce/commodity/wasde/latest.pdf

Vieira, R.D.; Penariol, A.L.; Perecin, D. & Panobianco, M. (2002). Condutividade elétrica e teor de água inicial das sementes de soja. *Pesquisa Agropecuária Brasileira*, Vol.37, No.9, (September 2002), pp. 1333-1338, ISSN: 0100-204X

Wilson, R.F.; Novitzky, W.P. & Fenner, G.P. (1995). Effect of fungal damage on seed composition and quality of soybeans. *Journal of the American Oil Chemists' Society*, Vol.72, No.12, (December 1995), pp. 1425-1429, ISSN 0003-021X

Zadernowski R.; Nowak-Polakowska H. & Rashed, A.A. (1999). The influence of heat treatament on the activity of lipo and hydrophilic components of oat grain. *Journal of Food Processing and Preservation*, Vol.23, No.3, pp. 177-191, (September 1999), ISSN 0145-8892

2

Soybean in the European Union, Status and Perspective

Yves Bertheau[1] and John Davison[2]
[1]Inra SPE, route de Saint Cyr, 78 026 Versailles cedex,
[2]Inra, route de Saint Cyr, 78 026 Versailles cedex(retired)
France

1. Introduction

Originating from China, soybean is currently the most important agricultural commodity traded around the world, both in terms of volumes and money. This crop also shows the most important changes over the last decades by the predominance of genetically modified (GM) crops, dominated by herbicide tolerance traits, and its worldwide cultivation. Due to its important protein content and the increasing demand for proteins in relation with the intensification of livestock production, the soybean surfaces have dramatically increased in several South American countries, such as Brazil, Argentina, Paraguay, Uruguay and Bolivia when compared to the surfaces of soybean in the USA, and at a less extent in Canada (James, 2011). China is still the main (non-GM) soybean producer but the main exporters of (GM) soybean are the USA, Brazil and Argentina. China which was, until the 1930s, the main worldwide exporter but was dethroned in the 1950s by USA exporting soybean as basis of feedstuffs and China is now the main importer with ca 60% of US soy.

The success of GM soybean can be explained by the ease of cropping due to GM trait in countries with large fields, particularly for farmers for whom weed management and soil erosion have always been an issue.

This segmentation of market between food and feed use is still prevalent into the international trade: Asia mostly cultivates and uses non-GM soybean as a food component while other regions of the world mostly use GM and non-GM soybean as feed component (Birthal et al., 2010). However, since a few years, this trend is changing with the improvement of the living standards of Asian countries such as China, which now imports huge quantities of soybean for livestock feeding due to an increased demand for meat. Currently, the soybean daily price is at its second highest peak after the 2007/2008 peak. Altogether, the soybean daily price increased of only 83% over the 3 last decades due to the current prices' peak (IndexMundi, International Monetary Fund. April 20, 2011[1]).

Despite the fact that several other sources of protein are available for livestock, the flexibility of soybean in feedstuff preparation, particularly for pork and poultry productions, drives the international trade. Its use in bovine production, meat or milk, can be more easily replaced by alternative protein sources, or simply be replaced by pasturage.

---

[1] http://www.indexmundi.com/commodities/?commodity=soybeans&months=360

Soybean has been first introduced in Europe during the XVIIth century as high-class food, however despite several scientific and popular reports during the XIXth century soybean was rarely cropped in Western Europe. The first massive importation of soybean in Europe started by the beginning of the XXth century for oil and meal production, declined during the 2 world wars, but with increasing imports between them. Since the 1950s, Europe dramatically increased its importations of soybean due to a new animal production scheme with highly concentrated livestock production. However, new dossiers in the pipeline of GMO approvals now consider cultivation in the EU. We thus examine in more depth this forthcoming issue in environmental surveillance.

Due to this increasing part of GM soybean in the international market and consumers' reluctance of several countries to accept these products, a new segmentation of the market appeared between GM and non-GM soybean linked to labeling of GM food, and feed in some countries, with an exemption of labeling below a threshold of fortuitous or technically unavoidable presence, ranging from 0.9% in the EU-27[2] and Russia to 3-5% in Korea, Taiwan and Japan. Generally speaking, the labeling thresholds are representative of the countries' dependence on feed and food imports.

2. GMO in the EU

The development of green biotechnology dates back to the 1970s and led to the development of Genetically Modified Organisms (GMOs) in the 1980s. On May 21, 1994, the genetically engineered FlavrSavr™ tomato was the first crop approved by the U.S. Food and Drug Administration for commercialization. Due to the controversy over GMOs, which started with the first arrivals in Europe of soybeans cargoes in 1996, and to its poor organoleptic qualities, this tomato was withdrawn from the market in 1998 (Bruening and Lyons, 2000).

2.1 European consumers

Today, around 148 million hectares of GM plants are grown and traded around the world annually, among which ca. 71% of GM soy according to the ISAAA lobbyist (James, 2011). Despite this development, the European public's perception of GM crops is still very negative as demonstrated by the recent results of the Eurobarometer surveys (Bonny, 2008; de Cheveigné, 2004; Gaskell et al., 2006; Gaskell et al., 2010; TNS Opinion & Social, 2010). However, this consumers' reluctance seems less pronounced in the eastern part of the EU-27 (Consumerchoice Consortium, 2008).

After several scandals in the 1990s' such as BSE, dioxin contaminations, the reluctance of consumers' and citizens to embrace GMOs has been considered by the EC and European Union Member States (EU-MS) which have implemented a legal framework enabling consumers to maintain their freedom of choice through both food and feed labeling (European Commission, 1997, 2000a, b, 2001, 2002d, 2003c, d). In counterpart, the freedom of choice of producers to cultivate GM or non-GM crops is considered through a set of coexistence rules to be implemented by EU-MS, according to the European principle of subsidiarity as recommended by the EC (European Commission, 2003a, 2010a). However, despite numerous requests from NGOs and consumers associations, animals reared with GM feed do not require labeling.

---

[2] Current European Union with 27 Member States

2.2 GMO approvals in the EU

In order to re-assure the European public on food safety and more particularly the question of GMOs, the European Community has developed a series of regulations (Table 1) to ensure GMO safety, detection, traceability and labeling.

Food safety assessment is the responsibility of the European Food Safety Authority (EFSA) which cooperates with EU-MS national advisory committees and covers food additives, animal welfare, plant health, allergies, mycotoxins, biological hazards, chemical and biological contaminants. It also assesses the safety of GMOs (seed, food, feed, and derivatives). EFSA is an independent scientific body providing advice on all aspects of food safety, and a positive EFSA assessment is necessary for authorization to place food on the European market. GMO dossiers can be notified to the European Commission either under the 2001/18 directive or the (EC) regulation 1829/2003. Although not implied in its name, EFSA also provides advice on GMO environmental issues.

Once a positive EFSA assessment has been obtained, and once validated GMO detection methods and control sample and reference materials are available (all being provided by the applicant company), the application is then sent to the EC. On the basis of the opinion of EFSA, in some instances amended on the basis of national advisory agencies and committees, the EC drafts a proposal for granting or refusing the authorization, which it submits to the Section on GM Food and Feed of the Standing Committee on the Food Chain and Animal Health. If this Standing Committee accepts the proposal, it is finally adopted by the EC. Otherwise, it is passed on to the Council of Ministers which has a time limit of 3 months to reach a qualified majority for, or against, the proposal. In the absence of such a decision (which is frequently the case), the EC adopts the proposal. Over the last years, all GMO approvals in the EU were accepted on that scheme basis with approvals for a renewable 10 years period.

In contrast to several claims against the "lengthy and costly" approval European procedure, it should be noted that the notifiers often use dossiers of previous approvals in third countries, such as USA, with thus very few changes and thus very low costs of compliance with the EU approval procedure. Secondly, the European theoretical approval duration is *per se* not very long; however dossiers are in numerous instances incomplete. In this case the clock of approval is stopped each time details are requested from the notifiers. Due to these several stop-and-go steps in such an approval, the effective duration of European approval may be rather long. The European procedure of safety assessment of GMOs is currently under review, for instance on the statistics to be used in comparing animal cohorts, the guidelines about environmental impact assessment, or the more important use of the "substantial equivalence" concept in the comparisons between GM and conventional plants.

Despite the relatively rather strict European approval procedure, several EU-MS introduced national bans on GMO, be these for import and transformation such as Austria, for baby food in Italy or for cultivation as in France, Austria or Bulgaria.

However, there are currently ca. 50 GMOs in the pipeline of approval or approved for import and transformation, including several stacked GMOs and a few modified flowers. For soybean, 11 transformation events or stacked genes are in the European approval process with 2 GM soybean as fully approved and the first approved one (MON GTS 40-3-2) in the renewal process.

Due to the rather long European approval process, several reports outlined the possible shortage of soybean for the feed industry due to these "asynchronous approvals" (DG AGRI European Commission, 2007; Stein and Rodriguez-Cerezo, 2009, 2010a). The EC recently issued a proposal of modification of 2001/18 directive to allow Low Level Presence of EU unapproved GMOs, as also discussed in the *Codex Alimentarius* instance, for GMO already approved in a third country and whose dossiers are already under EFSA discussion for at least 3 months.

- **Directive 1990/219/EEC** covered the contained use of genetically modified organisms. Directive 1990/220/EEC was modified by Directive 98/81/EEC.
- **Directive 1990/220/EEC** covered the notification for a deliberate release and of the placing on the market of GMOs. Directive 1990/220/EEC was repealed by Directive 2001/18/EEC.
- **Regulation (EC) 258/1997** concerning novel foods and novel food ingredients, not heavily used in the EU before 1997 and establishing a compulsory labeling for these novel foods and ingredients, such as GMO, irradiated food, etc. Part of the current revision of food and feed legislation.
- **Regulation (EC) 1139/1998** laid down the compulsory indication on the labeling of foods and food ingredients produced from genetically modified soya (*Glycine max* L.) covered by Commission Decision 1996/281/EC and genetically modified maize (*Zea mays* L.) covered by Commission Decision 1997/98/EC, of particulars other than those provided for in Directive 1979/112/EEC.
- **Regulation (EC) 49/2000** amended the 1139/1998 EC regulation and established a 1% labeling threshold which was further decreased to 0.9% by regulation 1829/2003.
- **Regulation (EC) 50/2000** establishing a mandatory labeling of additives and flavorings that have been genetically modified or have been produced from genetically modified organisms.
- **Directive 2001/18/EEC** covers the deliberate release of GMOs in the environment (field trials and cultivation), in the absence of specific containment measures. It also regulates commercialization (importation, processing and transformation) of GMOs into industrial products. Finally, the Directive requests post-commercialization, case specific and general, surveillance plans on unforeseen effects of GMO on both health and environment.
- **Regulation (EC) 178/2002** resulted in the creation of EFSA and in a general obligation for traceability of at least one step forwards and one step backwards in the food chain.
- **Regulation (EC) 1946/2003** is concerned with the trans-boundary movement, and accompanying documentation, for LMOs (living modified organisms) destined for deliberate release, or for food and feed or for immediate processing, under the terms of the Cartagena Protocol on Biosafety.
- **Regulation (EC) 1829/2003** covers mainly the commercialization of food and feed. It facilitates GMO detection by obliging the providers of GMO plants to disclose methods for their detection (Regulation (EC) 1981/2006 provides for a fee to be paid by the applicant to the CRL for this service). These methods are then verified and validated by the EURL-GMFF, hosted by the DG JRC laboratory of Ispra (Italy) with the support of the ENGL, before being made public. This regulation imposes labeling for authorized GMOs above a threshold of 0.9%. Labeling is not required for conventional or organic food and feed containing the adventitious, or technically unavoidable, presence of authorized GMOs at levels less than 0.9%. Unauthorized GMO are not permitted entry in the EU, even at levels less that 0.9% (the so-called "zero tolerance").

> * **Regulation (EC) 1830/2003** concerns the traceability and labeling of genetically modified organisms and the traceability of food and feed products produced from genetically modified organisms and amending Directive 2001/18/EC. It imposes a specific traceability requirement on GMOs, over and above that of the general traceability regulation 178/2002. Traceability archives must be kept for five years.
> * **Regulation (EC) 65/2004** establishes a system for the development and assignment of unique identifiers for genetically modified organisms.
> * **Regulation (EC) 882/2004** on official controls performed to ensure the verification of compliance with feed and food law, animal health and animal welfare rules.
> * **Regulation (EC) 1981/2006** establishing a financial contribution on a flat-rate basis in order to contribute to supporting the costs incurred by the EURL-GMFF in the methods' validations.

Table 1. summarized overview of the European legislative frame on GMO.

After the commercial withdrawal of Event 176 maize, only 2 GM plants are currently approved for cultivation in the EU, namely the MON 810 maize and the Amflora® potato. However several other GMOs are in the pipeline for cultivation approvals, such as Bt11 maize or GTS 40-3-2 soybean. In this later case, Romania, which was cultivating GTS 40-3-2 soybean before its entrance in 2007 in the EU, is pushing hard for this approval.

In December 2008, the European council of ministers in charge of Environment asked for a reform of the EFSA approval process and for integrating socio-economic factors into the approval considerations. So far, only the French High Council of Biotechnologies integrates such considerations into its advice system through its Economic, Ethical and Social Committee[3] due to a recent law (République Française, 2008).

2.3 Labeling and traceability

According to (EC) 178/2002 regulation, traceability is mandatory in the EU for all food items, one step forward - one step backward, with additional specific requirements for GM products such as keeping traceability document for at least 5 years as described below and in Table 1.

The operation of GMO food control systems (e.g. detection, labeling and traceability methods) are not within EFSA's remit, and remain the responsibility of the EC, through the European Reference Laboratory for Genetically Modified Food and Feed (EURL-GMFF), and the Competent Authorities (CA) of individual EU-MS. It should be noted, since it is a source of frequent miscomprehension, that EC traceability and labeling regulations are not concerned with GMO safety, risk evaluation or risk management, since food that does not have a positive EFSA assessment does not reach the market. Traceability data on food and feed, including GMOs, may serve, however, to enable the re-call of products from the supermarkets in the case of unforeseen mishaps, such as the accidental or deliberate contamination of food chains. Traceability is a non-discriminatory and inexpensive requirement since most of the companies already have quality assurance protocols in place and since numerous analyses are routinely carried-out for multiple purposes, including vitamin or toxins contents. Quality assurance procedures offer several advantages to the companies such as specific market niches, efficient low-cost withdrawal of products and easier implementation of control procedures for future mandatory requirements (e.g.

---

[3] Comité économique, éthique et social (CEES)

traceability and labeling of allergens in food and feed). Fees incurred by the EURL-GMFF, for validating the detection methods by inter-laboratory trials, are on a flat-rate covered by a financial contribution of the notifying companies while a new network of National Reference Laboratories has been established beside the European Network of GMOs Laboratories (ENGL) (European Commission, 2004d, 2006a). The EC released several reports on traceability experience in the EU-MS (European Commission, 2006d, 2008b).

So far there is no European obligation of labeling animals, or their derived products, reared on GM feed despite several requests of NGOs[4] and consumers associations. However, this possibility of animal and derived products labeling has been recently introduced in Austria and Germany with a threshold of 0.9%. More generally speaking, GMO-free labeling has been introduced at 0.9% in Germany and Austria, while France is currently considering a definition of GMO-free products at 0.1% according to the recommendation of the Economic, Ethical and Social Committee[5] of its High Council of Biotechnology[6] (http://agriculture.gouv.fr/remise-de-l-avis-du-haut-conseil). GMO-free labeling is currently used by several German companies for e.g. milk which according to some claims increased their market shares (http://www.bund.net/bundnet/themen_und_projekte/gentechnik/verbraucherinnenschutz/kennzeichnung/nutzende_unternehmen/).

Since the first commercialization of GMOs in third countries, the EU has been facing a great number of alerts (Davison and Bertheau, 2007, 2008). In most of the cases, these alerts resulted from a misappropriate segregation of approved GMO (Starlink™ maize), or the seeds' commercialization of unapproved GMOs (US Bt10 instead of Bt11 maize), or the release of unapproved GMOs as in the case of US LL601 rice or Chinese Bt63 and Kefeng 6 rice. While the issues of presence of unapproved GMO in domestic markets were previously considered as an issue for countries with labeling policies, the recent increase of GMOs from emerging countries led to the reaction of the US agencies (APHIS News release, 2010; GAO, 2008). In several ways, it appears that USA will move toward a more surveying attitude similar to the EU (Davison, 2010).

The issue of domestic unapproved GMO in local market is the basis of the current work of *Codex Alimentarius* on the Low Level Presence (Codex Alimentarius, 2003). Asynchronous approvals of GMOs has been recently taken into consideration by the EC which proposed a 0.1% presence in feedstuffs of EU unapproved GMOs after a revision of EU legislative frame (Aramyan et al., 2009a; Reuter, 2010; Stein and Rodriguez-Cerezo, 2010b).

However, the recent results of the European research project Co-Extra (www.coextra.eu) show that supply chains operators already use a contractual threshold of ca. 0.1% for the 0.9% European labeling threshold (European Commission, 2010b). Together with the possible labeling of animals reared with GMOs, between 0.9 and 0.1%, and below 0.1%, such a situation will drastically impact on the availability of "GMO-free" products. The reaction of consumers toward this new European proposal remains currently unknown as *ex ante* studies appear very difficult for providing accurate results.

Generally speaking, the European traceability greatly improved over the last decade but with still several issues such as the sowing of EU unapproved Bt11 in France a few years ago. In these circumstances, the recent approval of a GM potato, specifically destined for

[4] Non Governmental Organization
[5] Comité économique, éthique et social
[6] Haut Conseil des Biotechnologies (HCB)

industrial uses, may lead to the same issue of inappropriate segregation as the US Starlink™ maize (Miller, 2010).

3. European GMO coexistence issues

3.1 General overview

The freedom to producers to either produce GM, conventional or organic crops is the counterpart and the necessary basis of consumers' freedom to choose, or not, GMO into their food. Accordingly, the EC released in 2003 and updated in 2010 a recommendation on the coexistence of GM, conventional and organic farming (European Commission, 2003a, 2010a). Practical implementation and rules is the responsibility of the EU-MS according to the European subsidiarity principle. In parallel, several European regions declared themselves as GMO-free (http://www.gmofree-euregions.net:8080/servlet/ae5Ogm) despite the fact that some do not have the administrative legality for such a positioning. The European Commission reported on implementation of coexistence rules in EU-MS which is far from being implemented in a harmonized way in all EU-MS (European Commission, 2006b, 2009e). COEX-NET is a network established to facilitate the exchange of information on coexistence issues between EU-MS CA.

Different national (French ANR-OGM, British Farm Scale Evaluation, German BMBF project, etc.) and European (SIGMEA, Transcontainer, Co-Extra) research programs were launched the last decade to establish the scientific bases of coexistence. To aid national Competent Authorities, the EC has recently created a new 'co-existence bureau' specific for co-existence issues, at JRC-IPTS[7], Seville, Spain, which recently released its first document on maize crops coexistence (Czarnak-Kłos and Rodríguez-Cerezo, 2010). If most of the current work focused on maize, currently the only plant sown, several other crops have been studied from a coexistence viewpoint, as for instance oilseed rape and sugar beet, two crops for which ferals and crossing with wild relatives are important in the EU (Colbach, 2009; Colbach et al., 2009; Darmency et al., 2009; Darmency et al., 2007; Gruber et al., 2008).

Up to now 2 trends can be distinguished in the European coexistence schemes, a flexible coexistence frame and one based on dedicated production areas, be these GMO or GMO-free.

Until now, coexistence research has mostly focused on flexible coexistence, that is to say, the individual choice of a farmer, with a minimum of *ex-ante* duties (such as isolation distances, buffer and/or discard zones) together with an information system, including, for instance, a public register of GM crops to provide information to non-GM growing neighbors coupled with some *ex-post* economic solutions such as compensation schemes for economic losses (Demont et al., 2010; Demont et al., 2009; Desquilbet and Bullock, 2010; Devos et al., 2009; Mésséan et al., 2006; Mésséan et al., 2009). Current EU best practice guidelines and companies' stewardships for coexistence measures in maize give effective measures for the European 0.9% threshold by requesting separation distances, buffer and discard zones, and staggered flowering times, but without taking into consideration the threshold of 0.1% used by companies due to measurements and sampling uncertainties (Bartsch et al., 2009; Bock et al., 2002; European Commission, 2010b).

Generally speaking, the proposed flexible coexistence solutions are based on the assumption that individual choices should prevail, but information systems need to be available to allow other producers to know what is being produced and where, such as those already

---

[7] European Commission's Joint Research Centre, Institute for Prospective and Technological Studies

deployed in Portugal. However, at the same time territory / landscape multi-functionality is requested by the EU, and there is a growing call from consumers, and society as a whole, for both more sustainable production and so called quality-oriented produce as shown by the current yearly increase of ca. 20% of organic and other signs-of-quality based farming (Laisney, 2011). This leads to a conflict as GM produce is not seen as organic, even if it can be produced without, or with less pesticide, although it can definitely be more sustainable with yield increases over conventional farming (Cardwell, 2003; Grossman, 2003; Laurent et al., 2010; Marsden, 2008). Moreover, territory organization, with Natura 2000 areas (protected environmental areas, for example), is not taken into account by the flexible coexistence scheme while their domino impact is highly recognized (Demont et al., 2008). This dichotomy needs to be addressed. Finally, the landscape is highly structured by downstream supply chains (Coléno, 2008; Hannachi et al., 2009; Le Bail et al., 2010; Petit, 2009).

Due to the several requests of EU-MS to take into consideration socio-economic aspects into GMO approval, the 2010 updated EC recommendation considers more favorably the possibility of GMO-free areas. However the EC would not accept that requests by EU-MS be based on scientific or environmental grounds which are already assessed by EFSA. This last restriction is currently actively fought at the European Parliament. As a first demonstration of EU policy change, the Portuguese Madeira archipelago was established as the first GMO-free area, though cultivation of maize is relatively scarce in Madeira (Kanter, 2010).

On the opposite side, dissemination over long distance of pollen as well as the practical effect of the contractual threshold of operators militates in favor of dedicated production areas (Brunet et al., 2011; European Commission, 2010b). However, the research work on technical, economic and societal issues raised by this solution are drastically missing and the subject of strong opposition (DEFRA, 2006; Devos, 2008; Dobbs, 2011; European Economic and Social Committee, 2011; Jank et al., 2006; Sabalza et al. 2011).

As soybean is mostly an autogamous plant, numerous issues raised by pollen dissemination should not hamper the soybean cultivation in the EU. Similarly, the absence of out-crossing to wild-relatives, ferals and volunteers should facilitate the cultivation of GM soybean into a coexistence frame. However, the predominant herbicide trait would probably cause the same problems of resistant weeds as observed in the USA (Brasher, 2010; Cerdeira and Duke, 2006, 2007; Roberson, 2010). Transportation of GMO was identified as the cause of several incidental releases in the EU and third countries, including the growth of GM plants around harbors (Kawata et al., 2009; Kim et al., 2006; Lee et al., 2009).

Thus, due to its biological properties and despite the different structure of European farms and territories, the coexistence in the EU of GM and non-GM soybeans should be one of the easiest to implement; as it is in several third countries exporting non-GM soybean despite important GM soybean cropping.

4. Surveillance plans

Post-market release monitoring of GMOs approved both for import and cultivation (European Commission, 2001, 2003c) is one of several requests included in European GMO approval. According to the pre-market risk assessment (RA), this monitoring can be divided into case-specific monitoring (CSM), which covers any identified risk, and general surveillance (GS) for all risks that might not have been identified during the RA. EFSA published a series of documents about RA, CSM, and GS (Bartsch et al., 2006a; EFSA GMO

Panel, 2004, 2006a, b, 2010). Several guidance documents and reports on implementation of the monitoring were then published (Bartsch et al., 2006b; Bartsch et al., 2007; EU working group, 2003; European Commission, 2002a, b, c, 2004b, 2008b, 2009a).

Monitoring of the GMOs post-market release should include both health and environmental effects, should be carried out by the GMO consent holders, i.e. the companies having received a grant for a commercial release of a GMO, and may be supported by additional independent actions of EU-MS. The CSM and GS shall cover both GMO and non-GMO cultivated areas (EFSA, 2008; EFSA GMO Panel, 2004, 2006a).

Up to now, most of the surveillance activities of GMOs approved for import and processing have been delegated by the consent holders to European professional unions of importers, transporters, and processors, namely COCERAL, UNISTOCK and FEDIOL. However, the content of agreements between consent holders and such unions remains unknown. Due to the lack of precision, in particular about the methodology used for monitoring imported GMOs, the accuracy of such monitoring plans remains undetermined for the EFSA, GMO national advisory committees, and CA in charge of GMOs (Beissner et al., 2006).

As noted above, health and environmental monitoring, which also means animal health, of predictable and unexpected effects of GMO cultivation is mandatory in the EU on both GMO and non-GMO cropped areas (European Commission, 2001, 2002a, b, 2003c).

4.1 Specific surveillance

In the EU, several GMO CSM protocols have been pursued by notifiers, scientists, and enforcement authorities. A decade after the first GMO cultivation in the EU, a number of guidelines, conceptual frameworks, and reports are available for GMO CSM (Bartsch et al., 2007; Bontemps et al., 2004; Bourguet, 2004; Chaufaux et al., 2002; EU working group, 2003; European Commission, 2009b; Monsanto Co., 2006, 2009a, b). The consent holders[8] published results of insect resistance monitoring, but only from GMO cropped areas despite the European rules (Monsanto Co., 2006, 2009a, b).

4.2 General surveillance

This part of the chapter focuses on the monitoring activities of unexpected effects of GMO cultivation, i.e. general surveillance.

GMO GS frameworks proposed by GMO consent holders in relation to EFSA guidelines include literature survey, development and /or use of existing monitoring surveillance, and specific trials as necessary (EFSA GMO panel, 2006b, 2010).

General surveillance is designed to detect unanticipated effects on general safeguarded subjects such as natural resources, which must not be adversely affected by human activities like GMO cultivation. Monitoring has to be appropriate for detecting direct and indirect effects, immediate and long-term effects, as well as unforeseen effects. In its 2006 opinion, the EFSA GMO panel outlined that: *"general surveillance cannot be hypothesis driven, but should, when possible, make use of existing monitoring systems in addition to more focused monitoring systems (e.g. farm questionnaires). Data quality, management and statistical analysis are of high importance in the design of general surveillance plans and comparison should be made with baseline data"* (EFSA GMO panel, 2006b). A public consultation on the 2010 version of EFSA GMO panel opinion on GS is currently ongoing.

---

[8] Notifiers having received European approval for GMO import and/or cultivation.

The 2010 draft version of EFSA guidelines shows a drastic change of paradigm in the principle of environment GS and still does not establish guidelines for health effects surveillance. This draft version particularly outlines the importance of baselines, use and assessment of indicators after field trials, less oriented biodiversity studies without *a priori*, etc. This difference between 2006 and 2010 version may represent both the change into the EFSA GMO panel composition as well as an attempt of EU-MS, of their enforcement agencies and of the EC to master and retrieve the leadership in a scientific, but also highly political, issue.

For several years now, important scientific conceptual and practical works have indeed been developed in several EU-MS along with reports from national committees in charge of GMO approvals (ACRE, 2004; Breckling and Reuter, 2006; Garcia-Alonso et al., 2006; Graef et al., 2005; Monkemeyer et al., 2006; Wilhelm et al., 2009; Wolt et al., 2010; Zughart et al., 2008).

Most of these scientific works focused on environmental effects, while the effects on human health are roughly "delegated" by the consent holders to national health monitoring networks (Bakshi, 2003; Cellini et al., 2004; Covelli and Hohots, 2003; D'Agnolo, 2005; EFSA GMO panel, 2006b; Filip et al., 2004; Hepburn et al., 2008; Wal et al., 2003). To provide an example of EU-MS, in France, the "Sentinelles" network, ANSES[9], and InVS[10] might form parts of such a general surveillance plan on human health in application of the WHO and European rules, directives, and regulations. Animal health is relevant to the OIE[11] and European rules, directives, and regulations. As for GMO CSM and GS, the French Ministry of Agriculture (DGAl directorate) is in charge of animal health surveillance. However, no GMO-related GS activities in human and animal health are clearly identified in the European activity reports.

Indeed, GS of human and animal health is also particularly important given that GMOs not dedicated to food and feed purposes will rapidly arrive on the market as exemplified by the recent European approval of Amflora® potato for cultivation. For this kind of split approval, we must remember the first such issue raised by the incorrect segregation between food and feed/ industry storage facilities of the USA-approved Starlink™ maize (Alderborn et al., 2010; Beckie et al., 2010; Miller, 2010). Despite the past European experience of segregating crops dedicated to industrial uses as part of a specific derogatory cultivar list, the additional recent request Modena GM potato cultivation approval in the EU can lead us expect that more and more GMO dedicated to industrial use will enter the food chain and raise new controversies about human health.

From a decade of GMO cultivation in the EU, several remarks can be made about environmental GS reported by consent holders, scientists, and enforcement authorities.

• The consent holders include a literature survey and questionnaires to GMO cropping farmers and collaboration with existing networks in their environmental GS, as was done in Germany after the German Competent Authorities (CA) approval. However, in that latter case, a great deal of imprecision remains about the content of agreements with existing networks, the network's possible training, and the surveyed locations, i.e. representativeness and accuracy of the GS, particularly in non-GMO areas. Moreover, no statistics are provided which might alert the CA to start more in depth monitoring.

---

[9] Agence Nationale de Sécurité Sanitaire
[10] Institut National de Veille Sanitaire
[11] Office International des Epizooties

- Despite the positive EFSA assessment of the consent holders' monitoring reports, we can observe that, to date, the consent holders do not include non-GMO areas that are in practice delegated to the responsibility of the EU-MS (Alacalde et al., 2007; Lecoq et al., 2007; Monsanto Co., 2006, 2009a, b; Tinland, 2008; Tinland et al., 2007; Wandelt, 2007; Windels et al., 2009). Accordingly, several EU-MS have already, or are planning to, launched GS research projects and networks even in those EU-MS with bans on GMOs (Bartsch et al., 2009; Breckling and Reuter, 2006; de Jong, 2010; Gathmann, 2009; Gathmann and Bartsch, 2006; Pascher et al., 2009). This survey of non-GM cropped areas is however of utmost importance as recently shown in China (Lu et al., 2010). Unfortunately, we can again observe that private interests and benefits are supported by public funding when general goods are concerned, as usual in the "Tragedy of Commons" frame (Hardin, 1968; Hardin, 1998).

Most current environmental GS plans focus on changes in *ex ante* baseline and / or biodiversity assessments, sometimes along with a general approach looking at the effect of agricultural practices (Hintermann et al., 2002; Monkemeyer et al., 2006; Sanvido et al., 2007a; Sanvido et al., 2007b, 2009a; Sanvido et al., 2009b; Schmidt et al., 2009). However, the conceptual framework for environmental GS is far from being both a consensus and a reality. This situation motivated the European Commission to launch a call for proposals in 2010 (KBBE.2011.3.5-01) that address environmental GS and possible standardization (ACRE, 2004; Beismann et al., 2007; EFSA GMO panel, 2006b; Finck et al., 2006; Ostergard et al., 2009; Pascher et al., 2009; Sanvido et al., 2005; Schiemann, 2007; Seitz et al., 2010; Wilhelm et al., 2003; Wilhelm et al., 2009; Wilhelm and Schiemann, 2006).

The main conclusion that can be drawn from the current situation is that, despite the mandatory involvement of GMO consent holders into GMO GS and the monitoring of non-GMO areas, the main effort appears to be supplied by the EU-MS.

However, a number of environmental monitoring procedures are already in place in the EU, several of which partly embrace – episodically or on a longer term - biodiversity, GMO CSM, "epidemio-surveillance", or more general effects of agricultural practices on agro-environment. In several instances, these monitoring schemes are carried out by citizens in a so-called participatory science. These trends might be correlated with another trend for observing territory from societal and economic viewpoints (Barzman et al., 2005; Bodiguel, 2003; Cardwell and Bodiguel, 2005; Henle et al., 2008). Networks of citizens and/or scientists, as well as enforcement authorities already working on these issues, all have in common (i) a need for long-term studies, (ii) different demands on space and changes over time, (iii) different indicators which (iv) generally have to be reported to national CA and EC, sometimes according to international treaties. But up until now, results appear fragmented, collated in different databases generally without quality assessment or direct connection through a unique Web-based portal or automatic novelty detection capacity (Haggett, 2008).

Nonetheless the environmental liability directive and the right of European citizens to have access to environmental information reinforce the need for gathering these fragmented data (Cardwell, 2010; Ebert and Lahnstein, 2008; European Commission, 1985, 2003b, 2004a, c, 2006c, 2007b).

In the case that GM soybean cultivation would be allowed in the future, there are thus numerous issues that should be fulfilled by consent holders, particularly for herbicide tolerant crops whose uncontrolled use in third countries leads to numerous herbicide resistant weeds and costly companies' based eradication programs (Adams, 2011; Brasher, 2010; Cerdeira and Duke, 2006, 2007).

5. Soybean in the EU

5.1 A rapid historic overview of the last decades

In the Dillon round of GATT negotiations (1960-62), the EEC[12] negotiated zero duties on soybeans and several other agricultural products. At the time, soybean was of little importance in international trade with ca 4 MT traded in 1961. Furthermore, there were no varieties of soy available at this time that could be grown in this EU-6. Thus the EU-6 had no producers to protect and found in their interest to keep borders open to soybeans and their products. At that time, pasturages, cereal and some domestic protein rich crops provided most of the necessary feed. That period was the beginning of a drastic change into the European livestock production.

However, the high European internal costs of feed grains forced livestock producers to substitute cheaper soybean meal. In addition to the competition between European feed grains and imported soybean meal, soybean oil competed with domestic vegetable oils such as sunflower oil, olive oil or rapeseed oil, and, when used in margarine, competes with butter. To compete with cheap soybean oil, local oils were subsidized, which was attacked under GATT in 1987. In 1973, a shortage in US soybean exports impacted most of the current EU-MS, including France which was considered as the least soybean dependent EU-MS (Berlan et al., 1977; Hasha, 2002). As maize and soybean compete for both feeding and surface, this kind of shortage is expected to come back with the growing use of maize for bio-ethanol production (Headey, 2011). Several EU-MS attempted to reduce their growing dependence from soybean by national protein plans – but up to now unsuccessfully.

The European cropping of soybean was, up to 2007, restricted to some EU-MS and aimed at food or a few feed specialties (e.g. organic) with most of the production being based in Italy. Among the several reasons why EU is not a soybean producer we can distinguish a relatively unfavorable climate with cool spring and drought early summer, with a Northern predominance, in the EU compared to third countries producing soybean and a relatively high population density with rather small farms and fields. However, several soy varieties are cultivated in Canada and thus soy cropping in the EU-27 would now be possible after appropriate selection of cultivars, provided the seed companies could find some benefit in that selection. This would be probably the case after GM soybean approval for European cultivation.

The EU was in 2007 still under construction and two new countries coming from the implosion of the former soviet bloc entered the EU-25. At its entrance into the EU, Romania officially stopped cropping GM soybean and came back to old varieties of non-GM soybean whose cropping was also not favored by the current European Common Agricultural Policy (CAP; Badea and Pamfil, 2009; Dinu et al., 2010). However, the interest in GM soybean was declining from 1996 to 2002 (Brookes, 2005).

5.2 Soybean use in the EU

As most of the crops choices made by farmers over the last decades, soybean cultivation is highly linked to the several changes in CAP. Among those more related to soy cultivation (i.e. linked to oilseed and protein rich crops) we can notice the changes due the Blair House agreement, in 1992, for duty free soybean importation and the Berlin agreement, in 1999, for decreasing aids to oilseeds and open widely the European market to global trade.

---

[12] Economic European Communities or EU-6

soybean imports	Soybean exports	Soybean crushing	Soybean oil and fats production	Soybean oil and fats imports	Soybean oil and fats exports	Soybean oil and fats consumption	Soybean meals production	Soybean meals imports	Soybean meals exports	Compound feedstuffs production (FEDIOL)	Compound feedstuffs production (FEFAC)	Compound feedstuffs production in 2009 (FEFAC)	Compound feedstuffs turnover' (FEFAC) in 2008, M€
NL (4,004)	NL (1,079)	DE (3,364)	DE (643)	FR (547)	NL (633)	DE (591)	DE (2,676)	NL (5,236)	NL (4,769)	FR (22,424)	FR (22,679)	SP (21,549)	FR (8,450)
DE (3,485)	BE (188)	SP (3,026)	NL (572)	DE (249)	SP (323)	FR (556)	SP (2,388)	FR (4,399)	DE (1,491)	DE (21,825)	SP (22,117)	FR (21,236)	SP (6,804)
SP (3,279)	SL (84)	NL (2,898)	SP (549)	NL (223)	DE (301)	IT (415)	NL (2,257)	DE (3,573)	BE (782)	SP (19,220)	DE (21,825)	DE (20,829)	IT (6,500)
IT (1,640)	DE (47)	PT (1,170)	IT (259)	UK (187)	BE (88)	SP (317)	IT (1,232)	SP (3,385)	SL (697)	NL (14,507)	NL (14,643)	NL (14,108)	DE (6,290)
PT (1,190)	IT (39)	GB (559)	PT (199)	BE (185)	PT (68)	UK (279)	PT (930)	UK (2,216)	PT (247)	UK (14,222)	IT (14,349)	IT (13,860)	NL (4,750)

With an important increase since 2006 due to the food crisis.

Table 2. 2008 figures of soybeans, oil fats and meals' import, production and processing (sources: Fediol, 2011 and FEFAC, 2011). BE: Belgium; DE: Germany; FR: France; IT: Italy; NL: The Netherlands; PT: Portugal; SP: Spain; SL: Slovenia; UK: United Kingdom.

In 2006, the EU imported mostly soy beans and also meal. Crushing capacities have been developed in the Netherlands and Germany. As The Netherlands, Germany and Belgium have important harbors to where the most important part of EU imports of soybeans and meals are discharged, a large part of their imports are re-exported either as beans or more generally as meals and oils. Indeed, The Netherlands are currently a net exporter of soy meal toward other EU-MS. However, other European oilseeds compete with soybean. Generally speaking the proportion of rapeseed crushing is in constant increase over the last years, rapeseed crushing having overtaken soy crushing in 2005 due to the use of oilseed rape in European bio-fuels production. The table 2 provides the figures in 2008 of the soy beans, oil, fats and meals production imports and exports for the first top 5 EU-MS.

The table 3 provides a figure of soy beans and meals imports in the EU from 1980 to 2008. As it can be seen, European soy beans and meals imports are relatively stable, when compared to prices and yearly weather variations, since 2004, i.e. with the integration of 10 and then 2 new Member States into the EU in 2004 and 2007, respectively.

	2008 EU-27	2007 EU-27	2006 EU-25	2005 EU-25	2004 EU-25	2003 EU-15	2002 EU-15	2001 EU-15	2000 EU-15	1990 EU-12	1980 EU-9
Soy beans	15,298	15,064	14,127	14,670	14,732	17,353	18,239	17,922	14,779	13,301	11,760
Soy meals	23,227	24,321	23,405	23,029	22,632	20,352	19,605	17,870	15,840	10,471	7,226

Table 3. Imports (× 1,000 T) of soy beans and meals in the EU (source: Fediol, 2011).

5.2.1 Non-food non-feed use

Compared to the other uses of soybean the use of soybean, with or without chemical changes, in printer inks, as antifoam agent to bio-fuels and cosmetics are currently rather anecdotic (Gelder et al., 2008; Roebroeck, 2002). For instance most of the European sources of bio-fuel rely on oilseed rape. This part will probably increase with the new CAP reform favoring sustainable and environment friendly agriculture and supporting renewable, low carbon emitting energies sources.

5.2.2 Food use

Only a few percent of soybean is used for food purposes (Gelder et al., 2008). Lecithin and oil are the main products used in food, the latter being also used in margarine production, together with some specialties such as some kind of yogurts, vegetarian steaks, or the usual Asian specialties such as Tofu, nato (Roebroeck, 2002). Soy milk is mostly imported from Canada, by some worldwide companies.

5.2.3 Feed use

Soybean is mostly imported in the EU-27 for compound feedstuffs production (Popp, 2008). With the end, in the 1990's, of European intervention on cereals, which were used with some soy meal for compound feedstuff production, together with the ban of meat-and-bone meal from most of the feedstuffs (in fact meat-and-bone meal, despite being mostly destroyed, continues to be used in feeding short-living animals such as chickens and fish), the European feedstuff industry was looking for another source of cheap source of protein.

However, despite the end of European aids on cereals the proportion of cereals into compound feed increased since 1995, while the proportion of meals of all origins fluctuated between 30 Mt and 40 Mt.

Compound feed consumption in the EU-27 represents ca. 147.6 Mt, a quantity thus similar to the US consumption of 149 Mt, with a nearly constant percentage of the global consumption over the past decade. This compound feed consumption was accompanied by an increasing production of pig meat and poultry to be compared to a constant beef and veal production, a difference which is mostly due to the entrance of new EU-MS in 2004 and 2007 (FEFAC, 2010; European Commission DG-Agri, 2010). However, the increase in European meat production is parallel to a general trend of decrease (beef, veal, pig) or stagnation (poultry) of European meat consumption. In 2009, the proportion of compound feed for animal rearing again decreased to ca. 30% of the total feedstuffs quantity which corresponds to a general change into the European livestock production schemes.

The soybean meal is used for all animal feeding, particularly since the ban in 2001 of meat-and-bone meals due to the mad-cow / BSE disease, with an exception for organic production or some animals growing under specific signs-of-quality. This important source of protein cannot be fully replaced by fish meal which was another reason for increasing the imports of protein rich commodities. Due to its high content of protein and relative poorness in fat, the soy meal is relatively difficult to replace in poultry, piglets and calves feeding. Alternative sources of protein such as sweet lupine, field pea or rapeseed meal are generally less palatable until the animals reach maturity. This explains the figure of compound feed mostly used for poultry and pig production (FEFAC, 2011). The issue of protein source is of less importance for mature animals and more particularly for cattle.

The origin of imported soybean may depend on EU-MS, for instance France mostly imports soy from Brazil while The Netherlands and Portugal are the top 2 importers of the US exports of soybean (Dahl and House, 2008). Up to 2008, EU was the first destination of exported soybean from USA, Brazil and Argentina. The European protein crops imports represented in 2009 ca. 20 Mha cultivated outside Europe.

However, the development of GM crops in the 3 main exporting countries definitely impeded exports, particularly in the US. Brazil, up to now, took into account the undesirable effect of asynchronous approvals of GM crops on its exports toward the EU (Aramyan et al., 2009a; Aramyan et al., 2009b; Boshnakova et al., 2009; DG AGRI European Commission, 2007; Dobrescu et al., 2009; Konduru, 2008; Stein and Rodriguez-Cerezo, 2009; Stein and Rodriguez-Cerezo, 2010b).

5.2.4 The animal labeling issue

As the animals fed with GMOs do not have to be labeled in the EU, most of the feedstuffs in the EU-27 is produced from GM soybean. However several NGO and consumers associations are requesting such animal labeling, a request supported by polls and experimental auctions studies (Kontoleon and Yabe, 2006; Noussair et al., 2004). The EU organic farming threshold of labeling is also of 0.9% (European Commission, 2007a, 2008a). However, this EU threshold can be superseded by national measures. More generally speaking the EU has numerous signs-of-quality, based, not only on brands as in third countries, but mostly on EC-approved processes or origins. The consumers' reluctance was thus taken into account in production procedures of most of these quality signs by eliminating GMO use into feedstuffs.

Germany and Austria recently introduced a legislative frame for GMO-free labeling. In these countries the GMO-free threshold complies with the European 0.9% threshold of fortuitous or technically unavoidable presence of GMO. This labeling is applicable to both vegetal produce and animals reared with "GMO-free" feed.

In another hand, several French producers, such as *Poulets de Loué*, or retailers, such as Carrefour, or quality signs producers such as *Comté* cheese, used non-GM (Identity Preserved, at 0.9%) soybean since the beginning of the XXIst century but without the possibility of retrieving profits of their efforts (Milanesi, 2008, 2009). In 2009, the French *Conseil de la Consommation* as well as the *Haut Conseil des Biotechnologies* (HCB) released recommendations for the creation of a GMO-free supply chain at 0.1%, with, in the latter advice, animal labeling according to 2 thresholds: below 0.1% and between 0.1 and 0.9%. Despite the fact that the French decree related to GMO-free labeling is so far not published, several producers and retailers took this opportunity, and the further policy change of the French Repression of Fraud services, to label their animals as being reared with less than 0.9%. If the latest HCB recommendation is followed up by the French government, 2 kinds of GMO-free animal labeling would thus prevail in France: "reared with GMO-free feedstuffs below 0.9%" and "reared with GMO-free feedstuffs below 0.1%". The HCB also requested into its recommendation that the French government should precede the implementation by an *ex ante* socio-economic analysis of the viability of such a GMO-free supply chain at 0.1%. A feasibility study on this request for a *ex ante* socio-economic analysis is currently ongoing.

However, the availability of non-GM (including GMO content below 0.9%, IP, and "GMO-free" at 0.1% or "hard IP") soybean is far from being sustainable. Up to 2008, Brazil was the most important exporter of non-GM and GMO-free soy toward EU with negotiated premiums. But the breeding of new soy GM varieties more appropriate to Brazilian climate induced a new increase of GM surfaces in Brazil, particularly in Matto-Grosso with the largest farms and fields (Fok, 2010). Despite the fact that Parana state dedicated a whole harbor to non-GM soy, this state also moved, toward GM soy, particularly for the most weedy fields and by the farmers the less experienced into weed management. The main source of European non-GM soy could thus disappear unless operators facilitate the maintenance of non-GM cropping.

One of the first issues, for maintaining the interest of non-GM cropping, is the rather low level of premium (ca 1/4th of the final one paid by final buyers) received by the Brazilian farmers. The second is that this non-GMO related premium is not discerned by the buyers, such as cooperatives, from other premiums, all premiums being thus provided into a non transparent package of several premiums. The incentive of producing non-GM soy is thus rather low in Brazil despite the fact that the tech fees imposed by the traits' providers may be ca 40% of the seeds prices (Bonny, 2009; Fok, 2010). According to ABRANGE, a Brazilian association of non-GM farmers, Brazil would be however currently providing 53% of non-GM soybean while India and China would be providing 18 and 17%, respectively (Milanesi, 2011). However, these claims are not in line with the observation of the 2009 increase of non-GM soybean in USA, after a decade-long decrease of non-GM soybean areas, due to both more incentive premiums for non-GM beans and increased production costs of GMOs (prices of seeds and herbicide) (Milanesi, 2011).

With premiums, long-term contracts are the second driving force for farmers for maintaining what several authors call market niches (Foster, 2007, 2010). Long-term contracts have thus been established by European producers, such as *Poulets de Loué*, with or without the support of the European GMO-Free regions and Brazilian producers such as the Brazilian ABRANGE association. Generally speaking, the European GMO-free regions' network supports their producers into their search for long-term supply of "cheap" non-GM soy.

The premiums ranged from an average of 16 US$ in 2004 to ca. 70 US$ in 2009 for US farmers (Foster, 2010; Milanesi, 2011). While it is generally difficult to determine the premiums fluctuations over the year due to the confidentiality of the contracts, the changes observed into the non-GM soybean market of Tokyo show a general trend of a 10% premium over the GM soybean, with of course important peaks up to 40% in 2008 due to both the food crisis and an decrease of US 2007/08 soybean cropped surface in that year (Foster, 2010; Headey, 2011). Compared to the "only" 83% of soybean price increase over 30 years, such premiums could thus be very incentive, particularly when linked to long-term contracts.

The increase of price of compound feed due to this non-GM soy would be of less than 3% (Gryson and Eeckhout, 2011). Since feed cost represents ca. 77% of price of chickens, this would induce a final small increase of some cents per chickens' kilo (Milanesi, 2008, 2009).

Most of the European imports of both GM and non-GM soy are through the main commodities traders namely ADM, Bunge, Cargill and Louis Dreyfus (Green and Hervé, 2006). However, as these traders advertised they were facing a shortage of non-GM soy, several new SMEs, such as Solteam, are currently developing their own import network to provide European feed producers with non-GM commodities. With the growing surfaces of GM soybean in Brazil, alternative sources of non-GM soy are actively looked for beside long-term contracts and premiums use for sustaining the availability of American non-GM soy.

However, the main forthcoming issue might be the availability of low cost non-GM soybean varieties both in third countries and EU-MS (Milanesi, 2011; Then and Stolze, 2010). As currently observed, the availability of non-GM seeds is decreasing with a few new varieties being commercially released. Accordingly, old non-GM seeds cannot compete with new GM varieties what can explain, together with a lack of support of oilseeds by the European CAP, the dramatic decrease in yields and total production observed in Romania at its entrance in 2007 in the EU-27 (Dinu et al., 2010). Despite the fact that new but small plant breeder and seeds sellers (such as eMerge a Cargill subsidiary) are appearing, their ability to access to soybean germplams is questionable as private sector is focusing on GM varieties and public research is focused on germplasms (Heisey et al., 2005; Heisey et al., 2001; Naeve et al., 2010; Orf, 2004). The availability by the big seeds companies of non-GM soy varieties will highly depend on their forward or backward breeding strategies (Milanesi, 2011).

It may worth noting that while traders such as Cargill are developing such seeds companies which will help them to maintain the non-GM flows towards several importing countries: competing on global commodities trade does not mean, for such companies, excluding higher added value market niches. In the meantime, participatory breeding of non-GM soy varieties is also developing as observed for numerous other crops (Bellon and Morris, 2002; Desclaux et al., 2008; Smale, 1998). As noted by several authors, this availability of several kind of varieties is necessary for developing the segmented markets requested by farmers, retailers and consumers (Elbehri, 2007).

5.3 Soybean cultivation in the EU-27

As noticed by a recent motion of the European parliament, protein rich crop production occupies only 3% of EU-27 arable land and supplies only ca 30% of protein crops consumed as animal feed (LMC International, 2009a, b; Häusling, 2011). Table 4 provides figures on some EU-MS surfaces of soy cultivation (Eurostat, 2011).

With the *Agenda 2000* CAP reform, aid to European farmers became decoupled, i.e. aid were no longer received for oilseeds production, nor related to yields. There is thus no more European intervention for buying, export subsidizing or other market support available for oilseeds in the EU-27. Moreover, agricultural aid is now rather linked to environment preservation and sustainable agriculture, the second pillar of the new CAP, together with social criteria (Krautgartner et al., 2010b). These drastic changes into the European CAP could lead to drastic changes during the next years in the European cultivations' schemes.

	2010	2009	2008	2007	2006	2005	2004	2003	2002	2001	2000	1999
Italy	159	134.7	107.8	130.3	177.9	152.3	150.4	152.1	152	233.5	252.6	246.5
Romania	65.2	48.8	49.9	133.2[13]	190.8	143.1	121.3	128.8	71.8	44.8	117.	99.8
France	50.9	43.7	21.8	32.4	45.3	57.4	58.6	80.7	74.8	120.9	77.7	98.2
Austria	34.4	25.3	18.4	20.2	25	21.4	17.9	15.5	14	16.3	15.5	18.5
Hungary	33.5	31.5	29	32.9	35.9	33.6	27.3[14]	30.3	25	20.6	22.2	32.2

Table 4. Surface (× 1,000 ha) of soybean cultivation in the top 5 EU-MS (Eurostat, 2011).

In the Western part of the EU-27, soy cultivation attempts to satisfy the needs of high added values supplies such as food and meat production under signs-of-quality. However, in spite of several "protein plans", soy is still not considered as an important European crop.

While Western Europe was poorly considering soybean cultivation yet started to import soybean since ca. the second decade of the XXth century, the former USSR developed a soybean breeding institute since the beginning of this last century. As a result of Russian research, several soybean varieties were developed for the former Soviet bloc. An important area of soybean cropping is thus done in the eastern part of Europe, around the southern Danube basin, in particular in Romania, but also in Bulgaria and Hungary. As another example of such Soviet soybean production, and thus of varieties adapted to the European climate, Ukraine and Russia were cropping in 2006 725,000 and 810,100 ha, respectively (Otiman et al., 2008). However, Romania was the only country to extensively grow GM soybean over ca 137,000 ha for feedstuff production but with yield per ha nearly 2/3rd of the ones of USA, Brazil and Argentine. While irrigation is important for increasing the yield, it is also highly probable that the GM varieties were not fully adapted to the European eastern conditions.

Since its entrance in the EU in 2007, Romania stopped cultivating of GM soybean, but started with the MON 810 GM maize (Badea and Pamfil, 2009). Since that time Romanian farmers claim that after a period of self-sufficiency in feed, they had to import again soybean for livestock (Otiman et al., 2008). However, the decrease in soybean cultivation is more probably linked to the absence of European subsidies to soybean cultivation and a return to old, less productive, non-GM soy varieties. Romanian farmers are thus among those pushing to force the European approval of GM soybean cultivation, whose dossier is currently in the European approval pipeline. This may also be explained by the existence, in several Central and Eastern European Countries (CEEC), of large, corporate, farms up to 20,000 ha, inherited from the reforms after the Soviet bloc implosion, which face the same

---

[13] Integration into the EU
[14] Integration into the EU

weed management issue as the large farms of third countries (Csaki and Lerman, 1997; Eurostat. European Commission, 2010; Pouliquen, 2001).

Coexistence between GM and non-GM soy would probably not be an issue as soy is mostly autogamous as soon as the European seeds' threshold for non-GM seeds is defined (Roebroeck, 2002). However, case specific monitoring will be an important and costly workload for herbicide tolerant soybean cropping in order to avoid the issues of herbicide resistant weeds observed in the USA and eradication programs paid by companies (Brasher, 2010; Owen et al., 2010; Owen, 2009; Roberson, 2010).

As observed by most of the scholars and policy makers, European farming is highly dependent on CAP (Carlier et al., 2010; Cavaillès, 2009). Accordingly, European soybean cropping is currently only driven by global market prices. As finally observed by a recent EC sponsored study, a soybean shortage, such as the last US one in 2007, and thus an increase into soybean prices might induce reallocation of European arable surfaces toward soybean cultivation and probably allow several EU-MS to become self-sufficient. In this way, soybean might be considered as an opportunistic crop by European farmers driven by global soy prices, particularly for its non-GM counterpart. However, the European farmers have to "internalize" soybean cultivation into their agricultural practices and productions particularly in the EU-MS of Western Europe where soybean is not a familiar crop. For instance, farmers of Alsace region in France recently introduced a maize / soybean rotation as a tool to fighting Western corn rootworm. This "internalization" of rotation with soybean into maize culture may be rapid as the French government recently issued a decree making rotation mandatory.

Several recent changes in the CAP have to be kept in mind. In particular due to substantial reductions of aid in several agricultural sectors, European farmers have a closer look on the impact of the global commodities market on their sales prices with an increased trend toward crops' futures markets. In the meantime, farmers need to have new considerations toward the environment; the reduction into available chemicals for pest fighting, the carbon footprint, the multi-functionality of agriculture, etc. (Commission, 2006; European Commission, 2009d; Kaditi and Swinnen, 2006).

It is thus predictable that soybean cropping will differ from East to West among the EU-27, with probably GM crops in the eastern part, which is comparable to the gradient of sensitivity to GMO issues as observed for consumers (Consumerchoice Consortium, 2008).

5.4 Alternatives to imported soybean

Two considerations structure the soy importation issue: the first considering feedstuff production with GM soy and the second considering the use of non-GM soybean. Indeed alternative protein-rich crops did not succeed in the previous national or European "protein plans" and are less palatable to poultry and young pigs.

5.4.1 The issue of asynchronous approvals

As previously said, EU generally takes more time for approving GMO than several third countries, in part due to incomplete dossiers but probably also because of the EU-MS unclear economic interest of GM crops (European Commission, 2011).

The import of GM soybean is thus affected by this approval status as reported by numerous reports (ADAS ltd (for DEFRA feed import project), 2008; Aramyan et al., 2009a; Aramyan et al., 2009b; DG AGRI European Commission, 2007; Nowicki et al., 2010; Stein and Rodriguez-

Cerezo, 2009; Stein and Rodriguez-Cerezo, 2010b; Stein and Rodríguez-Cerezo, 2009; Tallage, 2010). All these reports concluded there is no alternatives to imported soybeans and meals, and thus recommended establishing a specific Low Level Presence threshold for EU unapproved GMOs to avoid any shortage in soybean which could hamper European livestock competitiveness, a recommendation recently officially taken in consideration by the EC by establishing a LLP threshold for EU unapproved GMOs dedicated to feed.

However, how reliable are these converging predictions? It might be helpful to have some insights on some recent reports on such issues of feed shortage in the EU-27. Are there some biases in those studies which are almost all based on modeling of feed use?

Models are clearly needed for simplifying complex situation and decision taking. In that way, it is thus understandable that models are used for forecasting international trade and soybean use in feedstuff production. However the choice of model or postulates, such as linear regression and "general equilibrium" instead of alternative is not neutral. Besides this essential questioning, common to all modeling issues, we will just examine some contextual questions.

- The first observation we can make is that those studies were carried out with limited funding in short time; thus impeding long studies and collective, contradictory expertise. Another problem is the use of very recently developed models, used by the EC without having been clearly in depth validated, or used out their scope, e.g. to foresee future trends while developed to analyze the past (Harrell, 2001). Some of these models were "validated" by discarding some crops in some parts of the evaluation but taking them into account into other evaluation parts. As these crops are used as adjustment variables in substitution strategies for low-price compound feedstuffs production, the validity of such models is highly questionable. The same issue of validating models applies to models attempting to merge ecological and CAP issues. It is finally rather surprising to read in a report about the development of a model, that one of the main goals was to simplify the yearly feeding of the model due to some lack of personal in the corresponding European Commission service. Avoiding complexity and simplifying the life of European personal does not help make sound policy.

- The large use of modeling is the expression of a general disinterest of economists for empirical studies and a preference toward modeling. This disinterest of economist scholars or consultants is due to (i) the difficulties to retrieve accurate information from companies and interviews, (ii) the duration necessary for establishing their own data-bases, together with (iii) a higher ranking in peer-reviewed journals for models, rather than for empirical data. There is thus a fundamental lack of sound, scholarly-established empirical data, i.e. not provided only by the companies in charge of feed production, before founding policy on models.

- A "business-as-usual" trend, i.e. a relative poorness in investigated scenarios and generally speaking in perspectives and alternatives. All considered scenarios take as read the requests of feed producers, i.e. the need for soybean and more generally proteins imports; just as previously cereals and then meat-and-bone were supposed to supply all needed proteins. This "business-as-usual" trend may in great part be explained by the power of lobbies, some kind of blindness, i.e. lack of prospective. However, it is generally recognized that a mass market is always turning into a market of niches and that EU is among the largest provider of market niches (Anderson, 2006). Such models are thus *inter-alia* not referring to market differentiation, European landscape use, consumers welfare, region competitiveness and ecological issues as requested by the second pillar of the new CAP (Hermans et al., 2010; Kissinger and Rees, 2010; Konefal and Busch, 2010; van Delden et al., 2010). As outlined by Konefal and Bush, maize and soybean market standardization also introduced a multiplicity of segmented markets which were not taken into these models.

- Short and over-simplified studies, as generally the policy makers need rapid results and the academic community is rather slow to mobilize for participating in such applied work. Thus calls for this type of studies are generally awarded by consultants' cabinets or by the few scientists having already worked on that issue and thus able to "reinvest" their initial work. Work is thus mostly desk-search with several biases, such as a more difficult access to the scientific literature, a large use of "Google" which highlights URL according to a Google ranking algorithm mostly based on the number of external links or sponsorship, thus a way of working which favors industry and lobbies reports and sites.

- Group "consanguinity": such a strong relationship between sponsors, for instance a technical officer in charge of supervising an European study, originating from a European institute, whose recent reports all biased in the same way the effect of EU unapproved GMOs on feed availability. This first type of consanguinity is then reinforced by the tenders who have been chosen after a call for tenders. In most of the instances, the scientists have published reports in the same way, e.g. the dramatic effect of EU unapproved GMOs on feed availability. Reinvesting initial work is clearly not the best way for sound prospective in comparison of a collective and contradictory expertise.

- Influence of working environment. Indeed several studies were carried out in an EU country highly depending on feedstuffs ingredients importations, and further re-exportations, which may hamper the independence of viewpoint of the scientists and criteria retained for the scenarios. This may also be linked to previous studies funded by the feed industry which may influence the viewpoints and future results. Again, it would be necessary to amplify the panel of viewpoints, e.g. with scientists from countries with different production schemes. By not taking into consideration the socio-economic context and history of some scientists, the EU is decreasing its chances to find a systemic and long-term solution.

Taken altogether, these considerations of the studies and modeling context show the limits of the available data and predictions. This militates for more scholar-driven, long-term, multidisciplinary and with people sharing different viewpoints about futures of agriculture into collective expertise using also different modeling and postulates bases. Rapid, biased studies for a very complex matter highly influenced by both uncontrolled events, (such as seasonal incidents, or policy controlled issues, such as the disappearance of fallows in the new CAP, further cultivable areas reallocations and integration of new players), are not the best conditions for forecasting the future of European agriculture needs.

The users, i.e. policy makers, should be aware of the strengths and weaknesses of models used by the technical officers, what is generally not the case in the reports provided to the policy makers or media (van Tongeren et al., 2001). To conclude, the over-simplified models, developed moreover under detrimental contextual influences, have drastic limitations for forecasting trade and supply chains trends but are routinely used and dramatically impact the European policy without sound "scientific" ground-bases. These observations together with other not reported in this paper show that the EC was in fact over the last years attempting to "scientifically" legitimate previous political decisions for "smoothing" global trade issues.

5.4.2 Perspectives

In spite of a careful survey of European scientific and grey literature on alternatives to GM soy, almost no one Western EU stakeholder involved into meat production is currently considering soybean cultivation in the EU as a solution. Beside some recommendations to come back to pasturage for cattle, the general trend in compound feedstuff production is a larger incorporation of European non-GM rapeseed meal as it can be seen in the statistics of Fediol as a by-product of European bio-fuel production. Substitution is thus generally retained in national schemes for non-GM soy use. However new Eastern EU-MS have a long tradition of soy cultivation with some very large farms which might, in the 'business-as-usual" trend benefit from GM, or non-GM, soy cultivation.

5.4.2.1 Domestic substitution to imported soybean

Two ways of substitution of imported soybean have to be considered: firstly, the European cultivation of soy, as this protein-rich feed is difficult to substitute in feedstuff of several young animals and, secondly, the replacement of soybean by some other protein-rich crop for adults or some young animals.

As observed in several studies the trend over the last decade to use low-price soybean and soy meal induced a clear disaffection of plant breeding companies for leguminous fodder crops (alfalfa, clover, etc.) and several protein-rich crops (field pea, sweet lupine, etc.) due to their small volumes and a constant decrease of cultivation over the last decade (European Parliament. Directorate-General for Internal Policies, 2010; LMC International, 2009a, b; Häusling, 2011). Moreover, public research programs on such European substitutes to soybean declined over the 3 last decades. As a example of such general decline, domestic leguminous crops to be incorporated into feed dropped in France from 11% in 1991 toward 2.5% in 2006, despite CAP subsidies of field pea, field bean and lupine (European Commission, 2009c). If some studies are currently ongoing, for instance on the use of lupine and pea for poultry, there is a lack of European research on substituting soybean by domestic protein-rich crops which however present the interest of currently being non-GM (Laudadio et al., 2009; Häusling, 2011).

Several changes in the CAP such as the "20-20-20 in 2020" objectives (reduction of 20% of emissions from 1990, 20% share of energy consumption from renewable sources and 20% improvement in energy efficiency by 2020) conducted to an increase of rapeseed oil production for bio-fuels and thus of alternative meal, at least for some livestock.

Under the Blair House agreement, oilseeds plantings were limited to an adjusted Maximum Guaranteed Area for producers benefiting from crop specific oilseeds payments. This limited the EU oilseeds production area and penalized overproduction till the 2003 renegotiation of Blair House agreement. Finally, with the *Agenda 2000*, the CAP relies on 2 pillars: the market and income policy (first pillar) and the sustainable development of rural areas (second pillar). Since 2010, the producers are free of the hectare limits set out by Blair House agreement. Additionally, the disappearance of European mandatory fallows is freeing new arable surfaces for, current or new, long-term or opportunistic cultivations.

As the 2003 CAP reform (linked to renegotiation of Blair House agreement) brings greater consideration to environmental integration we may expect several changes, in particular about soybean whose production in Brazil is criticized due to deforestation, an important use of chemicals and social impacts (Carlier et al., 2010). The recent reduced European interest for biofuels production in the EU due to a contrasted carbon footprint, as well as a possible effect on food prices could also free agricultural surfaces for soy opportunistic cultivation. However, studies on soybean use in cow feeding show that soy might have less environment impacting than rapeseed (Lehuger et al., 2009). The new CAP which embraces more environmental considerations may thus face new issues in the balance of environmental footprint and might let European prefer importing soy.

After a ban of about a decade, several lobbyists are pushing the reintroduction of meat-and-bone meal, probably the richest protein source. For instance in 2002, 220,000T were estimated equivalent to 503,000 T of soy meal. The European dependence onto imported soy could thus be dramatically decreased if meat-and-bone is safely re-incorporated into feedstuffs. However, the acceptance of European consumers of such a reintroduction is far from being obvious following the 1990s' mad-cow disease scandal.

CAP Health check in 2008 reduced again aids to cereals opening the opportunity to grow more oilseeds including soybean despite a previous EU support to protein-rich crops such as field pea, field bean and lupine.

Surprisingly, in all alternatives to soybean imports even though by NGO or organic farmers, no one proposed cultivation of soybean as protein sources alternatives, at least for conventional livestock (Billon et al., 2009; Confédération paysanne, 2002). However, production of European soybean showed in 2009 a 12.4% increase which demonstrates the opportunism of European farmers in front of high trade prices of the 2008/2009 food crisis (Krautgartner et al., 2010a). Such an alternative of soy cropping instead of imports should be more effectively considered in EU-MS, even though it looks difficult to dedicate ca. 20 Millions of European hectares to soy, the equivalent in surfaces of currently imported soy (see above).

The recent entrance of several new EU-MS, with a past of soybean crops and some very large corporate farms, could also accompany this trend of growing more soybean in the EU. Interviewed Spanish representatives agreed that Spain could grow soybean, be these GM or not, as soon as the prices would be of interest. Additionally, the fight against the Western corn rootworm (Diabrotica virgifera virgifera) in French maize monocultures was successful when introducing soybean into a newly implemented rotation, a choice of crop made in function of cropping practices and apparatus compatibility. Together with the environmental and economic interest of introducing a leguminous plant into rotations and the more general European request of reducing chemicals in cultivation, the interest of maize monoculture is questionable when we consider that the infected area[15] covers most of the Central and Eastern European Countries. Finally, the current trend of increase of petrol and thus of nitrogen fertilizer prices also favors reintroduction of leguminous crops into rotation.

In addition to an increasing part of local, pasture based and on-farm production for both "conventional" and under signs-of-quality meat production, the European soybean dependence might thus drastically decrease at least for bovine animals. Soybean imports would then mostly depend on intensive livestock production such as poultry and probably pig.

Altogether, these several observations show a balanced approach of European farmers toward the global market and an important dependence of EU farmers to CAP. As long as soybean and soy meal prices are low, there is no interest for European farmers to enter the

---

[15] http://extension.entm.purdue.edu/wcr/images/pdf/2010/EUROPEMap2010.pdf

very competitive commodities market. But all occasions can be taken to improve their niche markets of soy be this GM or non-GM.

5.4.2.2 Alternative sources of non-GM soy

With the growing trend to label animals reared on non-GM feed, availability of non-GM soy is of a growing interest for livestock producers. With the increased cropping of GM soy in exporting countries, alternative sources of non-GM soy are thus actively looked for by European importers.

Since the 2008 issue of Chinese organic soy meal spiked with melamine, The Peoples' Republic of China is no longer considered as a reliable source of non-GM soy despite recent claims of its interest for this country (Anonymous, 2010; Hansen et al., 2007; Takada, 2010). China is the most important importer of soybean and this expected to continue (Taylor and Koo, 2010). As numerous GM crops are under development in China together with a growing request of soy for livestock production, we may expect this country may rapidly cease to be a putative exporter of non-GM soybean. Indeed, India is currently a new source of non-GM soy for certain European traders and has been identified as such by US surveys (Ash, 2011).

This current relative shortage of non-GM (<0.9%) or GMO-free (<0.1%) soybean could be an opportunity for European soy producers, provided they find more incentives to grow soy. The new CAP trend considering more environmental issues might favor such changes into the European farmers practices. Integrating crops rotations with leguminous crops, for decreasing the use of costly fertilizers and fighting some pests such as the expending Western corn root worm, would be additional causes of such practices' changes with premiums and long-term contracts for non-GM productions,.

Beside a new consideration of soybean into crop rotation in Western EU or an increase of soy surfaces in EU-MS cultivating soy for market niches, the European soy status may also change by the integration in 2007 of Central and Eastern European countries such as Romania with a past and a future wish of soybean cultivation. The move of these countries, some cases having very large corporate farms, toward GM or non-GM soy cultivation will greatly depend on non-GM demand, premiums and long-term contracts as well as the volatility of GM soy commodities' prices.

6. Conclusion

Europe is so far highly dependent on protein importations for compound feedstuffs production particularly of soy for young animal production, poultry and pigs. However, several factors may lead to an increased soybean production in the EU over the next decade. Among the several reasons for such an increase are societal considerations such as carbon footprint of imported soybean, development of market niches - be these or not GMO-free due to animal labeling - entrance of new EU-MS with a past of soybean cropping as well as a general increase of American exports toward China inducing tensions on prices particularly for poultry and pig feeding.

Soybean cropping would however probably be considered as an opportunistic European crop due to e.g. rotation for fighting corn rootworm which is prevalent in Central and Eastern EU and extending into the Western area, continuous rises in nitrogen fertilizer prices, a persistent ban of meat-and-bones meal as well as an absence of alternative European protein-rich crops.

As the cultivations are rapidly adjusted to the market requests, there is thus not an issue of soybean supply *per se* but an issue of importation of soybean at the lowest prices for intensive livestock production. Such issue of competitiveness of European livestock drove the EC to introduce a "technical solution" to EU unapproved GMOs, i.e. a LLP threshold, for feed, which might impact food supply chains since segregating of food and feed is difficult.

At the same time, the need for a larger use of bio-fuels increased the production of oilseed rape in Europe which in counterpart decreased the imports of soybeans. It is currently difficult to determine what would be the future of such by-products of bio-fuels as the European policy bio-fuels appears to be changing due to new calculations of their carbon footprint and the need for "feeding the world". Such trend to develop domestic bio-fuels will probably impact European soy imports and cultivations.

After several shortages in the 1970s and 2000s, the current increases in feed and food commodities' prices after the 2007-2008 food crisis militates for an European alimentary sovereignty due in particular to the impact of the increasing living standards of emerging countries and thus of protein-rich feedstuffs.

Environmental, sustainability and social criteria newly incorporated into the European agricultural aids frame will probably push domestic oilseed production, including soybean and jeopardize oilseeds imports. The main driver of European livestock production and soy imports will also depend on the possible extension among EU-MS of the labeling of animals reared with GMO-free feed.

By the different past histories of Western and Eastern parts of the EU-27, it is also to be expected that the soy cropping strategies, i.e. the choice between GM and non-GM soy cultivation, will differ between the two. The lowest sensitivity of Eastern consumers to GM food and cultivation could facilitate the implementation of GM soy in the Eastern part of the EU, while non-GM soy might develop in the Western part of the EU. Such search for a European "sovereignty" is in line with the development of numerous markets niches, a usual counterpart to a more and more standardized and global trade.

Generally speaking, the recent initiative of the EC for establishing a LLP threshold for feed did not take into consideration the change of paradigm i.e. the European move from an "economy of offer" toward an "economy of demand" nor the difficulties to segregate food and feed commodities.

7. Note added in proof

As this chapter was in proof reading, EFSA published its final version of PMEM (EFSA GMO Panel, 2011) and the EC published the LLP related regulation (European Commission, 2011).

8. References

ACRE (2004). ACRE Guidance Note 16. Guidance on best practice in the design of post-market monitoring plans in submissions to the Advisory Committee on Releases to the environment. Guidance for applicants seeking permission to release genetically modified crops into the environment (under Directive 2001/18/EC), 21 p.

Adams, S. (2011). Dow AgriSciences, MU researcher develop a way to control "superweed". Available at:
http://munews.missouri.edu/news-releases/2011/0121-dow-agrisciences-mu-researcher-develop-a-way-to-control-%E2%80%9Csuperweed%E2%80%9D/

ADAS Ltd. (for DEFRA feed import project) (2008). What is the potential to replace imported soya, miaze and maize by-products with other feeds in livestock diets?, ADAS. Ltd, ed. (DEFRA, UK), 30 p.

Alacalde, E., Amijee, F., Blache, G., Bremer, C., Fernandez, S., Garcia-Alonso, M., Holt, K., Legris, G., Novillo, C., Schlotter, P., *et al.* (2007). Insect resistance monitoring for Bt maize cultivation in the EU: Proposal from the industry IRM working group. *Journal Fur Verbraucherschutz Und Lebensmittelsicherheit-Journal of Consumer Protection and Food Safety, 2*, pp. 47-49.

Alderborn, A., Sundstrom, J., Soeria-Atmadja, D., Sandberg, M., Andersson, H.C., and Hammerling, U. (2010). Genetically modified plants for non-food or non-feed purposes: Straightforward screening for their appearance in food and feed. *Food and Chemical Toxicology 48*, pp. 453-464.

Anderson, C. (2006). The long tail: why the future of business is selling less of more (Random House / Hyperion).

Anonymous (2010). Soybeans to go non-GM. In *China Business News*. Available at: http://cnbusinessnews.com/soybeans-to-go-non-gm/

APHIS News release (2010). USDA announces availability of biotechnology quality management system audit tandard and evaluation of comments. Available at: http://www.aphis.usda.gov/biotechnology/news_bqms.shtml

Aramyan, L.H., van Wagenberg, C.P.A., and Backus, G.B.C. (2009a). Impact of unapproved genetically modified soybean on the EU feed industry (Wageningen University, NL), 14 p.

Aramyan, L.H., Wagenberg, C.P.A.v., and Backus, G.B.C. (2009b). EU policy on GM soy. Tolerance threshold and asynchronic approval. Report 2009-052, 46 p.

Ash, M. (2011). Oil Crops Outlook 2011. In A report from the Economic Research Service, U. ERS, ed., 16 p.

Badea, E.M., and Pamfil, D. (2009). The status of agricultural biotechnology and biosafety in Romania. *Bulletin of University of Agricultural Sciences and Veterinary Medicine Cluj-Napoca Animal Science and Biotechnologies, 66*, pp. 8-15.

Bakshi, A. (2003). Potential adverse health effects of genetically modified crops. Journal of Toxicology and Environmental Health-Part B-Critical Reviews 6, pp. 211-225.

Bartsch, D., Bigler, F., Castanera, P., Gathmann, A., Gielkens, M., Hartley, S., Lheureux, K., Renckens, S., Schiemann, J., Sweet, J., *et al.* (2006a). Concepts for general surveillance of genetically modified (GM) plants: the EFSA position. Journal fur Verbraucherschutz und Lebensmittelsicherheit 1, pp. 15-20.

Bartsch, D., Bigler, F., Castanera, P., Gathmann, A., Gielkens, M., Hartley, S., Lheureux, K., Renckens, S., Schiemann, J., Sweet, J., *et al.* (2006b). The EFSA opinion on post-market environmental monitoring of GM plants. The 9th International Symposium on the Biosafety of Genetically Modified Organisms, Jeju Island, Korea, 24-29 September, 2006: biosafety research and environmental risk assessment, pp. 159-167.

Bartsch, D., Buhk, H.-J., Engel, K.-H., Ewen, C., Flachowsky, G., Gathmann, A., Heinze, P., Koziolek, C., Leggewie, G., Meisner, A., *et al.* (2009). BEETLE. Long-term effects of genetically modified crops on health and the environment (including biodiversity): Prioritisation of potential risks and delimitation of uncertainties. Executive summary and main report (Berlin, DE), 133 p.

Bartsch, D., Gathmann, A., Hartley, S., Hendriksen, N.B., Hails, R., Lheureux, K., Kiss, J., Mesdagh, S., Neemann, G., Perry, J., *et al.* (2007). First EFSA experiences with monitoring plans. Journal Fur Verbraucherschutz Und Lebensmittelsicherheit-Journal of Consumer Protection and Food Safety 2, pp. 33-36.

Barzman, M., Caron, P., Passouant, M., and Tonneau, J.P. (2005). Observatoire Agriculture et Territoires. Etude pour la définition d'une méthode de mise en place d'observatoires, 64 p. Available at : http://agriculture.gouv.fr/IMG/pdf/observatoire_rapport_final.pdf

Beckie, H.J., Hall, L.M., Simard, M.J., Leeson, J.Y., and Willenborg, C.J. (2010). A framework for postrelease environmental monitoring of second-generation crops with novel traits. *Crop Science*, 50, pp. 1587-1604.

Beismann, H., Finck, M., and Seitz, H. (2007). Standardisation of methods for GMO Monitoring on a European level. *Journal Fur Verbraucherschutz Und Lebensmittelsicherheit-Journal of Consumer Protection and Food Safety*, 2, pp. 76-78.

Beissner, L., Wilhelm, R., and Schiemann, J. (2006). Current research activities to develop and test questionnaires as a tool for the General Surveillance of important crop plants. *Journal fur Verbraucherschutz und Lebensmittelsicherheit-Journal of Consumer Protection and Food Safety*, 1, pp. 95-97.

Bellon, M.R., and Morris, M.L. (2002). Linking global and local approaches to agricultural technology development: the role of participatory plant breeding research in the CGIAR. In *Economics working paper* 02- 03, 30 p.

Berlan, J.-P., Bertrand, J.-P., and Lebas, L. (1977). The growth of the American 'soybean complex'. *European Review of Agricultural Economics*, 4, pp. 395-416.

Billon, A., Neyroumande, E., and Deshayes, C. (2009). Vers plus d'indépendance en soja d'importation pour l'alimentation animale en Europe - cas de la France, W.W.F. ENESAD, ed. (Dijon, France, ENESAD, WWF France), 49 p. Available at : http://www.wwf.fr/content/download/2687/20819/version/1/file/rapport+sub stitution+au+soja+complet+2.pdf

Birthal, P.S., Rao, P.P., Nigam, S.N., Bantilan, M.C.S., and Bhagavatula, S. (2010). Groundnut and soybean economies of Asia. Facts, trends and outlook, I.I.C.R.I.f.t.S.-A. Tropics), ed. (Patancheru 502 324, Andra Pradesh, India, ICRISAT), 92 p. Available at: http://dspace.icrisat.ac.in/bitstream/10731/3648/1/rso-gn-sbean-asia.pdf

Bock, A.K., Lheureux, K., Libeau-Dulos, M., Nilsagard, H., and Rodriguez-Cerezo, E. (2002). Scenarios for co-existence of genetically modified, conventional and organic crops in European agriculture. EUR 20394EN, xi + 133 p. Available at: http://ftp.jrc.es/EURdoc/eur20394en.pdf

Bodiguel, L. (2003). A way to establish the multifunctional agriculture: the notion of territory. *Economie Rurale*, pp. 61-75.

Bonny, S. (2008). How have opinions about GMOs changed over time? The situation in the European Union and the USA. *CAB Reviews: Perspectives in Agriculture, Veterinary Science, Nutrition and Natural Resources*, 3, pp. 1-17.

Bonny, S. (2009). Taking stock of the world market in transgenic seeds. In 13th ICABR conference *The emerging bio-economy* (Ravello, Italy).

Bontemps, A., Bourguet, D., Pelozuelo, L., Bethenod, M.T., and Ponsard, S. (2004). Managing the evolution of Bacillus thuringiensis resistance in natural populations of the European corn borer, Ostrinia nubilalis: host plant, host race and pheretype of adult males at aggregation sites. *Proceedings of the Royal Society of London Series B-Biological Sciences*, 271, pp. 2179-2185.

Boshnakova, M., Dobrescu, M., Flach, B., Henard, M.-C., Krautgartner, R., Lieberz, S., and the group of FAS oilseeds specialists in the EU. (2009). EU-27. Oilseeds crop update - U.S. soybean exports to EU threatened. In EU-27, U.F.A. Service, ed. (Vienna, Austria, USDA GAIN), 11 p. Available at:
http://gain.fas.usda.gov/Recent%20GAIN%20Publications/Oilseeds%20Crop%20Update%20%20U.S.%20Soybean%20Exports%20to%20EU%20Threatened_Vienna_EU-27_8-12-2009.pdf

Bourguet, D. (2004). Resistance to Bacillus thuringiensis toxins in the European corn borer: what chance for Bt maize? *Physiological Entomology*, 29, pp. 251-256.

Brasher, P. (2010). New weed strategies needed, scientists say. *CheckBiotech*. Available at: http://checkbiotech.org/node/29440/edit

Breckling, B., and Reuter, H. (2006). General surveillance of genetically modified organisms - the importance of expected and unexpected environmental effects. *Journal für Verbraucherschutz und Lebensmittelsicherheit*, 1, pp. 72-74.

Brookes G. (2005). The farm-level impact of herbicide-tolerant soybeans in Romania. *AgBioForum*, 8, 5, 235-241.

Bruening, G., and Lyons, J.M. (2000). The case of the FLAVR SAVR® tomato. *California Agriculture*, 54, pp. 6-7.

Brunet, Y., Dupont, S., Delage, S., Garrigou, D., Guyon, D., Dayau, S., Tulet, P., Pinty, J.-P., Lac, C., Escobar, J., *et al.* (2011). Long-distance pollen flow in large fragmented landscapes. In GM and non-GM supply chains coexistence and traceability, Y. Bertheau, ed. (Wiley Publishing), *In Press*.

Cardwell, M. (2003). Multifunctionality of agriculture: a European community perspective. In Agriculture and international trade: law, policy and the WTO, Cardwell, M. N. Grossman and M. R., Rodgers, C. P. (eds), CPL Press. pp. 131-164.

Cardwell, M. (2010). Public participation in the regulation of genetically modified organisms: a matter of substance or form? *Environmental Law Review*, 12, pp. 12-25.

Cardwell, M., and Bodiguel, L. (2005). Agriculture definition trend for advanced agriculture - comparative European Union/Great Britain/France approach. *Revue du Marche Commun et de l'Union Européenne*, 490, pp. 456-466.

Carlier, L., Rotar, I., and Vidican, R. (2010). The EU policy regulates and controls the farming practices. *Bulletin UASVM Agriculture*, 67, 1-11.

Cavaillès, E. (2009). La relance des légumineuses dans le cadre d'un plan protéine : quels bénéfices environnementaux ? In Etudes et documents, Commissariat général au développement durable – Service de l'économie, de l'évaluation et de l'intégration du développement durable, ed. (Paris, France, Commissariat général au développement durable – Service de l'économie, de l'évaluation et de l'intégration du développement durable), 44 p. Available at : http://www.developpement-durable.gouv.fr/IMG/pdf/E_D15.pdf

Cellini, F., Chesson, A., Colquhoun, I., Constable, A., Davies, H.V., Engel, K.H., Gatehouse, A.M.R., Karenlampi, S., Kok, E.J., Leguay, J.J., *et al.* (2004). Unintended effects and their detection in genetically modified crops. *Food and Chemical Toxicology*, 42, 1089-1125.

Cerdeira, A.L., and Duke, S.O. (2006). The current status and environmental impacts of glyphosate-resistant crops: A review. *Journal of Environmental Quality*, 35, pp. 1633-1658.

Cerdeira, A.L., and Duke, S.O. (2007). Environmental impacts of transgenic herbicide-resistant crops. *CAB Reviews: Perspectives in Agriculture, Veterinary Science, Nutrition and Natural Resources*, 2, 033, 14 p.

Chaufaux, J., Micoud, A., Delos, M., Naibo, B., Bombarde, F., Eychennes, N., Pagliari, C., Marque, G., and Bourguet, D. (2002). Transgenic maize and non target insects: what effects? *Phytoma*, pp. 13-16.

Codex Alimentarius (2003). Guideline for the conduct of food safety assessment of foods derived from recombinant-DNA plants. CAC/GL 45-2003. Annex 3: food safety assessment in situations of low-level presence of recombinant-DNA plant material in food. Available at:
http://www.codexalimentarius.net/download/standards/10021/CXG_045e.pdf

Colbach, N. (2009). How to model and simulate the effects of cropping systems on population dynamics and gene flow at the landscape level: example of oilseed rape volunteers and their role for co-existence of GM and non-GM crops. *Environmental Science and Pollution Research*, 16, pp. 348-360.

Colbach, N., Monod, H., and Lavigne, C. (2009). A simulation study of the medium-term effects of field patterns on cross-pollination rates in oilseed rape (Brassica napus L.). *Ecological Modelling*, 220, pp. 662-672.

Coléno, F.C. (2008). Simulation and evaluation of GM and non-GM segregation management strategies among European grain merchants. Journal of Food Engineering 88, 306-314.

Commission, E. (2006). Regulation (EC) No 1907/2006 of the European parliament and of the Council of 18 December 2006 concerning the Registration, Evaluation, Authorisation and Restriction of Chemicals (REACH), establishing a European Chemicals Agency, amending Directive 1999/45/EC and repealing Council Regulation (EEC) No 793/93 and Commission Regulation (EC) No 1488/94 as well as Council Directive 76/769/EEC and Commission Directives 91/155/EEC, 93/67/EEC, 93/105/EC and 2000/21/EC. *Official Jounal of the European Union*, L 396, pp. 1-849.

Confédération paysanne (2002). Un plan protéine pour l'Europe (Bagnolet, France), 26 p. Available at :
http://www.confederationpaysanne.fr/images/imagesFCK/file/02/plan_proteines.pdf

Consumerchoice Consortium (2008). Do European consumers buy GM foods? European Commission: Framework 6, Project no. 518435. "Consumerchoice". Final report., 346 p. Available at:
http://www.kcl.ac.uk/schools/biohealth/research/nutritional/consumerchoice/downloads.html

Covelli, N., and Hohots, V. (2003). The health regulation of biotech foods under the WTO agreements. *Journal of International Economic Law*, 6, pp. 773-795.

Csaki, C., and Lerman, Z. (1997). Land reform and farm restructuring in East Central Europe and CIS in the 1990s: Expectations and achievements after the first five years. *European Review of Agricultural Economics*, 24, pp. 428-452.

Czarnak-Kłos, M., and Rodríguez-Cerezo, E. (2010). European Coexistence Bureau (ECoB). Best Practice Documents for coexistence of genetically modified crops with conventional and organic farming. 1. Maize crop production. EUR 24509 EN. In JRC Scientific and Technical Reports, DG-JRC-IPTS, ed. (Seville, SP, DG-JRC-IPTS), 72 p. Available at: http://ecob.jrc.ec.europa.eu/documents/Maize.pdf

D'Agnolo, G. (2005). GMO: Human health risk assessment. *Veterinary Research Communications*, 29, pp. 7-11.

Dahl, E., and House, L. (2008). United States – European Union agricultural trade flows. In 110th EAAE Seminar 'System Dynamics and Innovation in Food Networks' (Innsbruck-Igls, Austria), 9 p. Available at: http://ageconsearch.umn.edu/bitstream/49841/2/Dahl-House.pdf

Darmency, H., Klein, E.K., De Garanbe, T.G., Gouyon, P.H., Richard-Molard, M., and Muchembled, C. (2009). Pollen dispersal in sugar beet production fields. *Theoretical and Applied Genetics*, 118, pp. 1083-1092.

Darmency, H., Vigouroux, Y., De Garambe, T.G., Richard-Molard, M., and Muchembled, C. (2007). Transgene escape in sugar beet production fields: data from six years farm scale monitoring. *Environmental Biosafety Research*, 6, pp. 197-206.

Davison, J. (2010). GM plants: science, politics and EC regulations. *Plant Science*, 178, pp. 94-98.

Davison, J., and Bertheau, Y. (2007). EU regulations on the traceability and detection of GMOs: difficulties in interpretation, implementation and compliance. *CAB Reviews: Perspectives in Agriculture, Veterinary Science, Nutrition and Natural Resources*, 2, 14 p.

Davison, J., and Bertheau, Y. (2008). The theory and practice of European traceability regulations for GM food and feed. *Cereal Foods World*, 53, pp. 186-196.

de Cheveigné, S. (2004). Quand l'Europe mesure les représentations de la science : une analyse critique des Eurobaromètres. In *Sciences, Médias et Société*, J.L. Marec, and I. Babou, eds. (Lyon, F, Ecole normale supérieure. Lettres et Sciences Humaines. Laboratoire JE 2419. Communication, culture et société.), pp. 45-55.

de Jong, T.J. (2010). General surveillance of genetically modified plants in the EC and the need for controls. *Journal Fur Verbraucherschutz Und Lebensmittelsicherheit-Journal of Consumer Protection and Food Safety*, 5, pp. 181-183.

DEFRA (2006). Consultation on proposals for managing the coexistence of GM, conventional and organic crops (DEFRA, UK), 92 p. Available at: http://www.defra.gov.uk/environment/quality/gm/crops/documents/gmcoexist-condoc.pdf

Demont, M., Daems, W., Dillen, K., Mathijs, E., Sausse, C., and Tollens, E. (2008). Regulating coexistence in Europe: beware of the domino-effect! *Ecological Economics*, 64, pp. 683-689.

Demont, M., Devos, Y., and Sanvido, O. (2010). Towards flexible coexistence regulations for GM crops in the EU. *Eurochoices*, 9, pp. 18-24.

Demont, M., Dillen, K., Daems, W., Sausse, C., Tollens, E., and Mathijs, E. (2009). On the proportionality of EU spatial ex ante coexistence regulations. *Food Policy*, 34, pp. 508-518.

Desclaux, D., Nolot, J., Chiffoleau, Y., Gozé, E., and Leclerc, C. (2008). Changes in the concept of genotype × environment interactions to fit agriculture diversification and decentralized participatory plant breeding: pluridisciplinary point of view. *Euphytica*, 163, pp. 533-546.

Desquilbet, M., and Bullock, D.S. (2010). On the proportionality of EU spatial ex ante coexistence regulations: a comment. *Food Policy*, 35, pp. 87-90.

Devos, Y., Demont, M., Dillen, K., Reheul, D., Kaiser, M., and Sanvido, O. (2009). Coexistence of genetically modified (GM) and non-GM crops in the European Union. A review. *Agronomy for Sustainable Development*, 29, pp. 11-30.

Devos, Y.A. (2008). Transgenic crops: a kaleidoscope impact analysis. PhD Thesis. Department of plant production, faculty of bioscience engineering (Ghent, BE, Ghent University), 342 p. Available at:
http://www.criticalphilosophy.ugent.be/index.php?id=2&type=file

DG AGRI European Commission (2007). Economic impact of unapproved GMOs on EU feed imports and livestock production (Brussels, B, European Commission). 11 p. Available at:
http://ec.europa.eu/agriculture/envir/gmo/economic_impactGMOs_en.pdf

Dinu, T., Alecu, I.N., and Stoian, E. (2010). Assessing the economic impact and the traceability costs in the case of banning the cultivation of GM soybean in Romania. In *10th International Symposium* " Prospects of Agriculture and Rural Areas Development in the context of Global Climate Change ", Session "Management, Marketing, Accounting, Financial Analysis, Finance", M. Draghici, and M. Berca, eds. (20-21 May 2010 , Bucharest, Romania, RAWEX COMS Publishing House in co-editing with DO-MINOR Publishing House), pp. 62-67.

Dobbs, M. (2011). Legalising general prohibitions on cultivation of genetically modified organisms. *German Law Journal*, 11, pp. 1347-1372.

Dobrescu, M., Henard, M.-C., Krautgartner, R., Lieberz, S., and the group of FAS oilseeds specialists in the EU (2009). EU-27. Soybean imports from the United States still impeded. In EU-27, USDA GAIN, ed. (Vienna, Austria), 12 p. Available at:
http://gain.fas.usda.gov/Recent%20GAIN%20Publications/EU-27%20Soybean%20Imports%20from%20the%20United%20States%20Still%20Imped ed_Vienna_EU-27_11-3-2009.pdf

Ebert, I., and Lahnstein, C. (2008). GMO liability: options for insurers. In *Economic Loss Caused by Genetically Modified Organisms*, B.A. Koch, ed. (Springer Vienna), pp. 577-581.

EFSA (2008). Environmental risk assessment of genetically modified plants - Challenges and approaches. Paper presented at: EFSA Scientific Colloquium Summary Report (Parma, Italy, EFSA). Available at:
http://www.efsa.europa.eu/fr/colloquiagmoera/publication/colloquiagmoera.pdf

EFSA GMO Panel (2004). Guidance document of the scientific panel on genetically modified organisms for the risk assessment of genetically modified plants and derived food and feed. *The EFSA Journal*, 99, 1-94.

EFSA GMO Panel (2006a). Guidance document of the scientific panel on genetically modified organisms for the risk assessment of genetically modified microorganisms and their derived products intended for food and feed use. *The EFSA Journal*, 374, pp. 1-115. Available at:
http://www.efsa.europa.eu/fr/scdocs/scdoc/374.htm

EFSA GMO panel (2006b). Opinion of the scientific panel on genetically modified organisms on the post market environmental monitoring (PMEM) of genetically modified plants (Question No EFSA-Q-2004-061). *The EFSA Journal*, 319, 1-27.

EFSA GMO Panel (2010). Scientific opinion. Guidance on the environmental risk assessment of genetically modified plants. *The EFSA Journal*, 8, 1-111.

EFSA GMO Panel. (2011) Guidance on the Post-Market Environmental Monitoring (PMEM) of genetically modified plants. EFSA Journal 9:2316.

Elbehri, A. (2007). The changing face of the U.S. grain system differentiation and identity preservation trends. *Economic Research Report - Economic Research Service, USDA*, 39 p. Available at: http://www.ers.usda.gov/publications/err35/err35.pdf

EU working group (2003). Appendix 1: EU working group on insect resistance management: harmonised insect resistance management (IRM) plan for cultivation of Bt maize in the EU, 28 p. Available at:
http://ec.europa.eu/food/food/biotechnology/docs/IRM_plan.pdf

European Commission (1985). Council Directive 85/374/EEC of 25 July 1985 on the approximation of the laws, regulations and administrative provisions of the Member States concerning liability for defective products *Official Jounal of the European Communities*, L 210, pp. 29-33.

European Commission. (2011) Commission Regulation (EU) No 619/2011 of 24 June 2011 laying down the methods of sampling and analysis for the official control of feed as regards presence of genetically modified material for which an authorisation procedure is pending or the authorisation of which has expired (Text with EEA relevance). Official Jounal of the European Union L 166:9-15.

European Commission (1997). Regulation (EC) No 258/97 of the European Parliament and of the Council of 27 January 1997 concerning novel foods and novel food ingredients. *Official Journal of the European Communities*, L 043, pp. 1-6.

European Commission (2000a). Commission regulation (EC) No 49/2000 of 10 January 2000 amending Council Regulation (EC) No 1139/98 concerning the compulsory indication on the labelling of certain foodstuffs produced from genetically modified organisms of particulars other than those provided for in Directive 79/112/EEC. *Official Journal of the European Communities*, L 6, pp. 13-14.

European Commission (2000b). Regulation (EC) No 50/2000 of 10 January 2000 on the labelling of foodstuffs and food ingredients containing additives and flavourings that have been genetically modified or have been produced from genetically modified organisms. *Official Journal of the European Communities*, L 6, 15-17.

European Commission (2001). Directive 2001/18/EC of the European Parliament and the Council of 12 March 2001 on the deliberate release into the environment of genetically modified organisms and repealing Council Directive 90/220/EEC. *Official Journal of the European Communities*, L 106, 1-38.

European Commission (2002a). Council decision 2002/811/EC of 3 October 2002 establishing guidance notes supplementing Annex VII to Directive 2001/18/EC of the European Parliament and of the Council on the deliberate release into the environment of genetically modified organisms and repealing Council Directive 90/220/EEC. *Official Journal of the European Communities*, L 280, pp. 27-36.

European Commission (2002b). Council decision 2002/812/EC of 3 October 2002 establishing pursuant to Directive 2001/18/EC of the European Parliament and of the Council the summary information format relating to the placing on the market of genetically modified organisms as or in products. *Official Journal of the European Communities*, L 280, pp. 37-61.

European Commission (2002c). Council decision 2002/813/EC of 3 October 2002 establishing, pursuant to Directive 2001/18/EC of the European Parliament and of the Council, the summary notification information format for notifications concerning the deliberate release into the environment of genetically modified organisms for purposes other than for placing on the market. *Official Journal of the European Communities*, L 280, pp. 62-83.

European Commission (2002d). Regulation (EC) No 178/2002 of the European Parliament and of the Council of 28 January2002 laying down the general principles and requirements of food law, establishing the European Food Safety Authority and laying down procedures in matters of food safety. *Official Journal of the European Communities*, L 31, pp. 1-24.

European Commission (2003a). Commission Recommendation of 23 July 2003 on guidelines for the development of national strategies and best practices to ensure the coexistence of genetically modified crops with conventional and organic farming (notified under document number C(2003) 2624). (2003/556/CE). *Official Journal of the European Communities*, L 189, pp. 36-47.

European Commission (2003b). Directive 2003/4/EC of the European Parliament and of the Council of 28 January 2003 on public access to environmental information and repealing Council Directive 90/313/EEC. *Official Jounal of the European Union*, L 41, pp. 26-32.

European Commission (2003c). Regulation (EC) No 1829/2003 of the European Parliament and of the Council of 22 September 2003 on genetically modified food and feed. *Official Journal of the European Union*, L 268, pp. 1-23.

European Commission (2003d). Regulation (EC) No 1830/2003 of the European Parliament and of the Council of 22 September 2003 concerning the traceability and labelling of genetically modified organisms and the traceability of food and feed products produced from genetically modified organisms and amending Directive 2001/18/EC. *Official Journal of the European Union*, L 268, pp. 24-28.

European Commission (2004a). Commission decision No 2004/204/EC of 23 February 2004 laying down detailed arrangements for the operation of the registers for recording information on genetic modifications in GMOs, provided for in Directive 2001/18/EC of the European Parliament and of the Council. *Official Journal of the European Union*, L 065, pp. 20-22.

European Commission (2004b). Commission recommendation No 2004/787 of 4 october 2004 on technical guidance for sampling and detection of genetically modified organisms as or in products in the context of Regulation (EC) No 1830/2003. *Official Journal of the European Union*, L 348, pp. 18-26.

European Commission (2004c). Directive 2004/35/CE of the European Parliament and of the Council of 21 April 2004 on environmental liability with regard to the prevention and remedying of environmental damage. *Official Journal of the European Union*, L 143, pp. 56-75.

European Commission (2004d). Regulation (EC) No 882/2004 of the European Parliament and of the Council of 29 April 2004 on official controls performed to ensure the verification of compliance with feed and food law, animal health and animal welfare rules. *Official Jounal of the European Union*, L 165, pp. 1-141.

European Commission (2006a). Commission Regulation (EC) No 1981/2006 of 22 December 2006 on detailed rules for the implementation of Article 32 of Regulation (EC) No 1829/2003 of the European Parliament and of the Council as regards the Community reference laboratory for genetically modified organisms (Text with EEA relevance), *Official Journal of the European Union*, L 368, pp. 99-103.

European Commission (2006b). Communication from the Commission to the Council and the European Parliament. Report on the implementation of national measures on the coexistence of genetically modified crops with conventional and organic farming. {SEC(2006) 313}. COM/2006/0104 final, 10 p. Availabel at: http://ec.europa.eu/agriculture/coexistence/com104_en.pdf

European Commission (2006c). Regulation (EC) No 1367/2006 of the European Parliament and of the Council of 6 September 2006 on the application of the provisions of the Aarhus Convention on Access to Information, Public Participation in Decision-making and Access to Justice in Environmental Matters to Community institutions and bodies. *Official Journal of the European Union*, L 264, pp. 13-19.

European Commission (2006d). Report from the Commission to the Council and the European Parliament on the implementation of Regulation (EC) No 1830/2003 concerning the traceability and labelling of genetically modified organisms and the traceability of food and feed products produced from genetically modified organisms and amending Directive 2001/18/EC. 13 p. Available at: http://eur-lex.europa.eu/LexUriServ/LexUriServ.do?uri=COM:2006:0197:FIN:EN:PDF

European Commission (2007a). Council Regulation (EC) No 834/2007 of 28 June 2007 on organic production and labelling of organic products and repealing Regulation (EEC) No 2092/91. *Official Journal of the European Union*, L 189, pp. 1-23.

European Commission (2007b). Directive 2007/2/EC of the European Parliament and of the Council of 14 March 2007 establishing an Infrastructure for Spatial Information in the European Community (INSPIRE). *Official Journal of the European Union*, L 108, pp. 1-14.

European Commission (2008a). Commission Regulation (EC) No 889/2008 of 5 September 2008 laying down detailed rules for the implementation of Council Regulation (EC) No 834/2007 on organic production and labelling of organic products with regard to organic production, labelling and control. *Official Journal of the European Union*, L 250, pp. 1-84.

European Commission (2008b). Report from the Commission to the Council and the European Parliament on the implementation of Regulation (EC) No 1830/2003 concerning the traceability and labelling of genetically modified organisms and the traceability of food and feed products produced from genetically modified organisms and amending Directive 2001/18/EC. COM (2008) 560 final. 11 p. Available at: http://eur-lex.europa.eu/LexUriServ/LexUriServ.do?uri=COM:2008:0560:FIN:EN:PDF

European Commission (2009a). Commission decision No 2009/770/EC of 13 October 2009 establishing standard reporting formats for presenting the monitoring results of the deliberate release into the environment of genetically modified organisms, as or in products, for the purpose of placing on the market, pursuant to Directive 2001/18/EC of the European Parliament and of the Council. *Official Journal of the European Union,* L 275, pp. 9-27.

European Commission (2009b). Commission Decision of 13 October 2009 establishing standard reporting formats for presenting the monitoring results of the deliberate release into the environment of genetically modified organisms, as or in products, for the purpose of placing on the market, pursuant to Directive 2001/18/EC of the European Parliament and of the Council. (notified under document C(2009) 7680). *Official Journal of the European Union,* L 975, pp. 9-27.

European Commission (2009c). Council regulation (EC) No 73/2009 of 19 January 2009 establishing common rules for direct support schemes for farmers under the common agricultural policy and establishing certain support schemes for farmers, amending Regulations (EC) No 1290/2005, (EC) No 247/2006, (EC) No 378/2007 and repealing Regulation (EC) No 1782/2003. *Official Jounal of the European Union,* L 30, pp. 16-99.

European Commission (2009d). Regulation (EC) No 1107/2009 of the Euopean Parliamne and of the Council of 21 October 2009 concerning the placing of plant protection products on the market and repealing Council Directives 79/117/EEC and 91/414/EEC. *Official Jounal of the European Union,* L 309, pp. 1-50.

European Commission (2009e). Report from the Commission to the Council and the European Parliament on the coexistence of genetically modified crops with conventional and organic farming (SEK(2009) 408). European Commission, ed. 12 p. Available at:
http://eur-
lex.europa.eu/LexUriServ/LexUriServ.do?uri=COM:2009:0153:FIN:EN:PDF

European Commission (2010a). Commission Recommendation of 13 July 2010 on guidelines for the development of national co-existence measures to avoid the unintended presence of GMOs in conventional and organic crops (2010/C 200/01). *Official Journal of the European Union,* C 200, pp. 1-5.

European Commission (2010b). A decade of EU-funded GMO research (2001 - 2010). EUR 24473, European Commission, ed. (Brussels, Belgium, European Commission), 264 p. Available at: http://ec.europa.eu/research/biosociety/pdf/a_decade_of_eu-funded_gmo_research.pdf

European Commission (2011). Report from the Commission to the European parliament and the Council on socio-economic implications of GMO cultivation on the basis of Member States contributions, as requested by the Conclusions of the Environment Council of December 2008. , D.-S. European Commission, ed. (Brussels, Belgium), 11 p. Available at:
http://ec.europa.eu/food/food/biotechnology/reports_studies/docs/socio_econ omic_report_GMO_fr.pdf

European Economic and Social Committee (2011). Opinion of the European Economic and Social Committee on the 'Proposal for a Regulation of the European Parliament and of the Council amending Directive 2001/18/EC as regards the possibility for the Member States to restrict or prohibit the cultivation of GMOs in their territory' COM(2010) 375 final — 2010/0208 (COD). (2011/C 54/16). *Official Jounal of the European Union,* C 54, pp. 51-57.

European Parliament. Directorate-General for Internal Policies (2010). The evaluations of the impact of reforms that have affected the sector and the needs of European livestock system. In Policy Department Structural and cohesion policies B, E.P.D.f.I. Policies, ed. (Brussels, Belgium), 31 p. http://www.europarl.europa.eu/activities/committees/studies/download.do?lan guage=en&file=32695

Eurostat. European Commission (2010). Agricultural statistics. Main results — 2008–09. In Eurostat pocketbooks, E. Commission, ed. (Eurostat), 186 p. Available at: http://epp.eurostat.ec.europa.eu/cache/ITY_OFFPUB/KS-ED-10-001/EN/KS-ED-10-001-EN.PDF

Filip, L., Miere, D., and Indrei, L.L. (2004). Genetically modified foods. Advantages and human health risks. *Revista medico-chirurgicala a Societatii de Medici si Naturalisti din Iasi*, 108, pp. 838-842.

Finck, M., Seitz, H., and Beismann, H. (2006). Concepts for general surveillance: VDI proposals standardisation and harmonisation in the field of GMO-monitoring. *Journal fur Verbraucherschutz und Lebensmittelsicherheit*, 1, pp. 11-14.

Fok, M. (2010). Un état de coexistence du soja transgénique et conventionnel au Paraná (Brésil). *Economie rurale*, 320, 53-68.

Foster, M. (2007). Challenges for agricultural markets, coexistence, segmentation of grain markets, costs and opportunities. ABARE CONFERENCE PAPER 07.9. Paper presented at: GMCC07 Third International Conference on Coexistence between Genetically Modified (GM) and non-GM based Agricultural Supply Chains. Seville, Spain, DG-JRC-IPTS (ed.). Available at: http://www.abare.gov.au/publications_html/conference/conference_07/agric_m arkets.pdf

Foster, M. (2010). Evidence of price premiums for non-GM grains in world markets. In Australian Agricultural and Resource Economics Society (Adelaide, Australia, ABARE, Australia), 16 p. Available at: http://www.abare.gov.au/publications_html/conference/conference_10/AARES _4.pdf

GAO (2008). Agencies are proposing changes to improve oversight, but could take additional steps to enhance coordination and monitoring. In Report to the Committee on Agriculture, Nutrition, and Forestry, US Senate (U.S. Government Accountability Office), 109 p. Available at: http://www.gao.gov/products/GAO-09-60

Garcia-Alonso, M., Jacobs, E., Raybould, A., Nickson, T.E., Sowig, P., Willekens, H., Van der Kouwe, P., Layton, R., Amijee, F., Fuentes, A.M., *et al.* (2006). A tiered system for assessing the risk of genetically modified plants to non- target organisms. *Environmental Biosafety Research*, 5, pp. 57-65.

Gaskell, G., Allansdottir, A., Allum, N., Corchero, C., Fischler, C., Hampel, J., Jackson, J., Kronberger, N., Mejlgaard, N., Revuelta, G., *et al.* (2006). Europeans and Biotechnology in 2005: Patterns and Trends. Eurobarometer 64.3. A report to the European Commission's Directorate-General for Research. In Eurobarometer, European Commission, ed. (Brussels, Belgium), 85 p. Available at: http://www.cibpt.org/docs/2006-jul-eurobarometro-bio-2nd-ed.pdf

Gaskell, G., Stares, S., Allansdottir, A., Allum, N., Castro, P., Esmer, Y., Fischler, C., Jackson, J., Kronberger, N., Hampel, J., *et al.* (2010). Europeans and biotechnology in 2010. Winds of change? A report to the European Commission's Directorate-General for Research. EUR 24537 EN. In Research*eu Studies and reports (Brussels, Belgium, European Commission), 176 p. Available at: http://ec.europa.eu/public_opinion/archives/ebs/ebs_341_winds_en.pdf

Gathmann, A. (2009). National implementation plan for MON810 monitoring in Germany – A way forward to improve General Surveillance? Journal fur Verbraucherschutz und Lebensmittelsicherheit 3, 50-50.

Gathmann, A., and Bartsch, D. (2006). National coordination of GMO monitoring - a concept for Germany. *Journal fur Verbraucherschutz und Lebensmittelsicherheit*, 1, pp. 45-48.

Gelder, J.W.v., Kammeraat, K., and Kroes, H. (2008). Soy consumption for feed and fuel in the European Union. A research paper prepared for Milieudefensie (Friends of the Earth Netherlands), 22 p. http://www.foeeurope.org/agrofuels/FFE/Profundo%20report%20final.pdf

Graef, F., Zughart, W., Hommel, B., Heinrich, U., Stachow, U., and Werner, A. (2005). Methodological scheme for designing the monitoring of genetically modified crops at the regional scale. *Environmental Monitoring and Assessment*, 111, 1-26.

Green, R., and Hervé, S. (2006). IP - Traceability and grains traders: ADM, Bunge, Cargill, Dreyfus. *INRA Les cahiers d'ALISS*, 03, 86 p. Available at: http://www.prodinra.inra.fr/prodinra/pinra/data/2007/11/PROD200752c85e26 _20071115125832834.pdf

Grossman, M.R. (2003). Multifunctionality and non-trade concerns. In Agriculture and international trade: law, policy and the WTO, Michael N Cardwell, Margaret R Grossman and Christopher Rodgers (eds), CABI. pp. 85-129.

Gruber, S., Colbach, N., Barbottin, A., and Pekrun, C. (2008). Post-harvest gene escape and approaches for minimizing it. *CAB Reviews: Perspectives in Agriculture, Veterinary Science, Nutrition and Natural Resources*, 3, 17 p.

Gryson, N., and Eeckhout, M. (2011). Co-existence and traceability in supply chains: a case study on Belgian compound feed. In GM and non-GM supply chains: their coexistence and traceability, Y. Bertheau, ed. (London, UK, Wiley Ltd.). In Press.

Haggett, S. (2008). Towards a multipurpose neural network approach to novelty detection (University of Kent), 236 p. Available at: http://kar.kent.ac.uk/24133/1/Towards_Detection.pdf

Hannachi, M., Coléno, F.C., and Assens, C. (2009). Collective strategies and coordination for the management of coexistence: the case studies of Alsace and western South of France. Paper presented at: Fourth International Conference on Coexistence between Genetically Modified (GM) and non-GM based Agricultural Supply Chains (Melbourne, AU). Available at: http://assens.perso.neuf.fr/GMO2009.pdf

Hansen, J., Lin, W., Tuan, F., Marchant, M.A., Kalaitzandonakes, N., Zhong, F., and Song, B. (2007). Commercialization of herbicide-tolerant soybeans in China: perverse domestic and international trade effects. In American Agricultural Economics Association Annual Meeting (Portland, OR, USA), 33 p. Available at: http://ageconsearch.umn.edu/bitstream/9906/1/sp07ha03.pdf

Hardin, G. (1968). Tragedy of commons. *Science*, 162, 1243-1248.

Hardin, G. (1998). Extensions of "The Tragedy of the Commons". *Science,* 280, 682-683.

Harrell, F.E. (2001). Regression modeling strategies. With applications to linear models, logistic regression, and survival analysis. Springer verlag.

Hasha, G. (2002). Livestock feeding and feed imports in the European Union—A decade of change. FDS-0602-01. In *Electronic Outlook Report from the Economic Research Service*, ERS, USDA, ed., 28 p. Available at: http://m.usda.mannlib.cornell.edu/usda/ers/FDS/2000s/2002/FDS-07-03-2002_Special_Report.pdf

Häusling M., Rapporteur of the Committee on Agriculture and Rural Development of the European Parliament (2011). The EU protein deficit: what solution for a long-standing problem? (2010/2111(INI)). In Report, European Parliament, ed. Available at: http://www.europarl.europa.eu/sides/getDoc.do?pubRef=-//EP//NONSGML+REPORT+A7-2011-0026+0+DOC+PDF+V0//EN&language=EN

Headey, D. (2011). Rethinking the global food crisis: the role of trade shocks. *Food Policy*, 36, pp. 136-146.

Heisey, P.W., King, J.L., and Rubenstein, K.D. (2005). Patterns of public-sector and private-sector patenting in agricultural biotechnology. *AgBioForum*, 8, 73.

Heisey, P.W., Srinivasan, C.S., and Thirtle, C. (2001). Public sector plant breeding in a privatizing world. In *Agriculture Information Bulletin*, E.R.S. Resource economics Division, USDA, ed. (Washington, DC, USA, Resource economics Division, Economic research Service, USDA). Available at: http://www.ers.usda.gov/publications/aib772/aib772.pdf

Henle, K., Alard, D., Clitherow, J., Cobb, P., Firbank, L., Kull, T., McCracken, D., Moritz, R.F.A., Niemelä, J., Rebane, M., *et al.* (2008). Identifying and managing the conflicts between agriculture and biodiversity conservation in Europe - A review. *Agriculture Ecosystems & Environment*, 124, pp. 60-71.

Hepburn, P., Howlett, J., Boeing, H., Cockburn, A., Constable, A., Davi, A., de Jong, N., Moseley, B., Oberdorfer, R., Robertson, C., *et al.* (2008). The application of post-market monitoring to novel foods. *Food and Chemical Toxicology*, 46, pp. 9-33.

Hintermann, U., Weber, D., Zangger, A., and Schmill, J. (2002). Biodiversity monitoring in Switzerland BDM - interim report. Environmental series. No. 342, Forest and Landscape, Swiss Agency for the Environment, Berne, Switzerland, ed. Available at: www.umwelt-schweiz.ch/buwal/shop/files/pdf/phpamJ0T6.pdf

James, C. (2011). Global status of commercialized biotech / GM crops: 2010. In ISAAA Brief, ISAAA, ed. (Ithaca, NY, USA).

Jank, B., Rath, J., and Gaugitsch, H. (2006). Co-existence of agricultural production systems. *Trends in Biotechnology*, 24, pp. 198-200.

Kaditi, E., and Swinnen, J.F.M. (2006). Trade Agreements, Multifunctionality and EU Agriculture. The Centre for European Policy Studies, Brussels, Belgium. 323 p.

Kanter, J. (2010). E.U. signals big shift on genetically modified crops. In *The New-York Times* (New-York, USA). Available at: http://www.nytimes.com/2010/05/10/business/energy-environment/10green.html?_r=1&src=busln

Kawata, M., Murakami, K., and Ishikawa, T. (2009). Dispersal and persistence of genetically modified oilseed rape around Japanese harbors. *Environmental Science and Pollution Research*, 16, pp. 120-126.

Kim, C.-G., Yi, H., Park, S., Yeon, J.E., Kim, D.Y., Kim, D.I., Lee, K.-H., Lee, T.C., Paek, I.S., Yoon, W.K., *et al.* (2006). Monitoring the occurrence of genetically modified soybean and maize around cultivated fields and at a grain receiving port in Korea. *Journal of Plant Biology*, 49, pp. 218-223.

Konduru, S.P. (2008). Three essays on the potential economic impacts of biotech crops in the presence of asynchronous regulatrory approval. PhD Thesis. Faculty of the Graduate School, Agricultural Economics (Columbia, U.S.A., University of Missouri-Columbia), 145 p. Available at:
https://mospace.umsystem.edu/xmlui/handle/10355/6642

Konefal, J. and Busch, L. (2010). Markets of multitudes: how biotechnologies are standardising and differentiating corn and soybeans. *Sociologia Ruralis*, 50, 4, pp. 409-427

Kontoleon, A., and Yabe, M. (2006). Market segmentation analysis of preferences for GM derived animal foods in the UK. *Journal of Agricultural & Food Industrial Organization*, 4, Article 8.

Krautgartner, R., Henard, M.-C., Lieberz, S., Boshnakova, M., and the group of FAS oilseeds specialists in the EU (2010a). EU-27. Oilseeds - Increased domestic soybean & soybean meal production. In EU-27, USDA GAIN, ed. (Vienna, Austria), 14 p. Available at:
http://gain.fas.usda.gov/Recent%20GAIN%20Publications/Oilseeds%20%20Incre ased%20Domestic%20Soybean%20and%20Soybean%20Meal%20Production_Vienn a_EU-27_11-30-2010.pdf

Krautgartner, R., Henard, M.-C., Lieberz, S., Dobrescu, M., Flach, B., Wideback, A., Guerrero, M., Bendz, K., and the group of FAS oilseeds specialists in the EU (2010b). EU-27. Oilseeds and products annual. 2010. In EU-27, USDA GAIN, ed. (Berlin, Germany), 33 p. Available at :
http://gain.fas.usda.gov/Recent%20GAIN%20Publications/Oilseeds%20and%20P roducts%20Annual_Berlin_EU-27_4-19-2010.pdf

Laisney, C. (2011). L'évolution de l'alimentation en France. Tendances émergentes et ruptures possibles. Futuribles, pp. 5-23.

Laudadio, V., Tufarelli, V., Dario, M., Cazzato, E., and Modugno, G.d. (2009). Evaluation of dehulled and micronized lupin (*Lupinus albus* L.) and pea (*Pisum sativum* L.) seed meal as an alternative protein source in diets for pre-laying hens. World Poultry Science Association (WPSA), 2nd Mediterranean Summit of WPSA, Antalya, Turkey, 4-7 October 2009, pp. 495-498.

Laurent, C., Berriet-Solliec, M., Kirsch, M., Labarthe, P., and Trouvé, A. (2010). Multifunctionality of agriculture, public policies and scientific evidences: some critical issues of contemporary controversies. *Applied Studies in Agribusiness and Commerce – APSTRACT*, pp. 53-58.

Le Bail, M., Lecroart, B., Gauffreteau, A., Angevin , F., and Mésséan, A. (2010). Effect of the structural variables of landscapes on the risks of spatial dissemination between GM and non-GM maize. *European Journal of Agronomy*, 33, pp. 12-23.

Lecoq, E., Holt, K., Janssens, J., Legris, G., Pleysier, A., Tinland, B., and Wandelt, C. (2007). General surveillance: roles and responsibilities, the industry view. *Journal Fur Verbraucherschutz Und Lebensmittelsicherheit-Journal of Consumer Protection and Food Safety*, 2, pp. 25-28.

Lee, B., Kim, C.G., Park, J.Y., Park, K.W., Kim, H.J., Yi, H., Jeong, S.C., Yoon, W.K., and Kim, H.M. (2009). Monitoring the occurrence of genetically modified soybean and maize in cultivated fields and along the transportation routes of the Incheon Port in South Korea. *Food Control*, 20, pp. 250-254.

Lehuger, S., Gabrielle, B., and Gagnaire, N. (2009). Environmental impact of the substitution of imported soybean meal with locally-produced rapeseed meal in dairy cow feed. *Journal of Cleaner Production*, 17, pp. 616-624.

LMC International (2009a). Evaluation of measures applied under the Common Agricultural Policy to the protein crop sector. Case study monographs (Report to the European Commission), LMC International, ed., 231 p. Available at: http://ec.europa.eu/agriculture/eval/reports/protein_crops/case_studies_en.pdf

LMC International (2009b). Evaluation of measures applied under the Common Agricultural Policy to the protein crop sector. Main Report (Report to the European Commission.), L. International, ed., pp. vIII, ii, 169. Available at: http://ec.europa.eu/agriculture/eval/reports/protein_crops/fulltext_en.pdf

Lu, Y., Wu, K., Jiang, Y., Xia, B., Li, P., Feng, H., Wyckhuys, K.A.G., and Guo, Y. (2010). Mirid bug outbreaks in multiple crops correlated with wide-scale adoption of Bt cotton in China. *Science*, 328, pp. 1151-1154.

Marsden, T. (2008). Agri-food contestations in rural space: GM in its regulatory context. *Geoforum*, 39, pp. 191-203.

Messéan, A., Angevin, F., Gomez-Barbero, M., Menrad, K., and Rodriguez-Cerezo, E. (2006). New case studies on the coexistence of GM and non-GM crops in European agriculture. EUR22102 EN (Seville, Spain, DG-JRC-IPTS), 116 p. Available at: http://www.jrc.es/home/pages/eur22102enfinal.pdf

Messéan, A., Squire, G., Perry, J., Angevin, F., Gomez, M., Townend, P., Sausse, C., Breckling, B., Langrell, S., Dzeroski, S., et al. (2009). Sustainable introduction of GM crops into european agriculture: a summary report of the FP6 SIGMEA research project. *OCL - Oleagineux, Corps Gras, Lipides*, 16, pp. 37-51.

Milanesi, J. (2008). Analyse des coûts induits sur les filières agricoles par les mises en culture d'organismes génétiquement modifiés (OGM) - Etude sur le maïs, le soja et le poulet Label Rouge, C.d.R.e.G. (CREG). ed. (Pau, France, Université de Pau et des Pays de l'Adour), 123 p. Availabel at :

Milanesi, J. (2009). Quel avenir pour les filières animales « sans OGM » en France? Illustration par le poulet Label Rouge. In 3èmes journées de recherches en sciences sociales INRA SFER CIRAD (Montpellier, France). Available at : http://hal.archives-ouvertes.fr/hal-00521222/en/

Milanesi, J. (2011). Current and future availability of non-genetically modified soybean seeds in the U.S., Brazil and Argentina. In GM and non-GM supply chains: their coexistence and traceability, Y. Bertheau, ed. (Wiley Publishing Ltd.), *In Press*.

Miller, H.I. (2010). Split approvals and hot potatoes. *Nature Biotechnology*, 28, pp. 552-553.

Monkemeyer, W., Schmidt, K., Beissner, L., Schiemann, J., and Wilhelm, R. (2006). A critical examination of the potentials of existing German networks for GMO monitoring. *Journal fur Verbraucherschutz und Lebensmittelsicherheit*, 1, 67-71.

Monsanto Co. (2006). Monitoring report. Mon810 cultivation. Czech republic, France, Germany, Portugal and Spain. 2005, 236 p. Available at: https://yieldgard.eu/en-us/YieldGardLibraryGrower/2005%20YieldGard%20Monitoring%20Report.pdf

Monsanto Co. (2009a). Annual monitoring report on the cultivation of MON 810 in 2008. Czech Republic, Germany, Portugal, Slovakia, Poland, Romania and Spain, 388 p. Available at: https://yieldgard.eu/en-us/YieldGardLibraryGrower/2008%20YieldGard%20Monitoring%20Report.pdf

Monsanto Co. (2009b). Monitoring Report. MON 810 cultivation. Spain. 2003-2004, 57 p. Available at: https://yieldgard.eu/en-us/YieldGardLibraryGrower/20032004%20YieldGard%20Monitoring%20Report.pdf

Naeve, S.L., Orf, J.H., and Miller-Garvin, J. (2010). 2010 analysis of the U.S. non-GMO food soybean variety pipeline. In Japan Soy Food Summit (Tokyo, Japan). Available at: http://soybase.org/meeting_presentations/soybean_breeders_workshop/SBW_2011/Naeve.pdf

Noussair, C., Robin, S., and Ruffieux, B. (2004). Revealing consumers' willingness-to-pay: A comparison of the BDM mechanism and the Vickrey auction. *Journal of Economic Psychology*, 25, pp. 725-741.

Nowicki, P., Aramyan, L., Baltussen, W., Dvortsin, L., Jongeneel, R., Domínguez, I.P., Wagenberg, C.v., Kalaitzandonakes, N., Kaufman, J., Miller, D., *et al.* (2010). Study on the implications of asynchronous GMO approvals for EU imports of animal feed products. Final report (Contract N° 30-CE-0317175/00-74). Executed on behalf of Directorate-General for Agriculture and Rural Development, European Commission. Available at: http://ec.europa.eu/agriculture/analysis/external/asynchronous-gmo-approvals/full-text_en.pdf

Orf, J.H. (2004). Overview of recent genetic improvement in public and private breeding programs in the USA. pp. 220-227.

Ostergard, H., Finckh, M.R., Fontaine, L., Goldringer, I., Hoad, S.P., Kristensen, K., van Bueren, E.T.L., Mascher, F., Munk, L., and Wolfe, M.S. (2009). Time for a shift in crop production: embracing complexity through diversity at all levels. *Journal of the Science of Food and Agriculture*, 89, pp. 1439-1445.

Otiman, I.P., Badea, E.M., and Buzdugan, L. (2008). Roundup Ready soybean, a Romanian story. *Bulletin of University of Agricultural Sciences and Veterinary Medicine Cluj-Napoca Animal Science and Biotechnologies*, 65, pp. 352-357.

Owen, M., Dixon, P., Shaw, D., Weller, S., Young, B., Wilson, R., and Jordan, D. (2010). Sustainability of glyphosate-based weed management: the benchmark study. *ISB News Report August 2010*, 3.

Owen, M.D.K. (2009). Herbicide-tolerant genetically modified crops: resistance management. In Environmental impact of genetically modified crops, Ferry, N. and Gatehouse, A. M. R. (eds). *CABI*. pp. 115-164.

Pascher, K., Moser, D., Dullinger, S., Sachslehner, L., Gros, P., Traxler, A., Sauberer, N., Frank, T., and Grabherr, G. (2009). Establishment of an Austrian monitoring design to identify potential ecological effects of genetically modified plants. *In* Fourth International Conference on Coexistence between Genetically Modified (GM) and Non-GM based Agricultural Supply Chains (GMCC09) (Melbourne, Australia,

GMCC09), 11 p. Available at: http://www.gmcc-09.com/wp-content/uploads/poster_03_82-pascher_poster.pdf

Petit, S. (2009). The dimensions of land use change in rural landscapes: Lessons learnt from the GB Countryside Surveys. *Journal of Environmental Management*, 90, pp. 2851-2856.

Popp, J. (2008). The future of GM crops. *Hungarian Agricultural Research*, 2/3, pp. 17-20.

Pouliquen, A. (2001). Competitiveness and farm incomes in the CEES agri-food sectors. Implication before and after accession for EU markets and policies., 95 p. Available at: http://ec.europa.eu/agriculture/publi/reports/ceeccomp/fullrep_en.pdf

République Française (2008). Loi n° 2008-595 du 25 juin 2008 relative aux organismes génétiquement modifiés. NOR: DEVX0771876L. *Journal Officiel de la République française*, pp. 10218-10228.

Reuter (2010). E.U. considers easing rules on biotech crops in animal feed. In *The New-York Times, October 8, 2010*. Available at: http://www.nytimes.com/2010/10/09/business/global/09gmo.html?_r=3&ref=g enetic_engineering

Roberson, R. (2010). Super weeds put USDA on hotseat, *CheckBiotech*, 2 p.

Roebroeck, L. (2002). Factsheet soybean, T.N. Biotechnology and Consumer Foundation, ed. (The Hague, The Netherlands), 44 p. Available at: http://www.voedingscentrum.nl/resources2008/Factsheet_sojapdf.pdf

Sabalza M., Miralpeix B., Twyman R.M., Capell T. and Christou P. (2011). EU legitimizes GM crops exclusion zones. Nature Biotechnology, 29, 4, pp. 315-317.

Sanvido, O., Aviron, S., Romeis, J., and Bigler, F. (2007a). Challenges and perspectives in decision-making during post-market environmental monitoring of genetically modified crops. *Journal Fur Verbraucherschutz Und Lebensmittelsicherheit-Journal of Consumer Protection and Food Safety*, 2, pp. 37-40.

Sanvido, O., Romeis, J., and Bigler, F. (2007b). Ecological impacts of genetically modified crops: Ten years of field research and commercial cultivation. *Advances in Biochemical Engineering/Biotechnology*, 107, pp. 235-278.

Sanvido, O., Romeis, J., and Bigler, F. (2009a). An approach for post-market monitoring of potential environmental effects of Bt-maize expressing Cry1Ab on natural enemies. *Journal of Applied Entomology*, 133, pp. 236-248.

Sanvido, O., Romeis, J., and Bigler, F. (2009b). Monitoring or Surveillance? Balancing between theoretical frameworks and practical experiences. *Journal fur Verbraucherschutz und Lebensmittelsicherheit*, 3, pp. 4-7.

Sanvido, O., Widmer, F., Winzeler, M., and Bigler, F. (2005). A conceptual framework for the design of environmental post-market monitoring of genetically modified plants. *Environmental Biosafety Research*, 4, 13-27.

Schiemann, J. (2007). Proceedings of the international workshop "Post Market Environmental Monitoring of Genetically Modified Plants: Harmonisation and Standardisation - a Practical Approach". 26-27 April, 2007, Berlin, Germany. *Journal fur Verbraucherschutz und Lebensmittelsicherheit*, 2, 91 p.

Schmidt, K., Mönkemeyer, W., Böttinger, P., Wilhelm, R., and Schiemann, J. (2009). Use of existing networks for post-market monitoring? *Journal fur Verbraucherschutz und Lebensmittelsicherheit*, 3, 13-13.

Seitz, H., Zughart, W., Finck, M., Beismann, H., Berhorn, F., and Eikmann, T. (2010). Standardization of methods for monitoring the environmental effects of genetically modified plants, development of VDI guidelines: a final report an R & D project (FKZ 804 67 010) on behalf of the Federal Office for Nature Conservation. BfN - Skripten (Bundesamt fur Naturschutz), 68 p. Available at: http://www.bfn.de/fileadmin/MDB/documents/service/Skript267.pdf

Smale, M., ed. (1998). Farmers, gene banks and crop breeding: economic analyses of diversity in wheat, maize, and rice (Springer Verlag). 288 p.

Stein, A.J., and Rodriguez-Cerezo, E. (2009). The global pipeline of new GM crops. EUR 23486 EN. In JRC scientific and technical reports, JRC-IPTS, ed. (Seville, SP, JRC-IPTS), 114 p. Available at: http://ftp.jrc.es/EURdoc/report_GMOpipeline_online_preprint.pdf

Stein, A.J., and Rodriguez-Cerezo, E. (2010a). International trade and the global pipeline of new GM crops. *Nature Biotechnology*, 28, pp. 23-25.

Stein, A.J., and Rodriguez-Cerezo, E. (2010b). Low-level presence of new GM crops: an issue on the rise for countries where they lack approval. *AgBioForum*, 13, pp. 173-182.

Stein, A.J., and Rodríguez-Cerezo, E. (2009). What can data on GMO field release applications in the USA tell us about the commercialisation of new GM crops? JRC52545. In JRC Technical notes, JRC-IPTS, ed., 16 p. Available at: http://ftp.jrc.es/EURdoc/JRC52545.pdf

Takada, A. (2010). China to compete with U.S., Canada in Asian market for soybean foods. In Bloomberg News. Available at: http://www.bloomberg.com/news/2010-11-19/china-s-non-modified-soybeans-to-compete-with-u-s-canada-in-asian-market.html

Tallage, Agri market forecasting (2010). Study on modelling of feed consumption in the European Union. Tender No. AGRI-2008-EVAL-09, Tallage, ed. (European Commission, DG-AGRI), 73 p. Available at: http://ec.europa.eu/agriculture/analysis/external/feed/fulltext_en.pdf

Taylor, R.D., and Koo, W.W. (2010). 2010 outlook of the U.S. and world corn and soybean industries, 2009-2019, C.f.A.P.a.T. Studies, ed. (Fargo, North Dakota, USA, Department of Agribusiness and Applied Economics. North Dakota State University), 35 p. Available at: https://ageconsearch.umn.edu/bitstream/92003/2/AgReport665%20-%20Corn_and_soybean.pdf

Then, C., and Stolze, M. (2010). Economic impacts of labelling thresholds for the adventitious presence of genetically engineered organisms in conventional and organic seed. Seed purity: costs, benefits and risk management strategies for maintaining markets free from genetically engineered plants (IFOAM EU Group), 60 p. Available at: http://www.ifoam.org/about_ifoam/around_world/eu_group-new/positions/publications/pdf/IFOAMEU_GMO-freeSeedStudy.pdf

Tinland, B. (2008). Implementation of post-market monitoring of insect-protected maize MON 810 in the UE. AFPP – 8ème Conference Internationale sur les Ravageurs en Agriculture, Montpellier SupAgro, France, 22-23 Octobre 2008, 165. Available at : http://www.cabdirect.org/abstracts/20093081604.html

Tinland, B., Delzenne, P., and Pleysier, A. (2007). Implementation of a post-market monitoring for insect-protected maize MON 810 in the EU. *Journal Fur*

Verbraucherschutz Und Lebensmittelsicherheit-Journal of Consumer Protection and Food Safety, 2, 7-10.

TNS Opinion & Social (2010). Eurobaromètre 72. L'opinion publique dans l'Union Européenne (Brussels, B).

van Tongeren, F., van Meijl, H., and Surry, Y. (2001). Global models applied to agricultural and trade policies: a review and assessment. *Agricultural Economics*, 26, pp. 149-172.

Wal, J.M., Hepburn, P.A., Lea, L.J., and Crevel, R.W.R. (2003). Post-market surveillance of GM foods: applicability and limitations of schemes used with pharmaceuticals and some non-GM novel foods. *Regulatory Toxicology and Pharmacology*, 38, pp. 98-104.

Wandelt, C. (2007). Implementation of general surveillance for amflora potato cultivation - Data management. *Journal Für Verbraucherschutz Und Lebensmittelsicherheit-Journal of Consumer Protection and Food Safety*, 2, pp. 70-71.

Wilhelm, R., Beissner, L., and Schiemann, J. (2003). Concept for the realisation of a GMO-monitoring in Germany. *Nachrichtenblatt des Deutschen Pflanzenschutzdienstes*, 55, pp. 258-272.

Wilhelm, R., Sanvido, O., Castanera, P., Schmidt, K., and Schiemann, J. (2009). Monitoring the commercial cultivation of Bt maize in Europe – conclusions and recommendations for future monitoring practice. Environmental Biosafety Research, 8, 4, pp. 219-225.

Wilhelm, R., and Schiemann, J. (2006). Does the baseline concept provide appropriate tools for decision making? *Journal fur Verbraucherschutz und Lebensmittelsicherheit*, 1, pp. 75-77.

Windels, P., Alcalde, E., Lecoq, E., Legris, G., Pleysier, A., Tinland, B., and Wandelt, C. (2009). General surveillance for import and processing: the EuropaBio approach. *Journal fur Verbraucherschutz und Lebensmittelsicherheit*, 3, pp. 14-16.

Wolt, J.D., Keese, P., Raybould, A., Fitzpatrick, J.W., Burachik, M., Gray, A., Olin, S.S., Schiemann, J., Sears, M., and Wu, F. (2010). Problem formulation in the environmental risk assessment for genetically modified plants. Transgenic Research, 19, pp. 425-436.

Zughart, W., Benzler, A., Berhorn, F., Sukopp, U., and Graef, F. (2008). Determining indicators, methods and sites for monitoring potential adverse effects of genetically modified plants to the environment: the legal and conceptional framework for implementation. *Euphytica*, 164, pp. 845-852.

3

High Pressure Treatments of Soybean and Soybean Products

Kashif Ghafoor[1], Fahad Y.I. AL-Juhaimi[1] and Jiyong Park[2]
[1]Department of Food and Nutrition Sciences, King Saud University, Riyadh,
[2]Department of Biotechnology, Yonsei University, Seoul 120-749,
[1]Saudi Arabia
[2]South Korea

1. Introduction

Soybean (*Glycine max*) is an industrial crop extensively cultivated for its oil and protein content. The global demand for soybean has increased dramatically over the last few years. Since the application of high hydrostatic pressure (HHP) to different food systems in the late 1800s (Bridgman, 1914), many researchers today are again applying this promising technology to the processing of foods. There is increasing worldwide interest in the use of HHP because of the advantages of this technology over other methods of processing and preservation. HHP offers homogeneity of treatment at every point in the product due to the fact the applied pressure is instantaneously and uniformly distributed within the HHP chamber. Therefore, processing time is not a function of sample size. Important advantages in using this technology are (1) significant or total inactivation of microorganisms (Knorr, 1993), and (2) better functional and nutritional retention of ingredients in the processed products, with improved food quality parameters (Hayashi, 1989). In addition, there is significant energy economy in comparison to thermal stabilization techniques, because once the desired pressure is reached, it can be maintained without the need for further energy input. Recently, processing of foods with HHP and low to moderate temperatures (less than 70 °C) was introduced as an alternative to thermal preservation. However, it was not until the late 1980s that researchers began investigating ways to commercialize high pressure treatment of foods (Hayashi, 1989).

The main uses of soybeans can be categorized into three groups: industrial, human food and livestock feed. Soybeans for human consumption are processed in many forms. Of major importance in Asian countries are soy foods such as tofu, soy sauce, miso, soy sprouts, and soymilk. These soy foods were recently introduced to the American market as were soy flour and soymilk previously. Soy flour is often mixed with other flours to increase the protein content, and soymilk provides an alternative source of protein for people allergic to the protein in cow's milk (Riaz, 1988). In addition to its versatility, the soybean is a commodity of unique chemical composition. On a mean dry matter basis, soybeans contain about 40% protein and 20% oil. Soybeans contains the highest protein content among food crops, and are second highest with respect to oil content among all the food legumes. Thus, the composition of soy products range in protein content from about 40% for full-fat flours to 95% or more for protein isolates (Wolf & Cowan, 1975).

2. High pressure processing of soybean

Due to the great variety of foods obtained from soybeans, different processing methods are required. Traditional methods include two types of processes: (1) fermentation, which uses fungi to produce fermented products; and (2) soaking and grinding of the soybeans to make bean curd and soymilk. During fermentation, protein is digested into peptides and later into amino acids for increasing digestibility of protein by the human body. Soaking and grinding are usually combined with thermal treatment to inactivate biologically active compounds such as trypsin inhibitors, lipoxygenase, and hemagglutinins, while increasing digestibility of proteins. In addition, thermal treatment (continued steaming) helps to diminish the characteristic beany flavor of raw soy products due to the volatilization of monocarbonyl compounds, which results from oxidation of fatty acids by the enzyme lipoxygenase. However, excessive heating may destroy certain amino acids that are sensitive to heat such as lysine, with losses of possibly more than 50% (Estrada-Giron et al., 2005).

Until recently, little research was being done on the effects of HHP on soybean grains and their sub-products. This is because of the recent interest in the use of HHP as a potential technology to improve the quality of cereals and textured products. These studies include reduction in the microbial population of soymilk curd, commonly known as tofu, to obtain a product with longer shelf life and to avoid secondary contamination. The solubilization of protein from whole soybean grains subjected to different treatments of pressure, time, and temperature was also reported. Additional information is available in a more extensive context about the effects of this technology on the inactivation of pure soybean lipoxygenase and lipoxygenase from some legumes (Estrada-Giron et al., 2005).

2.1 Microbial inactivation

The effects of high hydrostatic pressure on microbial inactivation depend on several factors such as type of microorganism, extent and duration of the high pressure treatment, temperature, and composition of suspension media or food. Therefore, suitable pressure treatment should be applied taking into account these factors to assure microbial inactivation of pathogenic, spoilage, and vegetative cells present in foods. Prestamo et al. (2000) reported that the microbial population of tofu pressurized at 400 MPa and 5 °C for 5, 30, and 45 min decreases from an initial microorganism count of 5.54×10^4 cfu/g to 0.31, 1.56, or 2.38 log units, respectively. Prestamo et al. (2000) also postulated that the effectiveness of HHP treatment to reduce microbial population at 400 MPa largely depends on the exposure time (Fig. 1). In the same study, after HHP treatment of tofu, psycrotrophs were reduced 2 log units from an initial population of 1×10^3 cfu/g. Mesophilic microorganisms were reduced 1 log unit from an initial number of 1.6×10^3 cfu/g, whereas yeast and molds decreased from an initial population of 2.64×10^3 cfu/g to 1×10^2 cfu/g. Other microorganisms such as *Pseudomonadaceae*, *Salmonella*, and Gram-negative bacteria (confirmed before HHP treatment) were not detected after HHP treatment of tofu. *Yersenia enterocolitica* and *Listeria monocytogenes*, which are more resistant to high pressure, were not found before and after HHP treatment. *Hafnia halvei* and *Bacillus cereus* remained active after high pressure treatment of tofu.

In addition to temperature and the extent and duration of high pressure treatment, a factor that significantly influences the effectiveness of HHP treatment on the inactivation and consequently the reduction in microbial population is the medium composition in which microorganisms are dispersed.

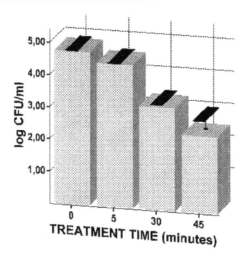

Fig. 1. Viable aerobic mesophilic population in tofu after treatment at 400 MPa and 5 °C for 5, 30, and 45 min (from Prestamo et al., 2000).

Food constituents such as sucrose, fructose, glucose, and salts affect the baro-resistance of microorganisms present in food (Oxen & Knorr, 1993). This effect is often observed since food constituents appear to protect microorganisms from the effects of high pressure. Therefore, a non-nutritive solution can reduce the microorganism's baro-tolerance. The presence of microorganisms such as *Hafnia halvei* and *Bacillus cereus* that remained active after HHP treatment could explain the baro-protective effect that food components exert over the extent on microbial reduction (Prestamo et al., 2000).

2.2 Proteins
Unlike allergenic proteins in cereals such as rice, soybeans contain a large number of proteins with important functional properties (Wolf & Cowan, 1975). Eighty percent of the proteins in soybeans are glycinin and β- and γ-conglycinin, which are globular salt-soluble proteins. On the basis of their sedimentation constants at pH 7.6 and ionic strength buffer of 0.5, the globulins are characterized as 11S or glycinin and 7S or β- and γ-conglycinin (Fukushima, 1991), with other less abundant globulins including 2S or α-conglycinin, 9S globulins, and 15S globulins. Functional properties associated with these kinds of soybean proteins:
1. Hydration properties such as swelling, solubility, and viscosity.
2. Protein–protein interactions resulting in precipitation and gelling.
3. Interfacial properties identified as surface tension and related foam/emulsion stability (Utsumi et al., 1998).
Therefore, the method of processing intact soybeans is important since the retention of proteins in the soybean seed is of special interest because of the high-quality vegetable protein, which also contains most of the essential amino acids (Steinke et al., 1992). When soybeans are immersed in hot water at 50–60 °C for 1 h, a considerable amount of protein solubilized from the soybean seeds is released to the surrounding water (Asano et al., 1989). Later studies identified these solubilized proteins as 7S globulins, which accounted for about 3% of the total protein in mature soybean seeds (Hirano et al., 1992).

In soybean seeds immersed in distilled water and treated at 300 MPa and 20 °C for 0–180 min, the solubilized proteins accounted for 0.5–2.5% of the total seed proteins. No apparent changes in shape, color, and size between treated and untreated soybeans were reported. The solubility of protein in surrounding water increased with increasing pressure, reaching a maximum value at 400 MPa (Fig. 2). Similar to what occurs when heat treatment is applied, SDS-PAGE patterns of high pressure treated seeds exhibited solubilization of 7S globulin, consisting of 27 and 16 KDa bands with staining intensity increasing as pressure increased to 400 MPa. The increase in staining intensity is indicative of the amount of release protein; thus, at higher intensity larger amounts of protein are released. At 700 MPa, 11S glycinin and 2S β-conglycinin also increased their staining intensity (Omi et al., 1996).

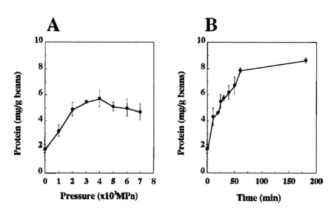

Fig. 2. Effect of high-pressure treatment on the release of proteins from soybean seeds. Water-immersed soybean seeds were pressurized at (A) 0-700 MPa and 20 °C for 25 min and at (B) 300 MPa and 20 °C for 0-180 min (From Omi et al., 1996).

2.3 Enzymes

At present, research regarding the inactivation of enzymes in intact grains and their sub-products is scarce. However, it is well known that high pressure modifies the activities of a whole range of unwanted food enzymes, which can result in a reduction in food quality or cause spoilage during storage. Recent investigations have reported the effect of combined pressure and temperature on soybean lipoxygenase. Lipoxygenase is one of the main anti-nutritional factors in soybean processing, which is also known to occur in other legume seeds, some cereal grains, and oil seeds. At least three types of lipoxygenase are well identified in soybeans as lipoxygenase I, II, and III. These enzymes catalyze the oxidation of unsaturated fatty acids in the presence of molecular oxygen. The presence of lipoxygenase can have detrimental effects on foods, for example:

1. Degradation of the essential fatty acids linoleic, linolenic, and arachidonic acid to yield fatty acid hydroperoxides.
2. Degradation of formed hydroperoxides, resulting in the formation of volatile compounds such as aldehydes, ketones, and alcohols, which cause the development of off-flavors.
3. Production of free radicals that can damage other compounds, including vitamins and proteins (Whitaker, 1972).

In soybean products, off-flavor development is highly dependent on the action of lipoxygenase since subsequent decomposition of the resulting hydroperoxides yields especially rancid flavor and beany aroma. Nevertheless, lipoxygenase is sensitive to heat and is destroyed at 82 °C when processed for 15 min (Baker & Mustakas, 1972). It is well known that thermal processing methods reduce considerably or completely inactivate unwanted enzyme activity, which limits or largely determines the conditions of storage needed to extend shelf life of food products. Although the behavior of enzymes under the influence of heat has been extensively studied, the effects of HHP treatment on enzyme inactivation are not clearly understood.

P/T treatment (MPa/°C)	Total time (min)	Cycling time	Activity retention
350/40	40	1×40	0.709
350/40	40	4×10	0.314
400/35	40	4×10	0.324
400/40	40	4×10	0.300
450/40	40	1×40	0.481
475/10	40	1×40	0.652
475/25	60	4×15	0.160
475/30	60	4×15	0.373
500/15	30	1×30	0.367
500/15	30	3×10	0.018
525/25	20	1×20	0.099
525/25	20	2×10	0.110

Adapted from Ludikhuyze et al. (1998a).

Table 1. Influence of multi-cycling on the inactivation of lipoxygenase in Tris–HCl at pH 9

Thermal inactivation of enzymes at atmospheric pressure occurs in the temperature range 60–70 °C. In contrast, pressure–temperature inactivation occurs in the pressure range 50–650 MPa at temperatures between 10 and 64 °C. Also, depending on the objectives of the research, pressure treatment may be applied in a single cycle or multi-cycles. Multi-cycling is the multiple application of pressure alone or in combination with temperature for the same total treatment time but with various numbers of cycles. Ludikhuyze et al. (1998a) reported the multi-cycling application of pressure to inactivate lipoxygenase. These authors found that in the pressure range 350–525 MPa and thermal treatment at 10–40 °C, the use of multi-cycles exerted an additional inactivation effect on lipoxygenase, compared to single cycle treatments (Table 1). Furthermore, temperature treatments at 10 °C caused an enhanced inactivation of lipoxygenase because the temperature inside the vessel dropped below zero upon depressurization.

In crude green bean extract, irreversible lipoxygenase inactivation was reported in the temperature range 55–70 °C at ambient pressure, whereas at room temperature, pressures around 500 MPa were required to inactivate lipoxygenase. High pressure treatment at 200 MPa and 50 °C resulted in 10% inactivation, while at least 50%, lipoxygenase inactivation occurred at pressures greater than 500 MPa and thermal treatment between 10 and 30 °C (Indrawati et al., 1999). The effect of HHP on enzyme inactivation in food systems is different compared to its effects on pure components dissolved in buffer solutions. As an example, solutions of commercial soybean lipoxygenase type I (100 mg/ml) dissolved in 0.2 M citrate-phosphate (pH range of 4.0–9.0) and 0.2 M Tris buffer (pH range of 6.0–9.0) were subjected to pressures of 0.1, 200, 400, and 600 MPa for 20 min. Under these conditions, lipoxygenase in citrate-phosphate buffer lost more than 80% of its activity at alkaline pH, whereas it was completely inactivated at acidic conditions and pressure treatment of 400 and 600 MPa (Tangwongchai et al., 2000). In Tris buffer, lipoxygenase activity was significantly inactivated at pH 9.0 and 400 MPa and lost all activity at 600 MPa and all pH values. Similar results were observed by Seyderhelm et al. (1996) who reported that lipoxygenase in Tris buffer pH 7.0 was completely inactivated at 600 MPa and temperatures 45 and 50 °C for 10 min and 5–10 min, respectively.

3. Effect of HHP on technological properties of tofu

Tofu with 0, 2.5 or 5% trehalose was pressurized at 100–686 MPa and approximately -20 °C for 60 min to determine changes in temperature and sensory evaluation of high-pressure-frozen tofu as affected by trehalose.

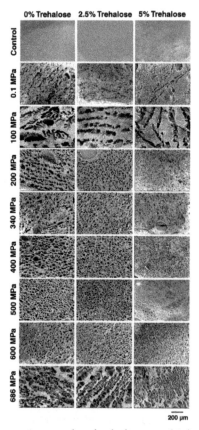

Fig. 3. Cryo-scanning electron micrographs of tofu frozen at high pressure. Tofu was pressurized at 100–686 MPa and approximately −20 °C for 60 min. After decompression, it was left in the pressure medium for 30 min then thawed at 20 °C. Control: unfrozen tofu (From Fuchigami et al., 2002).

Tofu froze during pressurization at 100 or 686 MPa; conversely, tofu did not freeze between 200 and 600 MPa and -20 °C, but it froze rapidly when the pressure was released. It was found that tofu frozen at 0.1, 100 or 686 MPa had larger ice crystals and was firmer (less like unfrozen tofu) than tofu frozen at 200–600 MPa. In the sensory evaluation, results showed that mouth feel (texture. of tofu frozen at 400 MPa) was more like the control when 2.5% trehalose was added (Fuchigami et al., 2002).

The micro structure of the tofu gel network high pressure frozen at 686 MPa was compared with untreated tofu (Fig. 3). Tofu (0% trehalose) frozen at 0.1–500 MPa maintained a comparatively coarse network (data not shown), but tofu gel frozen at 686 MPa was compressed. Compression of the protein gel network might have occurred above 600–686 MPa; however, the gel network in tofu frozen at 686 MPa became coarse with the addition of trehalose. This indicates that trehalose with high-pressure-freezing appears to protect against compression (effects of concentration of protein and coagulants on frozen tofu) (Fuchigami et al., 2002).

4. High pressure inactivation of soybean lipoxygenase

The high pressure inactivation of lipoxygenase in soy milk and crude soybean extract was studied in the pressure range 0.1–650 MPa with temperature varying from 5 to 60 °C.

T (°C)	Soy milk	Crude soybean extract
63	0.55 ± 0.02[a]	0.68 ± 0.03
	$r^2 = 0.993$	$r^2 = 0.995$
65	1.35 ± 0.05	1.54 ± 0.05
	$r^2 = 0.995$	$r^2 = 0.996$
67	3.57 ± 0.09	4.31 ± 0.12
	$r^2 = 0.998$	$r^2 = 0.997$
69	13.25 ± 0.50	14.43 ± 0.42
	$r^2 = 0.994$	$r^2 = 0.997$
71	47.72 ± 4.64	53.40 ± 2.74
	$r^2 = 0.972$	$r^2 = 0.987$
E_a (kJ/mol)	538.78 ± 29.04	526.94 ± 29.54
	$r^2 = 0.991$	$r^2 = 0.991$

[a] Standard error. (From Wang et al., 2008).

Table 2. Estimated inactivation rate constants ($\times 10^{-2}$ min^{-1}) for the isothermal inactivation of lipoxygenase in soy milk and in crude soybean extract

For both systems, the isobaric–isothermal inactivation of lipoxygenase was irreversible and followed a first-order reaction at all pressure–temperature combinations tested. In the entire pressure–temperature area studied, the lipoxygenase inactivation rate constants increased with increasing pressure at constant temperature for both systems; the rate constants were somewhat smaller in soy milk system than in crude soybean extract. At constant elevated pressure, lipoxygenase exhibited the greatest stability around 20 °C in both systems, indicating that the Arrhenius equation was not valid over the entire temperature range. For both systems, the temperature dependence of the lipoxygenase inactivation rate constants at high temperature decreased with increasing pressure, while the highest sensitivity of the lipoxygenase inactivation rate constants to pressure was observed at about 30 °C. The pressure–temperature dependence of the lipoxygenase inactivation rate constants was successfully described either using an empirical mathematical model or using a thermodynamic kinetic model for both systems. On a kinetic basis, neither the reaction order of inactivation nor the pressure and temperature sensitivities of the inactivation rate constants were influenced by the different levels of food complexity between the two systems (Wang et al., 2008).

In 63 to 71 °C temperature range, isothermal inactivation of soybean lipoxygenase followed first-order kinetics, allowing inactivation rate constants (k) to be determined from plots of the natural logarithm of relative residual activity, as a function of inactivation time. The estimated k values, together with standard errors and regression coefficients, are summarized in Table 2. Over the entire temperature domain studied, lipoxygenase was less thermostable in crude soybean extract than in soy milk and the temperature sensitivity of the rate constants for lipoxygenase inactivation in both systems could be estimated using the Arrhenius relation (Wang et al., 2008). First-order kinetics for thermal inactivation of soybean lipoxygenase has been frequently reported in the literature (Indrawati et al., 1999; Ludikhuyze et al., 1998b). Ludikhuyze et al. (1998a, 1998c) investigated the thermal inactivation kinetics of commercial soybean lipoxygenase in Tris–HCl buffer (0.01 M, pH 9) at two different concentrations (0.4 and 5 mg/ml) over the temperature range 60–70 °C (Table 3).

T (°C)	Lipoxygenase concentration	
	0.4 mg/ml[b]	5 mg/ml[c]
60		2.09 ± 0.13
		r^2 = 0.978
62	2.02 ± 0.09[a]	4.86 ± 0.21
	r^2 = 0.987	r^2 = 0.991
64	4.94 ± 0.16	10.8 ± 0.61
	r^2 = 0.993	r^2 = 0.984
66	9.18 ± 0.32	29.1 ± 3.37
	r^2 = 0.992	r^2 = 0.949
68	15.5 ± 0.52	
	r^2 = 0.992	
E_a (kJ/mol)	319.8 ± 27.3	408.2 ± 14.7
	r^2 = 0.986	r^2 = 0.997

[a] Standard error. [b] Ludikhuyze et al. (1998c). [c] Ludikhuyze et al. (1998a). (From Wang et al., 2008).

Table 3. Inactivation rate constants ($\times 10^{-2}$ min^{-1}) for the isothermal inactivation of commercial soybean lipoxygenase in 0.01 M, pH 9 Tris–HCl buffer

Lipoxygenase in soy milk or in crude soybean extract exhibited a higher thermal stability with the corresponding smaller inactivation rate constants. The two activation energy values derived from the plots of the natural logarithm of inactivation rate constants, as a function of the reciprocal of the absolute temperature were larger, pointing to higher temperature sensitivity of the k values. Likewise, kinetic inactivation of lipoxygenase from many different sources, such as green peas, green beans, potatoes, asparagus, wheat germ, and germinated barley, have also been studied (Bhirud & Sosulski, 1993; Ganthavorn et al., 1991; Guenes & Bayindirli, 1993; Hugues et al., 1994; Indrawati et al., 1999; Park et al., 1988; Svensson & Eriksson, 1974). Indrawati et al. (1999) reported that thermal inactivation of lipoxygenase in green bean juice could be described by a two-fraction first-order inactivation model, referring to the existence of two fractions (isozymes) with different thermal stability. However, in their study, this phenomenon was not observed.

Van Loey et al. (1999) studies soybean lipoxygenase inactivation [0.4 mg/mL in Tris-HCl buffer (0.01 M, pH 9)] quantitatively under constant pressure (up to 650 MPa) and temperature (-15 to 68 °C) conditions and kinetically characterized by rate constants, activation energies, and activation volumes. The irreversible lipoxygenase inactivation followed a first-order reaction at all pressure-temperature combinations tested. In the entire pressure-temperature area studied, LOX inactivation rate constants increased with increasing pressure at constant temperature. On the contrary, at constant pressure, the inactivation rate constants showed a minimum around 30 °C and could be increased by either a temperature increase or decrease. On the basis of the calculated rate constants at 102 pressure temperature combinations, an iso-rate contour diagram was constructed as a function of pressure and temperature. The pressure-temperature dependence of the LOX inactivation rate constants was described successfully using a modified kinetic model (Van Loey et al., 1999)

5. Immunoreactivity and nutritional quality of soybean products

Penas et al. (2011) reported that sprouts obtained from HHP-treated soybean seeds demonstrated an important reduction in immune-reactivity. Furthermore, they were a good source of proteins and essential amino acids, with Met and Cys corresponding to the limiting amino acids, as indicated by the chemical score (CS), and a high essential amino acid index (EAAI) (Table 4). These results suggested that HHP could constitute an important technological approach for the industrial production of hypoallergenic and nutritive soybean sprouts.

The HHP treatment of raw seeds (PRS) produced lower Gly and Cys levels than raw seeds (RS), while no significant differences ($P \leq 0.05$) were observed for total EAA (Table 4). The germination process resulted in a significant decrease in Glu, Trp and Cys, while no changes were observed in the other amino acids, compared to RS. The total EAA content showed a 4% reduction compared to RS. Pomeranz et al. (1977) found only minor differences in the amino acid composition of germinated and ungerminated soybean, while Mostafa et al. (1987) observed a marked increase in the relative contents of both EAA and NEAA after germination. The levels of sulphur amino acids in germinated soybean seeds remained almost constant, whereas Asp increased compared to raw seeds. Discrepancies between the data reported by other authors and those reported in the present work could be attributed to differences in the germination conditions and seed varieties.

The application of HHP treatment to seeds prior to germination (GPS) led to a reduction in Glu and Ala as NEAA and Trp, Met and Cys in EAA in comparison with GS, and also Pro, and Ile compared to PRS. GPS showed similar statistical ($P \leq 0.05$) values of total EAA content to GS (32 and 34 g/100 g protein, respectively), whilst significant ($P \leq 0.05$) differences were found compared to RS and PRS (34 g/100 g protein) (Table 4).

In another study, the effects of HPP on soybean cotyledon as a cellular biological material were investigated from the viewpoints of the cell structure and enzyme reaction system (Ueno et al., 2010). Damage to cell structure was evaluated by measuring dielectric properties using the Cole–Cole arc, the radius of which decreased as pressure level increased. Results suggested that cell structure was damaged by HPP. The distribution of free amino acids was measured after HPP (200 MPa) of soybean soaked in water or sodium glutamate (Glu) solution. HPP resulted in high accumulation of free amino acids in water-soaked soybean, due to proteolysis. HPP of soybean in Glu solution caused higher accumulation of γ-aminobutyric acid, suggesting that both proteolysis and specific Glu metabolism were accelerated by HPP. They concluded that HPP partially degraded cell structure and accelerated biochemical reactions by allowing enzyme activities to remain. These events were described as "high-pressure induced transformation" of soybean.

	RS	PRS	GS	GPS	Whole egg protein [B]
Protein (% d.w.)	45.0[a]	44.3[a]	47.3[b]	48.6[c]	
Non-essential amino acids					
Glu	15.9[c]	15.7[c]	14.8[b]	14.0[a]	
Asp	7.25[a]	7.26[a]	7.09[a]	6.80[a]	
Arg	5.44[ab]	5.60[b]	5.05[a]	5.27[ab]	
Pro	4.26[b]	4.25[b]	3.79[ab]	3.71[ab]	
Ala	4.05[b]	3.97[ab]	4.04[b]	3.71[a]	
Ser	3.87[a]	3.94[a]	3.92[a]	3.68[a]	
Gly	3.50[b]	3.07[a]	3.15[ab]	3.11[a]	
Essential amino acids					
Leu	6.10[a]	6.18[a]	6.52[a]	5.68[a]	8.6
Lys	4.32[a]	4.40[a]	4.18[a]	4.13[a]	7.0
Val	4.19[a]	4.26[a]	4.26[a]	4.11[a]	6.6
Phe	4.13[bc]	4.17[c]	3.70[ab]	3.90[abc]	9.3 (Phe+Tyr)
Ile	4.12[b]	4.22[b]	3.78[ab]	3.75[ab]	5.4
Tyr	2.91[c]	2.67[abc]	2.81[bc]	2.57[ab]	
Trp	2.26[c]	2.29[c]	1.99[b]	1.71[a]	1.7
Thr	2.89[b]	2.52[ab]	2.50[ab]	2.32[a]	4.7
His	2.03[a]	1.93[a]	1.96[a]	1.85[a]	2.2
Met	1.32[b]	1.36[b]	1.32[b]	1.16[a]	5.7 (Met+Cys)
Cys	1.24[c]	1.08[b]	1.06[b]	0.94[a]	
Total EAA	35.5[b]	35.1[b]	34.1[ab]	32.1[a]	
CS	45.0	42.8	41.8	37.0	
EAAI	72.4	70.9	68.5	65.6	

[A] Data are the mean of three independent results. Different superscripts in the same row mean statistically significant differences ($P \leq 0.05$). PRS: HHP-treated soybean seeds; RS: raw soybean seeds; GPS: germinated PRS, HHP-treated soybean seeds; GS: germinated soybean seed. EAA: essential amino acid; CS: chemical score; EAAI: essential amino acid index. [B] (FAO/WHO/UNU, 1985). (From Penas et al., 2011).

Table 4. Effect of HHP and/or germination on the total protein and amino acids (g/100 g protein) content of soybean seeds and sprouts during germination.[A]

6. Inactivation of soymilk trypsin inhibitors

Protease inhibitors (PIs) are generally considered the main anti-nutritional factors in soybeans. Soybean PIs belong to a broad class of proteins that inhibit proteolytic enzymes, such as trypsin and chymotrypsin. Both compounds are important animal digestive enzymes for splitting proteins to render dipeptides and tripeptides (Scheider, 1983). However, the specificity of these inhibitors is not necessarily restricted to trypsin and chymotrypsin but also to elastase and serine proteases for which serine constitutes the active site. Nevertheless, the literature reports two main types of soybean PIs, specifically called trypsin inhibitors (TIs). The Kunitz soybean inhibitor, with a molecular weight of 20,000 and two disulfide bridges, exhibits specificity to inhibit trypsin. The Bowman–Birk inhibitor, on the other hand, with a molecular weight ranging from 6000 to 10,000 and seven disulfide bonds, exhibits specificity to inhibit chymotrypsin (Liener, 1994).

Residual trypsin was measured in soymilk subjected to selected pressures, temperatures and holding times. Treatment combination at higher pressures and temperatures, for selected holding times resulted in an increased inhibition rate of trypsin inhibitors in soymilk. It was not possible to obtain inactivation rate parameters for treatments at 550 MPa and 80 °C because the data did not fit a first order kinetics model. However, a clear increase of residual trypsin was observed as treatment times increased. Soaking of soybeans in sodium bicarbonate solution, prior to preparation of soymilk, resulted in smaller inhibition rates of trypsin at the working selected pressures, combined with thermal treatment and holding times, than in soybeans soaked in distilled water. The use of sodium bicarbonate, as soaking medium of soybeans, did not result in a significant increase in the percentage of residual trypsin in soymilk treated at 550 MPa and 80 °C for the selected holding times (Fig. 4).

Ven et al. (2005) also evaluated HPP as an alternative for the inactivation of TIs in soy milk and also studied the effect of HPP on in whole soybeans and soy milk. For complete lipoxygenase inactivation either very high pressures (800 MPa) or a combined temperature/pressure treatment (60 °C/600 MPa) was needed. Pressure inactivation of TIs was possible only in combination with elevated temperatures. For TIs inactivation, three process parameters, temperature, time, and pressure, were optimized using experimental design and response surface methodology. A 90% TIs inactivation with treatment times of <2 min can be reached at temperatures between 77 and 90 °C and pressures between 750 and 525 MPa.

7. Conclusions

High hydrostatic pressure (HHP) processing is an innovative technology for processing of soybean which is an important food from nutritional point of view. HHP enables the inactivation of pathogenic bacteria at ambient temperatures. It also showed an increase in protein solubility and staining intensity due to the release of more protein after application of HHP. It also favored the inactivation of quality deteriorating enzymes such as lipoxygenase at room temperatures. Treatment of soybean with HHP improved the bio-availability of nutrients such as amino acids and the reduction of immune-reactivity. HHP also favored the activity of proteases, probably by reducing the activity of their inhibitors. It can be inferred that soybean and its products which are valuable food commodities can be effectively processed using this innovative processing technology, however more research needs to be done on HHP optimization and its effects on various physicochemical properties of soybean and different soy-foods.

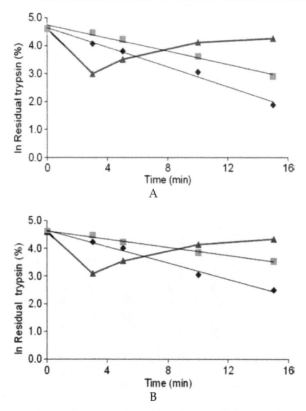

Fig. 4. Residual trypsin in HHP-processed (550 MPa) soymilk from soybeans previously soaked in water (A) and sodium bicarbonate solution (0.5%) (B). Temperatures: 19 (♦), 65 (■), and 80 °C (▲) (From Guerrero-Beltrán et al., 2009).

8. References

Asano, M., Okubo, K., & Yamauchi, F. (1989). Effect of immersing temperature on the behavior of exuding compounds from soybean. *Journal of the Japanese Society for Food Science and Technology*, 36, pp. 636–642, ISSN: 1344-6606

Baker, E. C., & Mustakas, G. C. (1972). Heat inactivation of trypsin inhibitor, lipoxygenase and urease in soybean: Effect of acid and base additive. *Journal of the American Oil Chemists Society*, 50, pp. 137–141, ISSN: 0003-021X

Bhirud, P. R., & Sosulski, F. W. (1993). Thermal inactivation kinetics of wheat germ lipoxygenase. *Journal of Food Science*, 58, pp. 1095–1098, ISSN: 1750-3841

Bridgman, P. W. (1914). The coagulation of albumen by pressure. *Journal of Biological Chemistry*, 19, pp. 511–512, ISSN: 0021-9258

Estrada-Giron, Y., Swanson, B. G., & Barbosa-Canovas, G. V. (2005). Advances in the use of high hydrostatic pressure for processing cereal grains and legumes. *Trends in Food Science and Technology*, 16, pp. 194–203, ISSN: 0924-2244

Fuchigami, M., Ogawa, N., & Teramoto, A. (2002). Trehalose and hydrostatic pressure effects on the structure and sensory properties of frozen tofu (soybean curd). *Innovative Food Science and Emerging Technologies*, 3, pp. 139-147, ISSN: 1466-8564

Fukushima, D. (1991). Recent progress of soybean protein foods: chemistry, technology and nutrition. *Food Review International*, 7, pp. 323–351, ISSN: 8755-9129

Ganthavorn, C., Nagel, C. W., & Powers, J. R. (1991). Thermal inactivation of asparagus lipoxygenase and peroxidase. *Journal of Food Science*, 56, pp. 47–49, 79, ISSN: 1750-3841

Guenes, B., & Bayindirli, A. (1993). Peroxidase and lipoxygenase inactivation during blanching of green beans, green peas and carrots. *LWT-Food Science and Technology*, 26, pp. 406–410, ISSN: 0023-6438

Guerrero-Beltrán, J. A., Estrada-Girón, Y., Swanson, B. G., & Barbosa-Cánovas, G. V. (2009). Pressure and temperature combination for inactivation of soymilk trypsin inhibitors. *Food Chemistry*, 116, pp. 676-679, ISSN: 0308-8146

Hayashi, R. (1989). Application of food processing and preservation: philosophy and development. In: *Engineering and Food*, W. E. L. Spiess, & H. Schuber, pp. (815–826), Elsevier Applied Science, ISBN: 1851664475, London

Hirano, H., Kagawa, H., & Okubo, K. (1992). Characterization of proteins released from legume seeds in hot water. *Phytochemistry*, 31, pp. 731–735, ISSN: 0031-9422

Hugues, M., Boivin, P., Gauillard, F., Nicolas, J., Thiry, J. M., & Richard-Forget, F. (1994). Two lipoxygenases from germinated barley – Heat and kilning stability. *Journal of Food Science*, 59, pp. 885–889, ISSN: 1750-3841

Indrawati, A., Van Loey, A. M., Ludikhuyze, L. R., & Hendrickx, M. (1999). Single, combined, or sequential action of pressure and temperature on lipoxygenase in green beans (*Phaseolus vulgaris* L.): A kinetic inactivation study. *Biotechnology Progress*, 15, pp. 273–277, ISSN: 1520-6033

Ludikhuyze, L. R., Indrawati, Van den Broeck, I., Weemaes, C. A., & Hendrickx, M. E. (1998a). High pressure and thermal denaturation kinetics of soybean lipoxygenase: A study based on gel electrophoresis. *LWT-Food Science and Technology*, 31, pp. 680–686, ISSN: 0023-6438

Ludikhuyze, L. R., Indrawati, Van den Broeck, I., Weemaes, C. A., & Hendrickx, M. A. (1998b). Effect of combined pressure and temperature on soybean lipoxygenase. 2. Modelling inactivation kinetics under static and dynamic conditions. *Journal of Agricultural and Food Chemistry*, 46, pp. 4081–4086, ISSN: 0021-8561

Ludikhuyze, L. R., Indrawati, Van den Broeck, I., Weemaes, C. A, & Hendrickx M. A. (1998c). Effect of combined pressure and temperature on soybean lipoxygenase. 1. Influence of extrinsic and intrinsic factors on isobaric–isothermal inactivation kinetics. *Journal of Agricultural and Food Chemistry*, 46, pp. 4074–4080, ISSN: 0021-8561

Mostafa, M. M., Rahma, E. H., & Rady, A. H. (1987). Chemical and nutritional changes in soybean during germination. *Food Chemistry*, 23, pp. 257–275, ISSN: 0308-8146

Omi, Y., Kato, T., Ishida, K., Kato, H., & Matsuda, T. (1996). Pressure-induced release of basic 7S globulin from cotyledon dermal tissue of soybean seeds. *Journal of Agricultural and Food Chemistry*, 44, pp. 3763–3767, ISSN: 0021-8561

Oxen, P., & Knorr, D. (1993). Baroprotective effect of high solute concentration against inactivation of Rhodotorula rubra. *LWT-Food Science and Technology*, 26, pp. 220–223, ISSN: 0023-6438

Park, K. H., Kim, Y. M., & Lee, C. W. (1988). Thermal inactivation kinetics of potato tuber lipoxygenase. *Journal of Agricultural and Food Chemistry*, 36, pp. 1012–1015, ISSN: 0021-8561

Penas, E., Gomez, R., Frias, J., Baeza, M. L., & Vidal-Valverde, C. (2011). High hydrostatic pressure effects on immunoreactivity and nutritional quality of soybean products. *Food Chemistry*, 125, pp. 423–429, ISSN: 0308-8146

Pomeranz, Y., Shogren, M. D., & Finney, K. F. (1977). Flour from germinated soybeans in high protein bread. *Journal of Food Science*, 42, pp. 824–828, ISSN: 1750-3841

Prestamo, G., Lesmes, M., Otero, L., & Arroyo, G. (2000). Soybean vegetable protein (tofu) preserved with high pressure. *Journal of Agricultural and Food Chemistry*, 48, pp. 2943–2947, ISSN: 0021-8561

Riaz, M. N. (1988). Soybeans as functional foods. *Cereal Foods World*, 44, pp. 88–92, ISSN: 0146-6283

Seyderhelm, I., Boluslawski, S., Michaelis, G., & Knorr, D. (1996). Pressure induced inactivation of selected food enzymes. *Journal of Food Science*, 61, pp. 308–310, ISSN: 1750-3841

Steinke, F. H., Waggle, D. H., & Volgarev, M. N. (1992). *New Protein Foods in Human Health: Nutrition, Prevention and Therapy*. CRC Press, Inc., ISBN: 0849369045, Florida

Svensson, S. G., & Eriksson, C. E. (1974). Thermal inactivation of lipoxygenase from peas (*Pisum sativum* L.). III. Activation energy obtained from single heat treatment experiments. *LWT-Food Science and Technology*, 7, pp. 142–144, ISSN: 0023-6438

Tangwongchai, R., Ledward, D. A., & Ames, J. M. (2000). Effect of high-pressure treatment on lipoxygenase activity. *Journal of Agricultural and Food Chemistry*, 48, pp. 2896–2902, ISSN: 0021-8561

Ueno, S., Shigematsu, T., Watanabe, T., Nakajima, K., Murakami, M., Hayashi, M., & Fujii, T. (2010). Generation of free amino acids and γ-aminobutyric acid in water-soaked soybean by high-hydrostatic pressure processing. *Journal of Agricultural and Food Chemistry*, 58, pp. 1208–1213, ISSN: 0021-8561

Utsumi, S., Matsuda, Y., & Mori, T. (1998). Structure-functions relationships of soy proteins. In: *Functional food*, G. Mazza (Ed.), pp. (46-51), Technomic Publishing Company Inc., ISBN: 1566764874 , Lancaster

Van Loey, I. A. M., Ludikhuyze, L. R., & Hendrickx, M. E. (1999). Soybean lipoxygenase inactivation by pressure at subzero and elevated temperatures. *Journal of Agricultural and Food Chemistry*, 1999, 47, pp. 2468–2474, ISSN: 0021-8561

Ven, C., Matser, & Berg, R. W. (2005). Inactivation of soybean trypsin inhibitors and lipoxygenase by high-pressure processing. *Journal of Agricultural and Food Chemistry*, 2005, 53, pp. 1087–1092, ISSN: 0021-8561

Wang, R., Zhou, X., & Chen, Z. (2008). High pressure inactivation of lipoxygenase in soy milk and crude soybean extract. *Food Chemistry*, 106, 603–611, ISSN: 0308-8146

Whitaker, J. R. (1972). *Principles of Enzymology for the Food Sciences*. Marcel and Dekker, ISBN: 0824791487, New York

Wolf, W. J., & Cowan, J. C. (1975). *Soybeans as a Food Source*. CRC Press, Inc., ISBN: 0878191127, Cleveland

4

Lipids, Nutrition and Development

Juliana M.C. Borba, Maria Surama P. da Silva
and Ana Paula Rocha de Melo
Federal University of Pernambuco
Brazil

1. Introduction

Lipids are classified as simple, compound and derived based on the hydrolysis, which result in breaking the fatty acids off, leaving free fatty acids and a glycerol, using up three water molecules. Simple lipids are esters of fatty acids with various types of alcohol. They are distinguished into fats and oils. Compound lipids contain an inorganic or organic group in addition to fatty acids and glycerol. They include phospholipids, glicolipids and lipoproteins. Finally, derived lipids are obtained by hydrolysis of simple and complex lipids. These lipids contain glycerol and other alcohols. They correspond to steroid hormones, ketone bodies, hydrocarbons, fatty acids, fatty alcohols, mono and diglycerides, terpenes and carotenoids. Sometimes they are present as waste products of metabolism. Lipids also can be classified, depending on its solubility or function, as polar or apolar compounds and as structural or reserve substances, respectively (Basso, 2007).

The main source of body energy comes from the triglycerides. These compounds are esters formed by one molecule of glycerol and three molecules of fatty acids. Fatty acids are carboxylic acids that usually have in its structure an unbranched carbon chain and one carboxyl. According to the saturation of the carbon chain, they can be classified as saturated, monounsaturated and polyunsaturated fatty acids (Basso, 2007).

In general, all mammals are able to synthesize saturated and monounsaturated fatty acids, but this ability is limited to polyunsaturated fatty acids (PUFAs), without them the organisms could not function properly. For this reason, these compounds are considered "essential" fatty acids. Thus, these fatty acids must be supplied by the diet. Linoleic acid 18:2 (n-6), a member of the n-6 family of fatty acids, was identified as the first "essential" fatty acid, whereas α-linolenic acid, 18:3 (n-3) represents the other essential fatty acid. These two essential fatty acids are the only sources for important longer chain fatty acids and physiological synthesis of complex lipids (Yehuda et al., 1999). Linoleic acid [18:2 (n-6)] is usually found in large quantities in soybean, corn, canola and safflower oil while α – linolenic [18:3 (n-3)] is easily found in green leafy vegetables, linseed and marine fish oil (Takahashi, 2005).

The main source of dietary lipids is obtained through the intake of triglycerides which can be found as fats or oils. The concept of fat or oil is based on the consistency and on the fatty acid present in the triglyceride molecule. At room temperature, oils are liquid because are constituted of tiacylglycerols containing a high proportion of mono and/or polyunsaturated fatty acids. These come from the vegetable sources such as soybean, corn, sunflower, olive or canola oil or from animal source such as fish oil. On the other hand, fats are solid or pasty at room temperature and contain a large proportion of saturated fatty acids and/or unsaturated with trans double bonds. Fats can be from animal source such as butter, beef tallow or pig and from vegetable source such as cocoa butter and hydrogenated fats. Trans fats are produced naturally at the rumen or from an industrial process adding hydrogen to unsaturated fatty acids found in vegetable oils (partial or total hydrogenation) (Basso, 2007). After hydrogenation, vegetable oils may be converted from liquids to solids, resulting in margarines and shortenings which have excellent culinary properties even though have detrimental health effects. The partial hydrogenation increases the melt pointing of the fats at a room temperature and the degree of hydrogenation controls the final consistency of the manufactured products. These fats are regularly found in foods such as ice cream, chocolates, biscuits, cookies, cakes, mass and margarines (Basso, 2007; Kinsella et al., 1981).

Changes in the structure of the lipids molecules have metabolic and nutritional repercussions (Kummerow, 2009), since, produce a significant loss of essential fatty acids (Martin et al., 2004). According to Martin et al. (2004), human nutrition has been moulded substantially along the modernization and industrialization process. Because the property of hydrogenated fats in increasing validity of food products and making them look like more crispy and less oily, food industry and "fast food"restaurants has been intensifying the use of trans fatty acids. On the other hand, since 1990´s it has been a substantial interest of the scientif community in investigating the adverse effects over health caused by the long lasting intake of trans fatty acids.

1.1 Effects of early and prolonged intake of hydrogenated fat for health

Once considered a problem only in high-income countries, overweight and obesity are now dramatically present in high rates in low- and middle-income countries, particularly in urban areas. World Health Organization predicts that by 2015, approximately 2.3 billion adults will be overweight and more than 700 million will be obese. According to World Health Organization/Food and Agriculture Organization (2003), obesity and overweight is caused by an energy imbalance between calories consumed and calories expended. Among several factors related to the present development of overweight and obesity, an increased intake of foods high in fat and sugar but low in vitamins, minerals and fiber is one of the worldwide problems of this century (WHO, 2003; Remig et al., 2010). Fats are important nutrients in diet and have wide chemical properties that drive diverse metabolic effects.

Although optimal dietary fat quantity has been keenly pursued over past decades, attention has recently centered on the value of dietary fat quality (Gillingham et al., 2011). Recent population studies have shown the important role of monounsaturated and polyunsaturated fats as key nutrients in preventing chronic diseases in modern societies. In addition, sufficient intakes of polyunsaturated fats during childhood are required for optimal growth and development (Carrillo-Fernandez et al., 2011). Consumption of a healthy diet, containing adequate rates of omega6:omega3 throughout the life is important to maintain cardiovascular and possibly also cognitive and immune health (WHO/FAO, 2003). In addition to PUFA, dietary monounsaturated also promote benefits to health in terms of blood lipid profile, blood pressure, insulin sensitivity and glycemic control. Due to existing and emerging research on health attributes of monounsaturated rich diets, and to the low prevalence of chronic disease in populations consuming monounsaturated rich Mediterranean diets, recommendations have been made to replace saturated fat acids with unsaturated fats (Gillingham et al., 2011).

Although several studies have been shown the importance for health in replacing saturated for unsaturated fats in diets, currently, food products consumed in a typical Western diet contain a significant proportion of fats that have been industrially altered. It is estimated that 2% to 8% of energy needs in the typical Western diet come from chemically modified lipid products (Craig-Schmidt, 2006). Trans fatty acids frequently present in partially hydrogenated vegetable oils have been consumed on large scale all over the world. In terms of consumption, trans fatty acids have long been used in food manufacturing for reasons that were described above. Increasing epidemiologic and biochemical evidence suggest that high consumption of trans fats is related to many metabolic alterations and also induce a significant risk factor for cardiovascular disturbs (de Oliveira et al, 2011). Recently, Mozaffarian et al. (2006) showed that a 2% absolute increase in energy intake from trans fat has been associated with a 23% increase in cardiovascular risk. In general, these cardiovascular disturbs are related to some of known mechanisms such as reduction of c-HDL concentration, increase of low density lipoprotein and triglycerides; disturbance in prostaglandin balance and they may also promote insulin resistance (Castro-Martínez et al., 2010). Since , the American Heart Association as well as the World Health Organization recommend limiting trans fats to <1% energy and many others health institutions in the United States of America all recommend limiting dietary trans-fat intake from industrial sources as much as possible.

The presence of an inflammatory process in the arteries seems to be another risk factor in heart disease, and studies show that hydrogenated trans fats increase the inflammation in the arteries and promote endothelial dysfunction (Sun et al., 2007; Lopez-Garcia et al., 2005). Endothelial cells are responsible for the regulation of local vascular tone by means of releasing relaxing and contracting factors synthesized mainly from arachidonic acid. Trans fat inhibits COX-2, an enzyme which converts arachidonic acid to prostacyclin that is needed to prevent blood clots in the coronary arteries (Kummerow, 2009).

In addition to the relation between high rates of trans fatty acids in diet and increased risk for developing cardiovascular disease, a detrimental relationship was found between trans fatty acids intake and depression risk. Rising secular trends in the incidence of depressive disorders have been paralleled by a dramatic change in the sources of fat in the Western diet. This change mainly consists in the replacement of polyunsaturated or monounsaturated fatty acids by saturated fats and trans-unsaturated fats (Pawels & Volterrani, 2008). These findings suggest that cardiovascular disease and depression may share some common nutritional determinants related to subtypes of fat intake (Sanchez-Villegas et al., 2011).

It seems that the exposure to trans fatty acids in utero has negative consequences early in life. How much the in utero environment dictates birth weight and the programming of long-term obesity related disorders is still unclear, especially when compared with that of early neonatal growth rate. The placental transfer of trans fatty acids is still contradictory, both in human and animals (Haggarty, 2010). However, experimental studies have demonstrated that placenta is not completely impermeable to these compounds, since a number of trans fatty acids cross this barrier and accumulates in the liver and in the total body lipids of the fetus. However, despite that the myelinogenesis process is not finished in fetus, the amount of trans isomers transferred to fetal brain was negligible in all studies. At least in animals, this finding suggests the brain might be protected from the trans fatty acids accumulation, but no data have yet been reported for human newborns (Haggarty, 2010).

1.2 The role of lipids on the development of nervous system

Among the various organic systems, the nervous system plays the main role in controlling several physiological processes. During development, this system presents a rapid growth spurt period or a vulnerable period which corresponds to the highest rate of cellular migration and differentiation, neurogenesis, synaptogenesis, myelinization and maturation of neurotransmitter pathways. Depending on the animal species, this critical period of intense neural development occurs at different time points early in life (Dobbing, 1968). In human, for example, it occurs during the last trimester of gestation until the second year of life, while in rats it corresponds the lactation period. During this period, the nutrition is one of the essential environmental factors to a normal development because it provides nutrients without them the neurodevelopment would be impaired (Walker, 2005).

The classical studies about malnutrition show that nutritional deficiency in macro or micronutrients has deleterious effects on the brain (Winick & Rosso, 1969). In rats, malnutrition induces functional and developmental failures, as well as reduction in brain size (Morgane et al., 1992;1993). In children, malnutrition showed influence both short and long term problems of cognition and behavior (Grantham-McGregor & Baker-Henningham, 2005; Benton, 2008).

Essential fatty acids are important constituents of structural lipids in nervous membranes cell and signaling molecules, as such, are involved in many brain functions. Around 30–40% of the total phospholipids in these structures are docosahexaenoic acid molecules (Young et al., 2000), which appear to be specifically concentrated in membranes surrounding synapses (Carlson, 2001). Changes in the quantity and quality of the dietary fatty acids are often associated with developmental and functional alterations in the nervous system.

At the cellular level, an α-linolenic deficient diet can induce less complex patterns of dendritic branching (Wainwright, 2002), smaller neurite growth in hippocampal neurons (Calderon & Kim 2004) and reduced neuronal soma size in some brain regions (Ahmad et al., 2002). Modification in the fatty acid composition of rat brain cell membranes of neurons, astrocytes, oligodendrocytes, and of subcellular fractions, such as myelin and synaptosomes, are also induced by a diet with reduced levels of n-3 fatty acids (Bourre et al., 1984). It has been shown that essential fatty acids imbalance as well as specific fatty acid deficiencies in the maternal diet can affect the neuromotor development of pups, including the ability to respond to environmental stimulation (Lamptey & Walker, 1976; Wainwright, 2002; Anselmo et al., 2006).

Fats and oils as they exist in nature must be processed before they are suitable for human consumption (González et al., 2007). On the other hand, only a few studies have described the effects of the ingestion of trans fatty acids early in life on the rat brain development. In this study we investigated the replacement of soybean oil in the diet by partially hydrogenated vegetable oil, rich in trans fatty acids, from the beginning of gestation through lactation on reflex ontogeny.

2. Methods

2.1 Animals and diets

Female pregnant *Wistar* rats weighing 200-250 g were obtained from the colony of Department of Nutrition of the Federal University of Pernambuco. Twenty-four hours after the birth of the entire mothers´ nestling, born on the same day, were contained in a large group and randomly each litter was culled to six males and two female pups. The litters were also randomly assigned to isocaloric diets containing as lipid source 7% soybean oil (control group-C; n=32) or 7% of hydrogenated vegetal fat (experimental group-E; n=39), since gestation until weaning (21 days old). Both diets (table 1) were formulated based on recommendations of the American Institute of Nutrition-AIN-93 (Reeves et al., 1993). The animals were kept on a 12:12-h light –dark photoperiod at 24o C temperature during the whole period. The animals were maintained according to recommendations from the National Institute of Health (USA) and approved by the Ethics Committee for Use Animal, Federal University of Pernambuco (CEUA, protocol.23076.020339/2010-24). After delivery, from P1 until 21d, the pups were weighted at 1, 7, 14 and 21days of age. Indicators of somatic maturation and reflex ontogeny were studied daily from P1 to P21 between 07:00 am and 09:00 am. Daily it was observed the occurrence of the reflex responses, being considered the consolidation day to be the first one, of a series of three consecutive days where the reply was verified. For each reflex it was established a maximum observation time of 10s according to the experimental model established by Smart & Dobbing (1971).

Ingredients	Control diet (g/100g)	Experimental diet (g/100g)
Casein	20.0	20.0
Cellulose	5.0	5.0
Corn starch	52.95	53.49
Sucrose	10.0	10.0
Soyabean oil	7.0	-
Hydrogenated vegetable fat	–	6.46
Vitamin mix [1]	1.0	1.0
Mineral mix [2]	3.5	3.5
L-Cystine	0.3	0.3
Choline Bitartrate	0.25	0.25

Table 1. Composition of the diets. Vitamin mixture (Rhoster Ind.Com. LTDA. SP. Brazil) containing (m%): folic acid (20); niacin (300); biotin (2); calcium pantothenate 160; pyridoxine 70; riboflavin 60; thiamine chloride 60; vitamin B_{12} 0.25; vitamin K_1 7.5. Additionally containing (UI%): vitamin A 40.000; vitamin D_3 10.000; vitamin E 750. 2 Mineral mixture (Rhoster Ind. Com. LTDA. SP. Brazil) containing (m%): $CaHPO_4$ (38); K_2HPO_4 (24); $CaCO_3$ (18.1); NaF (0.1); NaCl(7.0); MgO (2.0); $MgSO_4$ $7H_2O$ (9.0); $FeSO_4$ $7H_2O$ (0.7); $ZnSO_4$ H_2O (0.5); $MnSO^+$ H_2O (0.5); $CuSO_4$ $5H_2O$ (0.1); $Al_2(SO_4)_3K_2SO_4$ $24H_2O$ (0.02); Na_2SeO_3 $5H_2O$ (0.001); KCl (0.008).

2.2 Indicators of somatic maturation
The following indicators of somatic maturation, as illustrated in figure 1, were analyzed in order to study whether the replacement of soybean oil by vegetable hydrogenated fat in the diet influenced the development of physical features of the rat pups.

2.3 Indicators of reflex ontogeny
The following indicators of the reflex ontogeny investigated at the present study are described and illustrated below (figure 2):
Palmar Grasp (PG)- This reflex is present at birth and consists of a dorso flexion of the digits ('grasping') in response to the stimulation of the hand-palm with a small metallic stick. The

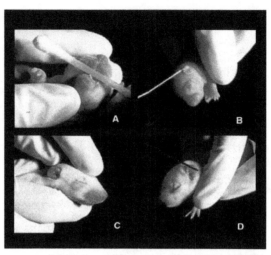

Fig. 1. Indicators of somatic maturation. A) Eruption of the Upper Incisors (EUI) and Eruption of the Lower Incisors (ELI); B) Ear Unfolding (EU); C) Eye Opening (EO); D) Auditory Conduit Opening (ACO).

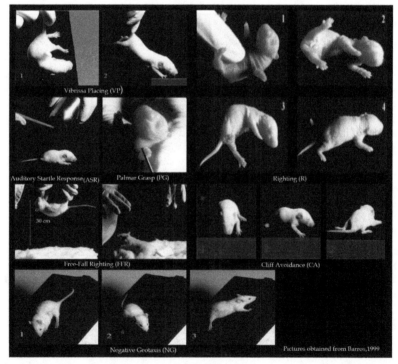

Fig. 2. The figure represents the indicators of reflex ontogeny observed during lactation period.

expected response is the disappearance of the palmar grasping response, as the organism matures.

Righting (R)- The newborn is placed on its back on a flat surface and the expected mature response is to turn over on the ventral surface, resting in the normal position with the four feet on the ground.

Vibrissa Placing (VP)- The pup is held by the tail, with the head facing the edge of a table and the vibrissae just touching the vertical surface of the table. The expected response is to lift the head and to extend the fore legs in direction of the table.

Cliff Avoidance (CA) - The newborn is placed on the edge of a 'cliff' (for instance, on the edge of a table), with the forepaws and face just over the edge. The expected response is to move away from the cliff, to avoid dropping.

Negative Geotaxis (NG)- The pup is placed on an inclined ramp (45° slope) with its head pointing to the ground. The expected mature response is to turn around and crawl up the slope.

Free-Fall Righting (FFR)- The pup is held with the back downwards 35cm above a cotton pad and dropped. The expected response is to turn in mid-air to land on its four paws.

Auditory Startle Response (ASR)- The newborn is exposed suddenly to a loud, sharp noise. The expected response is a prompt extension of the head and the limbs, followed by withdrawal of the limbs and a crouching posture.

2.4 Statistical analysis

The body weight was analyzed by Student's t-test. Results are presented as means±standard error of the mean (SEM). Differences were significant when $p < 0.05$. Results of the somatic maturation and reflex ontogeny were evaluated by Mann-Whitney test. Results are expressed as median ± interquartile. Statistical significance was set at $p < 0.05$.

3. Results

3.1 Body weight

The figure 3 shows that pups fed an experimental diet did not exhibited significant differences in the body weight at the 1st, 7th, 14th and 21th day, during the lactation period when compared to the controls. (C = 6.7 g ± 0.12; 16.3 g ± 0.49; 29.6 g ± 1.18; 46.4 g ±1.52; E = 6.6 g ± 0.11; 16.5 g ± 0.22; 30.3 g ± 0.48; 47 g ± 0.86).

3.2 Indicators of somatic maturation

The effect of the experimental diet on somatic development is shown in Figure 4, where the results are expressed in median (min. – max.) and compared with the control group. There was no difference between the number of days for ear unfolding, eye opening and the eruptions of superior and inferior incisors. However, the opening of the external auditory canal was significantly delayed in the experimental group as compared to the control (C: 12-1; E: 14-2.5; p<0.05).

3.3 Indicators of reflex ontogeny

As can be seen in Figure 5, the development of the early reflexes which appear in the first postnatal week such as righting, cliff avoidance and vibrissa placing did not differ between control and experimental pups. In the second postnatal week, the negative geotaxis was the only reflex which was delayed in the experimental group when compared to the control (C: 11-3; E: 9-1.5).

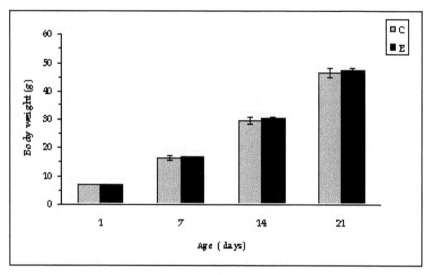

Fig. 3. Body weight in control (C) and experimental (E) offsprings from the 1st , 7th, 14th and the 21st day. Each bar represents the mean ± SEM. C = control group fed a diet containing 7% of soybean oil (n=32); E = experimental group fed a diet containing 7% of hydrogenated vegetal fat (n=39).

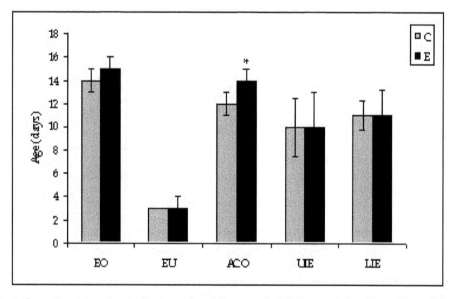

Fig. 4. Somatic maturation indicators of suckling rats fed the control diet (C; n=32) and the experimental diet (E; n=39). Each bar represents the median ± interquartile. EO = Eye Opening; EU = Ear Unfolding; ACO = Auditory Conduit Opening; UIE = Upper Incisors Eruption; LIE = Low Incisors Eruption. * p< 0.05 vs C group (Mann-Whitney Test).

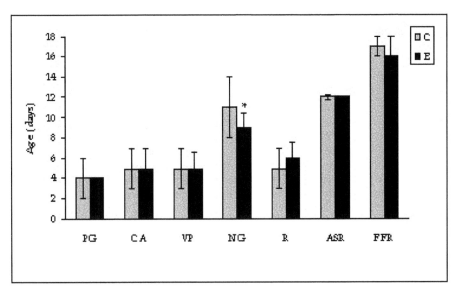

Fig. 5. Reflex ontogeny. Each bar represents the median + inertquartile. C = control group (n=32); E = experimental group (n=39). PG = Palmar Grasp; CA = Cliff Avoidance; VP = Vibrissa Placing; NG = Negative Geotaxis; R = Righting; ASR = Auditory Startle Response; FFR = Free-fall Righting. * p< 0.05 vs C group (Mann-Whitney Test).

4. Discussion

The currently hydrogenation of vegetable oils produces trans fatty acids which can be found in most manufactured products. The high consumption of these foods is related with the increase of health problems. When consumed, trans fat acids can be found in plasma and in the maternal milk (Carlson et al., 1997) and its concentration varies with daily mother intake. Circulating trans fat acids are also carried and incorporated into tissues such as the brain, liver, adipose tissue and spleen and its levels on these tissues depend on the amount ingested (Larqué et al., 2001).

In this study, we observed that offsprings fed a hydrogenated fat based diet did not exhibited differences in body weight from P0 until P21 when compared to the controls. One could explain this finding by the fact of the experimental drawing, herein used, kept the same proportion of lipids in the both diets, only replacing soybean oil for hydrogenated vegetal fat in the experimental diet. Although this dietary treatment has been offered during a period of an intense growth and body development, it did not compromise the body weight gain of the pups over the lactation period. These data are similar to those found in a previous study of our laboratory when rats fed a diet containing 5%coconut-oil or soybean-oil during pregnancy and lactation did not show significant changes in body weight gain from P0 to P21 when compared to the control group (Borba et al., 2010). Santillán et al., (2010) did not find differences in the body weight of mice fed a commercial diet enriched with soybean or sunflower oil diet over gestation and lactation when compared with those fed a commercial diet. On the other hand, mice known to be prone to obesity and insulin resistance when consumed a high fat diet during pregnancy and lactation exhibited a growth delay at the first week of life, but accelerated the growth in the subsequent two weeks (Kavanagh, et al., 2010). It is known that the fat content variation in human milk is clearly the result of different dietary, metabolic and physiologic controls (German, 2011). In humans, the fatty acid composition in maternal diet and in breastmilk during lactation may not affect the infant body composition in the early postpartum period but may be a factor in the development of childhood overweight later in life (Anderson et al., 2010).

The somatic maturation and reflex development from the 2nd over the 21st postnatal days of life are good indicators to understand how environmental factors can influence the functional maturity of the brain development (Smart & Dobbing, 1971; Gramsbergen, 1998). Among the somatic parameters herein investigated, the opening of the external auditory canal was delayed in the experimental group when compared to the control group. However, we did not observe this effect on the auditory startle response. These results suggest that the consumption of trans fat during the critical period of development did not influence the function of the auditory system since the hearing sensation was preserved. It has been known since the 19th century that hearing may occur through bone conduction; however the way how this physiologic pathway works is not completely understood. Some factors can contribute to the bone conduction, such as: sound radiated into the external ear canal, middle ear ossicle inertia, inertia of the cochlear fluids, compression of the cochlear walls and pressure transmission from the cerebrospinal. Of these five, inertia of the cochlear fluid seems the most important. The efficiency of the bone conduction is largely dependent on the skull bone where the skull acts as a rigid body at low frequencies and incorporates different types of wave transmission at higher frequencies (Stenfelt & Goode, 2005).

Regarding the others somatic maturation indicators no differences were observed between the group fed a trans fat based diet and the group fed a soybean oil based diet. When the reflex ontogeny was analyzed, we observed that there was a significant delayed in the negative geotaxis of the pups fed an experimental diet. These data suggested that there was a negative effect possibly induced by the lack of any essential fatty acid or by the trans fat *per se* on the development of the motor and the cerebellar system. The negative geotaxis reflex is stimulated by the abnormal position of the head and the body which are under control of the vestibular and postural systems (Adams et. al, 1985). This reflex requires a sequence of organized motor events (Ramirez & Spear, 2010) but this only occurs if the motor system is maturated. In rats, spinal cord descending projections develop relatively early. Projections from vestibulospinal and reticulospinal origin reach the cervical levels of the spinal cord at 13rd or 14 rd embrionary day. Around the same period, motoneurons in the ventral horn of thoracic and lumbosacral spinal cord segments start developing and two days thereafter, their axons invade the muscle mass of the caudal limb bud (Altman & Bayer, 1984; 1997; Gramsbergen, 1993).

The development of cerebellum occurs in the postnatal period, reaching its peak of development at the end of the first week (Smart & Dobbing, 1971). This period results of a number of events including: neuronal and glial proliferation, outgrowth of axons and dendrites, establishment of synaptical contacts, as well as myelination (Altman & Bayer, 1997). This late development makes the cerebellum a structure particularly vulnerable to insufficient supply of nutrients or to side and possible beneficial effects of pharmacological treatments (Gramsbergen, 2003). It has been shown that a restriction of daily food intake to dams delays the motor development and behavior associated with a disturbed cerebellar development of the offspring (Gramsbergen, 2003). On the other hand, Collucia et al. (2009) showed that omega-3 supplementation during gestation and lactation improved motor coordination in juvenile-adults rats.

One of the problems with the process of hydrogenation is the fact that it possibly produces a loss of essential fatty acids of the original vegetable oils. Hill et al. (1982) showed that rats fed a diet containing as a lipid source partially hydrogenated soybean oil showed a reduction of essential fatty acids levels in the liver and heart. It is evident that partially hydrogenated fats have excellent culinary properties, but from a nutritional point of view, the consumption of trans fatty acids represents a loss of essential fatty acids intake that may have a hazardous impact on health. This study is the first evidence that the consumption of hydrogenated vegetable fat during the critical period of development may compromise some parameters of the reflex and somatic development of rat pups.

5. Conclusion

Although the maternal intake of a diet containing trans fatty acid in replacement of soybean oil have not changed the body weight in the early postpartum period of the pups, it influenced negatively both somatic and reflex development. Recently, there is an increase in the level of interest in fatty acids and lipids. This interest is not limited to brain biochemistry, but also to the effects of levels and ratios of fatty acids on physiological and behavioral aspects. For these reasons, more research is warranted regarding the influence of maternal dietary on the fatty acid composition of the breast milk and their effects on body composition, the development of overweight and behavior changes later in life of rat pups.

6. Acknowledgements

This publication was made possible in part by support from the Universidade Federal de Pernambuco, Pró-Reitoria para Assuntos de Pesquisa e Pós-Graduação (PROPESQ). The authors gratefully acknowledge the invaluable assistance of Dr. Edeones França for the animal care.

7. References

Adams, J., Buelke-Sam, J., Kimmel, C.A., Nelson, C.J., Reiter, L.W., Sobotka, T.J., Tilson, H.A. & Nelson, B.K. (1985). Collaborative Behavioral Teratology Study: protocol design and testing procedures. *Neurobehav. Toxicol. Teratol.*, Vol.. 7, No. 6 (November-December 1985), pp. 579-586, ISSN 0275-1380

Ahmad, A., Moriguchi, T. & Salem N. (2002). Decrease in neuron size in docosahexaenoic acid-deficient brain. *Pediatr. Neurol.*,Vol. 26, No. 3, (March 2002), pp. 210-8, ISSN 0022-3166

Altman, J. & Bayer, S.A. (1984). The development of the rat spinal cord. *Adv. Anat. Embryol. Cell. Biol.*, Vol. 85, (1984), pp. 1-164, ISSN 0022-3077

Altman, J. & Bayer, S.A. (1997). Development of the cerebellar system in relation to its evolution, structure and functions, In: *The Central Nervous System, Structure and Function*, Brodal, P. (Ed.), pp. 1-783, Oxford University Press, ISBN 084-93-9490-2, New York, USA

Anderson, A.K., McDougald, D.M. & Steiner-Asiedu, M. (2010). Dietary trans fatty acid intake and maternal and infant adiposity. *European Journal of Clinical Nutrition*, Vol. 64, (November 2010) pp.1308-1315, ISSN 0954-3007

Anselmo, C.W., Santos, A.A., Freire, C.M., Ferreira, L.M., Cabral Filho, J.E., Catanho, M.T. & Medeiros, Mdo C. (2006). Influence of a 60 Hz, 3 micro T, electromagnetic field on the reflex maturation of Wistar rats offspring from mothers fed a regional basic diet during pregnancy. *Nutr. Neurosci.*, Vol. 9, No. 5-6, (October-December 2006), pp. 201-6, ISSN 1028-415X

Basso, R. (2007). Bioquímica e Metabolismo dos Lípides, In: *Tratado de Alimentação, Nutrição & Dietoterapia*, Silva, S.M.C.S. & Mura, J.D.P. (Ed.), pp. 55-73, Roca, ISBN 978-85-7241-678-8, São Paulo, Brazil

Benton, D. (2008). Sucrose and behavioral problems. *Crit. Rev. Food. Sci. Nutr.*, Vol. 48, No. 5, (May 2008), pp. 385-401, ISSN 1040-8398

Borba, J.M., Rocha-de-Melo, A.P., dos Santos, A.A., da Costa, B.L., da Silva, R.P., Passos, P.P. & Guedes, R.C. (2010). Essential fatty acid deficiency reduces cortical spreading depression propagation in rats: a two-generation study. *Nutr. Neurosci.*,Vol. 13, No. 3, (June 2010), pp. 144-50, ISSN 1028-415X

Bourre, J.M., Pascal, G., Durand, G., Masson, M., Dumont, O. & Piciotti, M. (1984). Alterations in the fatty acid composition of rat brain cells (neurons, astrocytes, and oligodendrocytes) and of subcellular fractions (myelin and synaptosomes) induced by a diet devoid of n-3 fatty acids. *J. Neurochem.*, Vol. 43, No. 2, (August 1984), pp. 342-8, ISSN 0022-3042

Calderon, F. & Kim, H.Y. (2004). Docosahexaenoic acid promotes neurite growth in hippocampal neurons. *J. Neurochem.*, Vol.90, No. 4, (August 2004), pp. 979-88, ISSN 0022-3042

Carlson, S.E., Clandinin, M.T., Cook, H.W., Emken, E.A. & Filer, L.J. Jr. (1997). Trans Fatty acids: infant and fetal development. *Am. J. Clin. Nutr.*, Vol.66, No. 3, (September 1997), pp.715S-36S, ISSN 0002-9165

Carlson, S.E. (2001). Docosahexaenoic acid and arachidonic acid in infant development. *Semin. Neonatol.*, Vol. 6, No. 5, (October 2001), pp. 437-449, ISSN 1744-165X

Carrillo-Fernández, L., Dalmau- Serra, J., Martínez Álvarez, J.R., Solà Alberich, R. & Pérez Jiménez, F. (2011). Dietary fats and cardiovascular health. *Anales de Pediatría (Barc)*, Vol.74, No. 03, (February 2011), e1-e16, ISSN 1695-4033

Castro-Martínez, M.G., Bolado-García, V.E., Landa-Anell, M.V., Liceaga-Cravioto, M.G., Soto-González, J. & López-Alvarenga, J.C. (2010). Dietary trans fatty acids and its metabolic implications. *Gac. Med. Mex.*, Vol. 146, No. 4, (July-August 2010), pp. 281-288, ISSN 0016-3813

Craig-Schmidt, M.C. (2006). World-wide consumption of trans fatty acids. *Atherosclerosis Supplements*, Vol. 7, pp. 1–4, ISSN 1567-5688

Coluccia, A., Borracci, P., Renna, G., Giustino, A., Latronico, T., Riccio, P. & Carratu, M.S. (2009). Developmental omega-3 supplementation improves motor skills in juvenile-adult rats. *Int. J. Devl. Neuroscience*, Vol. 27, (May 2009), pp. 599–605, ISSN 0736-5748

de Oliveira, J.L., Oyama, L.M., Hachul, A.C., Biz, C., Ribeiro, E.B., Oller do Nascimento, C.M. & Pisani, L.P. (2011). Hydrogenated fat intake during pregnancy and lactation caused increase in TRAF-6 and reduced AdipoR1 in white adipose tissue, but not in muscle of 21 days old offspring rats. *Lipids Health Dis.*, Vol. 10, No. 22, (January 2011), pp. 1-9, ISSN 1476-511X

Dobbing, J. (1968). The development of the blood-brain barrier. *Prog. Brain. Res.* Vol. 29, pp. 417-27, ISSN 0079-6123

German, J. B. (2011). Dietary lipids from an evolutionary perspective: sources, structures and function. *Maternal & Child Nutrition*, Vol. 7, No. 2, (2011), pp. 2-16, ISSN 1740-8709

Gillingham, L.G., Harris-Janz, S. & Jones, P.J. (2011). Dietary Monounsaturated Fatty Acids Are Protective Against Metabolic Syndrome and Cardiovascular Disease Risk Factors. *Lipids*, Vol. 46, No. 3, (March 2011), pp. 209-28, ISSN 0024-4201

González, C., Resa, J.M., Concha, R.G. & Goenaga, J.M. (2007). Enthalpies of mixing and heat capacities of mixtures containing acetates and ketones with corn oil at 25°C. *J. Food Eng.*, Vol., 79, pp. 1104-1109, ISSN 0260-8774

Gramsbergen, A. (1993). Posture and locomotion in the rat: independent or interdependent development? *Neurosci. Biobehav. Rev.*, Vol. 22, No. 4, (July 1998), pp. 547-53, ISSN 0149-7634

Gramsbergen A. (2003). Clumsiness and disturbed cerebellar development: insights from animal experiments. *Neural Plast.*,Vol. 10, No. 1-2, (2003), pp. 129-40, ISSN 1687-5443

Grantham-McGregor S & Baker-Henningham H. (2005). Review of the evidence linking protein and energy to mental development. *Public. Health. Nutr.*, Vol. 8, No. 7A, (October 2005), pp. 1191-201, ISSN 1368-9800

Haggarty, P. (2010). Fatty Acid Supply to the Human Fetus. *Annal Review of Nutrition*, Vol. 30, (May 2010), pp. 237-255, ISSN 0199-9885

Hill, E.G., Johnson, S.B., Lawson, L.D., Mahfouz, M.M. & Holman, R.T.(1982). Perturbation of the metabolism of essential fatty acids by dietary partially hydrogenated vegetable oil. *Proc. Natl. Acad. Sci.*, Vol. 79, No. 4, (February 1982), pp. 953-7, ISSN 0027-8424

Lamptey, M.S. & Walker, B.L. (1976). A possible essential role for dietary linolenic acid in the development of the young rat. *Journal of Nutrition*, Vol. 106, No. 1, (January1976), pp. 86-93, ISSN 0022-3166

Larqué, E., Zamora, S. & Gil, A. (2001). Dietary trans fatty acids in early life: a review. *Early Hum Dev.*, Vol. 65, Suppl 2, (November 2001), pp. S31-S41, ISSN 0378-3782

Lopez-Garcia, E., Schulze, M. B., Meigs, J. B., Manson, J.A. E. , Rifai, N., Meir, J. S., Willett , W. C. & Hu, F. B. (2005). Consumption of Trans Fatty Acids Is Related to Plasma Biomarkers of Inflammation and Endothelial Dysfunction. *Journal of Nutrition*, Vol. 135, No. 3, (March 2005), pp. 562-566, ISSN 0022-3166

Kavanagh, K., Sajadian, S., Jenkins, K.A., Wilson, M.D., Carr ,J.J., Wagner, J.D. & Rudel, L.L.(2010). Neonatal and fetal exposure to trans-fatty acids retards early growth and adiposity while adversely affecting glucose in mice. *Nutr. Res.*,Vol. 30, No. 6, (June 2010), pp. 418-26, ISSN 0271-5317

Kinsella, J.E., Bruckner, G., Mai, J. & Shimp, J. (1981). Metabolism of trans fatty acids with emphasis on the effects of trans, trans-octadecadienoate on lipid composition, essential fatty acid, and prostaglandins: an overview. *The American Journal of Clinical Nutrition*, Vol.34, (October 1981), pp. 2307-2318, ISSN 0002-9165

Kummerow, F.A. (2009). The negative effects of hydrogenated trans fats and what to do about them. *Atherosclerosis*, Vol. 205, No. 2, (March 2009), pp. 458–465, ISSN 0021-9150

Martin, C. A., Matshushita, M. & Souza, N. E. (2004). Ácidos graxos trans: implicações nutricionais e fontes na dieta. *Revista de Nutrição*, Vol. 17, No. 3, (July-September 2004), pp. 361-368, ISSN 1415-5273

Morgane, P.J., Austin-LaFrance, R., Bronzino, J., Tonkiss, J., & Galler, J.R. (1992). Malnutrition and development central nervous system. In: The Vulnerable Brain and Environmental Risks, Issacson, R.L., Jensen, K.F. (Ed.), pp. 2-42, Springer Us, ISBN 0306441489, New York, USA

Morgane, P.J., Austin-LaFrance, R., Bronzino, J., Tonkiss, J., Díaz-Cintra, S., Cintra, L., Kemper, T. & Galler, J.R. (1993). Prenatal malnutrition and development of the brain. *Neurosci. Biobehav. Rev.*,Vol. 17, No. 1, pp. 91-128, ISSN 0149-7634

Mozaffarian, D., Katan, M.B., Ascherio, A., Stampfer, M.J. & Willett, W.C. (2006). Trans Fatty Acids and Cardiovascular Disease. *N. Engl. J. Med.*, Vol. 354, (April 2006), pp. 1601-1613, ISSN 0028-4793

Pawels, E.K. & Volterrani, D. (2008). Fatty acid facts, Part I. Essential fatty acids as treatment for depression, or food for mood? *Drug News Perspect,*Vol. 21, No. 8, (October 2008), pp. 446-451, ISSN 0214-0934

Ramirez, R.L. & Spear, L.P. (2010). Ontogeny of ethanol-induced motor impairment following acute ethanol: assessment via the negative geotaxis reflex in adolescent and adult rats. *Pharmacol Biochem. Behav.,* Vol. 95, No.2, (April 2010), pp. 242-248, ISSN 0091-3057

Reeves, P. G., Nielsen, F. H. & Fahey, G. C. (1993). AIN-93 Purified Diets for Laboratory Rodents: Final Report of the American Institute of Nutrition. *Journal of Nutrition,* Vol. 123, pp. 1939-1951, ISSN 0022-3166

Remig, V., Franklin, B., Margolis, S., Kostas, G., Nece, T. & Street , J. C. (2010). Trans Fats in America: A Review of Their Use, Consumption, Health Implications, and Regulation. *Journal of the American Dietetic Association,* Vol. 110, No. 4 , (April 2010), pp. 585-592, ISSN 0002-8223

Sánchez-Villegas, A., Verberne, L., De Irala, J., Ruíz-Canela, M., Toledo, E., Serra-Majem, L. & Martínez-González, M.A. (2011). Dietary Fat Intake and the Risk of Depression: The SUN Project. *PLoS One,* Vol. 26, No. 1, (January 2011), e16268, ISSN 1932-6203

Santillán, M.E., Vincenti, L.M., Martini, A.C., de Cuneo, M.F., Ruiz, R.D., Mangeaud, A. & Stutz, G. (2010). Developmental and neurobehavioral effects of perinatal exposure to diets with different omega-6:omega-3 ratios in mice. *Nutrition,* Vol. 26, No. 4, (April 2010), pp. 423-31, ISSN 0899-9007

Smart, J.L. & Dobbing, J. (1971). Vulnerability of developing brain. II. Effects of early nutritional deprivation on reflex ontogeny and development of behavior in the rat. *Brain Research,* Vol. 28, pp. 85-95, ISSN 0006-8993

Stenfelt, S. & Goode, R.L. (2005). Bone-conducted sound: physiological and clinical aspects. *Otol. Neurotol.,*Vol. 26, No. 6, (November 2005), pp. 1245-61, ISSN 1531-7129

Sun, Q., Ma, J., Campos, H. & Hu, F.B. (2007). Plasma and erythrocyte biomarkers of dairy fat intake and risk of ischemic heart disease. *Am. J. Clin. Nutr.,* Vol. 86, No.4, (May 2007), pp. 929-937, ISSN 0002-9165

Takahashi, N.S. (October 2005). Importância dos ácidos graxos essenciais. 04.12.2010, Available from: <ftp://ftp.sp.gov.br/ftppesca/acidos_graxos.pdf>

Wainwright, P.E. (2002). Dietary essential fatty acids and brain function: a developmental perspective on mechanisms. *Proc Nutr Soc.,* Vol. 61, No. 1, (February 2002), pp. 61-9, ISSN 0029-6651

Walker, C.D. (2005). Nutritional aspects modulating brain development and the responses to stress in early neonatal life. *Prog. Neuropsychopharmacol. Biol. Psychiatry,* Vol. 29, No. 8, (December 2005), pp. 1249-63, ISSN 0278-5844

WHO/FAO (2003). Diet, Nutrition and the Prevention of Chronic Diseases: Report of a Joint WHO/FAO Expert Consultation. *WHO Technical Report Series,* No. 916, World Health Organization: Geneva

WHO. (2003). Diet, nutrition and the prevention of chronic diseases. In: Organization, WH, editor. WHO Technical Report Series. Geneva

Winick, M. & Rosso, P. (1969). The effect of severe early malnutrition on cellular growth of human brain. *Pediatric Research,* Vol. 3, (1969), pp. 181-184, ISSN 0031-3998

Yehuda S., Rabinovitz S. & Mostofsky D. I.(1999). Essential Fatty Acids Are Mediators of Brain. *Journal of Neuroscience Research,* Vol. 56, (1999), pp. 565-570, ISSN 0360-4012

Young, C., Gean, P.W., Chiou, L.C. & Shen, Y.Z. (2000). Docosahexaenoic acid inhibits synaptic transmission and epileptiform activity in the rat hippocampus. *Synapse,* Vol. 37, No. 2, (August 2000), pp. 90-94, ISSN 0887-4476

5

Inhibition of Soybean Lipoxygenases – Structural and Activity Models for the Lipoxygenase Isoenzymes Family

Veronica Sanda Chedea[1] and Mitsuo Jisaka[2]
[1]Laboratory of Animal Biology, National Research Development Institute
for Animal Biology and Nutrition (IBNA),
[2]Faculty of Life Sciences and Biotechnology, Shimane University,
[1]Romania
[2]Japan

1. Introduction

Lipoxygenases (EC 1.13.11.12, linoleate:oxygen, oxidoreductases, LOXs) which are widely found in plants, fungi, and animals, are a large monomeric protein family with non-heme, non-sulphur, iron cofactor containing dioxygenases that catalyze the oxidation of polyunsaturated fatty acids (PUFA) as substrate with at least one $1Z$, $4Z$-pentadiene moiety such as linoleic, linolenic and arachidonic acid to yield hydroperoxides (Gardner, 1991).

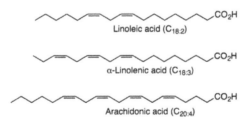

Fig. 1. Lipoxygenase substrates, linoleic, α- linolenic and arachidonic acid.

Theorell et al. (1947) succeeded in crystallizing and characterizing lipoxygenase (LOX) from soybeans and since then among plant LOXs, soybean lipoxygenase isozyme 1 (LOX-1) can be regarded as the mechanistic paradigm for these nonheme iron dioxygenases (Coffa et al., 2005; Minor et al., 1996; Fiorucci et al., 2008).

Designing agents to modulate activities of the variety of so closely homologous enzymes, such as different LOXs, require an intimate knowledge of their 3D structures, as well as information about metabolism of the potential xeno- or endobiotics. So far only the structures of soybean isozymes LOX-1 and LOX-3 have been determined for native enzymes, and several structures of their and rabbit 15-LOX (from reticulocytes) molecular complexes with inhibitors are known. Due to lack of sufficiently purified human enzymes most of the structural research has been done on soybean LOX (Skrzypczak-Jankun et al., 2003).

Understanding the mechanism of inhibition of LOXs can have profound effect in the development of many anti-cancer and anti-inflammatory drugs. On the basis of the available LOX data it was suggested that a combination of LOX modulators might be needed to shift the balance of LOX activities from procarcinogenic to anticancerogenic as a novel strategy for cancer chemoprevention (Skrzypczak-Jankun et al., 2003).

The aim of the present study is to present knowledge on different lipoxygenases having the soybean lipoxygenases as a structural and activity template for their inhibition by natural antioxidant compounds as theoretical approach for food biochemistry and medical applications.

2. Lipoxygenase structure and activity

The three-dimensional structure of soybean lipoxygenase-1 has been determined to 2.3 Å resolution by single crystal X-ray diffraction methods (Boyington et al., 1993). It is a two-domain, single-chain prolate ellipsoid of dimensions 90 x 65 x 60 Å with a molecular mass of 95 kDa. The 839 residues are organised in two domains: one 146 residue N-terminal domain (domain I), and a major, 693 residue C-terminal domain (domain II) (Prigge et al., 1997). Overall, the three-dimensional structure of lipoxygenase-1 shows a helical content of 38.0% and a β-sheet content of 13.9%. The structure of another crystal form of soybean lipoxygenase-1 determined to 1.4 Å resolution (Minor et al., 1996) showed very similar results. The structure of lipoxygenase-3, another soybean lipoxygenase isozyme (Skrzypczak-Jankun, 1997) shows that the lipoxygenase-3 isozyme is very similar in structure despite significant differences in sequence: 857 residues vs 839, deletions at 7 positions, insertions at 25 positions, and substitutions at 224 residues (72% identity).

Soybean seed isoenzymes are 94–97 kDa monomeric proteins with distinct isoelectric points ranging from about 5.7 to 6.4, and can be distinguished by optimum pH, substrate specificity, product formation and stability (Siedow, 1991; Mack et al., 1987). LOX-1 is the smallest in size (838 amino acids; 94 kDa), exhibits maximal activity at pH 9.0 and converts linoleic acid preferentially into the 13-hydroperoxide derivative. LOX-2 is characterized by a larger size (865 amino acids; 97 kDa), by a peak of activity at pH 6.8, and forms equal amounts of the 13-and 9-hydroperoxide compounds (Loiseau et al., 2001). LOX-2 oxygenates the esterified unsaturated fatty acid moieties in membranes in contrast to LOX-1 which only uses free fatty acids as substrates (Maccarrone et al., 1994). LOX-3 (857 amino acids; 96.5 kDa) exhibits its maximal activity over a broad pH range centred around pH 7.0 and displays a moderate preference for producing a 9-hydroperoxide product. It is the most active isoenzyme with respect to both carotenoid cooxidation and production of oxodienoic acids (Ramadoss, 1978).

3. Lipoxygenase reaction

The initial step of LOX reaction is removal of a hydrogen atom from a methylene unit between double bonds in substrate fatty acids (Fig. 2A). The resulting carbon radical is stabilized by electron delocalization through the double bonds. Then, a molecular oxygen is added to the carbon atom at +2 or –2 position from the original radical carbon, forming a peroxy radical as well as a conjugated *trans,cis*-diene chromophore. The peroxy radical is then hydrogenated to form a hydroperoxide. The initial hydrogen removal and the following oxygen addition occur in opposite (or antarafacial) sides related to the plane formed by the 1Z,4Z-pentadiene unit. In most LOX reactions, particularly those in plants, the resulting hydroperoxy groups are in S-configuration, while one mammalian LOX and some marine invertebrate LOXs produce R-hydroperoxides. Even in the reactions of such "R-LOXs", the antarafacial rule of hydrogen removal and oxygen addition is conserved.

In cases of plant LOXs, including soybean LOXs, the usual substrates are C18-polyunsaturated fatty acids (linoleic and α-linolenic acids), and the products are their 9S- or 13S-hydroperoxides (Fig. 2B). Most plant LOXs react with either one of the regio-specificity, while some with both. Therefore, based on the regio-specificty, plant LOXs are classified into 9-LOXs, 13-LOXs, or 9/13-LOXs.

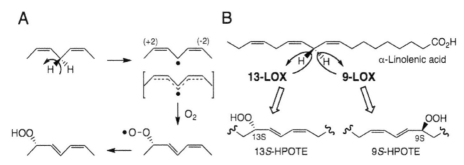

Fig. 2. LOX reaction showing the principal steps of LOX reaction (Panel A), and the actual reactions of plant LOXs and α-linolenic acid (Panel B). HPOTE: hydroperoxyoctadecatrienoic acid.

4. Biological and metabolic functions

4.1 In plants
Lipid peroxidation is common to all biological systems, both appearing in developmentally and environmentally regulated processes of plants (Feussner & Wasternack, 2002). The hydroperoxy polyunsaturated fatty acids, synthesized by the action of various highly specialised forms of lipoxygenases, are substrates of at least seven different enzyme families (Feussner & Wasternack, 2002). Signaling compounds such as jasmonates, antimicrobial and antifungal compounds such as leaf aldehydes or divinyl ethers, and a plant-specific blend of volatiles including leaf alcohols are among the numerous products. Thus, the lipoxygenase pathway becomes an initial step in the interaction of plants with pathogens, insects, or abiotic stress and at distinct stages of development (Feussner & Wasternack, 2002).

4.2 In humans
Besides polyunsaturated fatty acids, H_2O_2, fatty acid hydroperoxides, and synthetic organic hydroperoxides support the lipoxygenase-catalyzed xenobiotic oxidation the major reactions documented thus far including oxidation, epoxidation, hydroxylation, sulfoxidation, desulfuration, dearylation, and N-dealkylation (Kulkarni, 2001). It is noteworthy that lipoxygenases are also capable of glutathione conjugation of certain xenobiotics (Kulkarni, 2001). Available data suggest that lipoxygenases contribute to in vivo metabolism of endobiotics and xenobiotics in mammals (Kulkarni, 2001).

Recent reviews describe the role of lipoxygenase in cancer (Bhattacharya et al., 2009; Pidgeon et al., 2007; Moreno, 2009), inflammation (Duroudier et al., 2009; Hersberger, 2010) and vascular biology (Chawengsub et al., 2009; Mochizuki & Kwon, 2008) and for an extensive presentation of the role of eicosanoids in prevention and management of diseases the reader is referred to the review of Szefel et al. (2011).

5. Interaction of lipoxygenase with inhibitors as theoretical approach for food industry and medical applications

In terms of the structure and function, LOXs are unique, because their metal cofactor is a single ion bound by the side chains of the surrounding amino acids and the carboxylic group of the C-terminus, and their inhibitors bind to or near the Fe co-factor (Skrzypczak-Jankun et al., 2007). Lipoxygenases are inhibited by a large number of chemicals, some of which also serve as co-substrates (Kulkarni, 2001).

5.1 Importance of lipoxygenase inhibition for food industry

Besides their physiological role, plant lipoxygenases are of significant importance to the food industry, since these enzymes have been implicated in the generation of the flavour and aroma in many plant products. For instance, they are responsible for the undesirable 'beany', 'green' and 'grassy' flavours produced during processing and storage of protein products derived from legume seeds (Fukushima, 1994; Robinson et al., 1995) and the development of the stale flavour in beer during storage (Kobayashi et al., 1993). Lipoxygenases also play an important role in the baking industry. They are quite effective as bleaching agents, increase mixing tolerance and improve dough rheology (Nicolas & Potus, 1994; Larreta-Garde, 1995; Cumbee et al., 1997; Borelli[b] et al., 1999).

Freshly refined soybean oil is practically odourless and bland, but "green, grassy, fishy" off-flavors may develop quickly if the oil is heated or stored under conditions that expose it to light and oxygen or by contamination with pro-oxidant metals such as copper and iron (Berk, 1992). "Beany" flavour is the principal inconvenience of traditional soymilk and its products (e.g., tofu) and is caused by some ketones and aldehydes, particularly hexanals and heptanals, produced through LOX catalyzed oxidation (Berk, 1992).

Fish lipids are susceptible to oxidation owing to the high levels of polyunsaturated fatty acids (PUFA), even in frozen storage, and this can affect the flavour, texture, taste, aroma and shelf life of fish (Ke & Ackman, 1976). Since the direct interaction between oxygen and highly unsaturated lipids is kinetically hindered (Kanner et al., 1987), the enzymatic initiation of oxidation by enzymes such as lipoxygenase, peroxidases and microsomal enzymes has been gaining favour.

Green tea glazing was shown to improve the storage quality of frozen bonito fillets (Lin & Lin, 2005). In addition, hot water tea extract was shown to suppress the pro-oxidant activities of the dark meat and skin of blue sprat (Seto et al., 2005). Banerjee (2006) proposes that the improvement in the shelf life of fish by green tea polyphenols is at least in part due to inhibition of LOX resulting in delaying oxidation of fish lipids and because of that impregnation of muscle fillets in tea extract by itself or in combination with other natural inhibitors may improve the shelf-life and storage quality of fish fillets.

Besides its function of oxidizing the polyunsaturated fatty acids (linoleic, linolenic and arachidonic), the enzyme may also catalyse the co-oxidation of carotenoids, resulting in the loss of natural colorants and essential nutrients (Robinson et al., 1995). LOX have been implicated in the generation of the flavour and aroma in many plant products, in the decolourisation of pigments and in the potential of compromising the anti-oxidant status (Casey, 1999). In pasta the involvement of LOX in colour loss is demonstrated by positive correlation between the decrease of β-carotene content after pastification and LOX activities in semolina. In addition to this, the hydroperoxidation and bleaching activities of LOX are highly correlated demonstrating that the bleaching might be ascribable to a co-oxidative action by LOX (Borrelli[a] et al., 1999).

During pasta processing in which the maximal pigment degradation by LOX activity occurs (Borrelli[b] et al., 1999), it is shown that externally added β-carotene can act as inhibitor of the LOX-catalysed linoleate hydroperoxidation and an inverse relation between the % of carotenoid loss and the initial carotenoid content in semolina from durum varieties, showing similar LOX activity, was found (Trono et al., 1999).

The complete characterisation of lipoxygenase from pea seeds (*Pisum sativum var. Telephone L.*) gives possibility to avoid destructive influence during food processing and storage (Szymanowska et al., 2009) by the action of this enzyme.

5.2 LOX inhibition in cancer

Molecular studies of the well-known relationship between polyunsaturated fatty acid metabolism and carcinogenesis have revealed novel molecular targets for cancer chemoprevention and treatment (Lipkin et al., 1999; Willett, 1997; Klurfeld & Bull, 1997; Guthrie & Carroll, 1999).

The role of lipoxygenase in the development and progression of cancer is complex due to the variety of lipoxygenase genes that have been identified in humans, in addition to different profiles of lipoxygenase observed between studies on human tumor biopsies and experimentally induced animal tumor models (Pidgeon et al., 2007). The literature emerging on the role of lipoxygenases in tumor growth, for the most part, suggests that distinct lipoxygenase isoforms, whose expression are lost during the progression of cancer, may exhibit anti-tumor activity, while other isoforms may exert pro-tumorigenic effects and are preferentially expressed during the development of various cancers.

The involvement of 5-lipoxygenase and 12-lipoxygenase in human cancer progression is now supported by a growing body of literature. The involvement of 15-lipoxygenase-1 in colorectal cancer involves its implication in carcinogenesis having pro-carcinogenic as well as anti-carcinogenic roles (Bhattacharya et al., 2009). The co-localization of these enzymes and the similarities of their bioactions on cancer cell growth suggest that the simultaneous inhibition of these enzymes may represent novel and promising therapeutic approaches in selected cancer types (Pidgeon et al., 2007). Therefore, when targeting the regulation of arachidonic acid metabolism, blocking 5-lipoxygenase, 12-lipoxygenase and 15-lipoxygenase-1 without altering the expression of the anti-carcinogenic 15-lipoxygenase-2 may be the most effective, however at present no drug recapitulates these capabilities (Pidgeon et al., 2007).

5.3 Mechanisms of lipoxygenase inhibition

In general, lipoxygenase inhibitors can bind covalently to iron or form the molecular complexes blocking access to iron (Skrzypczak-Jankun et al., 2007). It was pointed out by Walther et al., that a course of inhibition, by the drug ebselen, (noncompetitive vs competitive) and its reversibility depend on the oxidation state of iron, i.e. whether the enzyme is catalytically silent with Fe^{2+} when it binds covalently, causing irreversible inhibition or preoxidized and active with Fe^{3+} in the presence of the fatty acid substrate (Walther et al., 1999). In both cases the enzyme's performance can be illustrated by a classic Lineweaver-Burk plot. Many inhibitors do not follow such a linear relation between velocity and the inhibitor's concentration showing a hyperbolic curve instead as observed by Skrzypczak-Jankun et al. (2002) for polyphenolic inhibitors (curcumin, quercetin, epigallocatechin gallate and epigallocatechin) interacting with soybean lipoxygenase-3. In general, the kinetic data are seldom reported (Skrzypczak-Jankun et al., 2007). Xenobiotic oxidation by soy lipoxygenase has been investigated and described, while human enzymes lack such thorough studies (Skrzypczak-Jankun et al., 2007). The in vivo susceptibility of lipoxygenases' inhibitors may depend not only on the source of lipoxygenase and its isozyme (Pham et al., 1998; Schewe et al., 1986) but also on the oxidation state of iron and the competition between peroxidase and co-oxidase activities of enzyme (Borbulevych et al., 2004).

The first mode to inhibit the lipoxygenase would be a direct reduction of iron to its inactive form. For soybean lipoxygenase, it has been demonstrated that nordihydroguaiaretic acid rapidly reduces the active ferric species of the enzyme to its inactive ferrous form, thus causing interruption of the catalytic cycle (Kemal et al., 1987). For the polyphenol inhibition of lipoxygenase it was firstly suggested that this molecules strongly complex of the ferric iron moiety of the lipoxygenase, thus preventing its reduction *via* the catalytic cycle as proposed for the action of 4-nitrocatechol on the soybean lipoxygenase-1 (Spaapen et al., 1980). The second observation that complexation of the flavanols with Fe^{3+} did not abolish the inhibitory effect may rule out a direct complexation of the iron moiety in ferric lipoxygenase by these catechol compounds. The X-ray analysis shows 4-nitrocatechol near iron with partial occupancy, blocking access to Fe but not covalently bound to it (Skrzypczak-Jankun et al., 2004). If a similar mode of action holds for the interaction of flavanols with mammalian lipoxygenases, the corresponding iron polyphenol complexes may retain their lipoxygenase-inhibitory effect.

A third conceivable mode of action of polyphenols is the effective reduction of hydroperoxides that are essential activators of lipoxygenase *via* conversion of the enzymatically silent ferrous species to the active ferric form. The observation of Schewe et al. (2001) that lowering of the hydroperoxide tone by glutathione plus glutathione peroxidase did not modulate the inhibitory effects of flavanols on 15-lipoxygenase-1 does not support the latter possibility.

In case of carotenoids, more specifically, β-carotene, the lipoxygenase was inhibited by keeping it in the inactive form of Fe(II) (Serpen & Gökmen, 2006). These authors suggest that β-carotene reacts with linoleyl radical (L•) at the beginning of the chain reaction, so it prevents the accumulation of conjugated diene forms (LOO•, LOO− and LOOH). Since L• transforms back to its original form of LH, the enzyme cannot complete the chain reaction and thus remains in the inactive Fe(II) form, which is not capable of catalyzing linoleic acid hydroperoxidation (Serpen & Gökmen, 2006).

Wu et al. (1999) have reported that β-carotene scavenges the linoleyl peroxy radical (LOO•) by a hydrogen transfer mechanism and the oxidation of β-carotene occurs during this action. In these conditions, it is absolutely clear that the amount of inactivated enzyme depends on the concentration of β-carotene present in the medium (Serpen & Gökmen, 2006).

According to Mahesha et al. (2007) the lipoxygenase inhibition by isoflavones follows the next mechanism: an electron donated by isoflavones is accepted by the ferric form (Fe^{3+}) of lipoxygenase, which is reduced to resting ferrous form (Fe^{2+}), thus inhibiting lipoxygenase.

Genistein is neither consumed nor does state change during the course of the reaction of lipoxygenase (Mahesha et al., 2007), while quercetin entrapped within lipoxygenase undergoes degradation (Borbulevych et al., 2004).

5.4 Inhibition of soybean lipoxygenase by different classes of polyphenols

Gillmor et al. (1997) obtained the structure of the rabbit reticulocyte enzyme as a complex with the inhibitor RS75091. Located in one of the hydrophobic channels of the enzyme, the inhibitor was found to be close but not biding the iron atom of the catalytic situs. These observations provided the first indications of how the native enzyme can interact with potential ligands (Pham et al., 1998).

Natural flavonoids don't affect only the lipoxygenase oxidation of its classical substrates but also the co-oxidation of xenobiotics by this enzyme. Epigallocatechin-gallate, quercetin and rutin proved to reduce the co-oxidation rate of guaiacol, benzidine, paraphenylenediamine and dimethoxybenzidine by soybean lipoxygenase-1 (Hu et al., 2006). This data suggest that flavonoids may have anticarcinogenic and antitoxic effect through inhibition of oxidative activation generated by lipoxygenase (Hu et al., 2006). Green tea polyphenols have potent free radical quenching and antioxidant activities (Wiseman et al., 1997) and have structural features that may specifically interfere with the arachidonic acid cascade, including the lipoxygenase pathway (Hong & Yang, 2003; Hussain et al., 2005). In addition, with growing concerns regarding the safety of synthetic antioxidants such as BHT and BHA, alternative mechanisms of antioxidant protection by the use of natural antioxidants have been in review over the years (Barlow, 1990).

Polyphenols, mainly flavonoids and phenolic acids, are abundant in a number of dietary sources such as certain cocoas, tea, wine, fruits and vegetables. More than 8000 different flavonoids of natural origin are known (Schewe & Sies, 2003). The flavonoids exist in nature as aglycons (free form) or conjugated (with O-glucosides or methylated). The aglycons can be subdivided in different subclasses (flavanols, flavanones, flavones, izoflavones, flavonols, anthocyanidines, aurones, chalcones) in function of how the B ring from their structure is linked to the heterocycle C, of the oxidation state and of the functional groups linked to the C ring (Beecher, 2003).

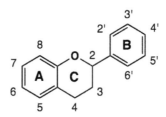

Fig. 3. The basic structure of flavonoids.

The basic structure of flavonoids is represented by the flavan nucleus containing 15 carbons structured in 2 benzene rings, named A and B and linked by a C_3 unit, which together with an oxygen atom forms the γ-pyronic or γ-pyranic ring, named the C ring as shown in Fig.3. A number of *in vitro* and *in vivo* studies as well as clinical trials suggest beneficial effects of flavonoids for health, counteracting the development of cardiovascular diseases, cancer and obesity. Bors et al. (1990) were the first to claim three partial structures contributing to the radical-scavenging activity of flavonoids: (a) an o-dihydroxyl structure in the B ring (catechol structure) as a radical target site providing good electron delocalization and stabilization of the phenoxy radical; (b) a 2,3-double bond with conjugation to the 4-oxo group which is necessary for delocalization of an unpaired electron from the B ring, (c) hydroxyl groups at the 3-and 5-positions, which are necessary for enhancement of radical scavenging activity, increasing the delocalisation of electrons across the flavonoid scaffold. The catechol group is essential for the radical-scavenging activity of flavan-3-ols and flavanones lacking 2,3-double bonds (Bors et al., 1990).

In parallel to the free radical-scavenging properties the following structural features were found to enhance the inhibitory potency: (i) presence of a catechol arrangement in the B or A ring, (ii) a carbonyl group together with a 2,3-double bond in the C ring (Schewe & Sies, 2003). Other structural features were opposite to the free radical scavenging potencies: (i) presence of a 3-OH group in the C ring diminished than reinforced the inhibition of lipoxygenases, (ii) in the absence of a catechol arrangement there was an inverse correlation to the total number of OH groups in the flavonoid molecule (Schewe & Sies, 2003). Although either reducing or ferric iron chelating properties are prerequisites for a lipoxygenase- inhibitory compound, and both of them are also inherent to flavonoids, the inhibitory effects cannot be ascribed solely to one of these mechanisms (Schewe & Sies, 2003). The inhibition of lipoxygenases by flavonoids appears to be of more complex nature (Schewe & Sies, 2003).

Inhibitors studies of lipoxygenase from pea showed that phenolic antioxidant components were effective and can be used to protect food lipids against oxidation (Szymanowska et al., 2009). The conducted research proved that activity of lipoxygenase from pea seeds could be effectively inhibit by some phenolic compounds. The most effective inhibitor is caffeic acid (about 57% of inhibition). Flavonoids like catechin and quercetin considerably inhibit the lipoxygenase activity. Inhibitors used for investigation in this study were placed in the following order: caffeic acid > quercetin > catechin > benzoic acid > ferulic acid > kaempferol (Szymanowska et al., 2009).

5.5 Lipoxygenase inhibition by quercetin

Quercetin is the most abundant among the flavonoid molecules and can be found in the fruits, vegetables, seeds, nuts, and flowers of many plants. Its documented impact on human health includes cardiovascular protection, anticancer, antiviral, anti-inflammatory activities, antiulcer effects and cataract prevention. Like other flavonoids, quercetin appears to combine both lipoxygenase-inhibitory activities and free radical-scavenging properties in one agent and thus belongs to a family of very effective natural antioxidants (Sadik et al., 2003). Quercetin is a flavonol that can be easily oxidized in an aqueous environment, and in the presence of iron and hydroxyl free radicals (Borbulevych et al., 2004).

The inhibition of rabbit 15-lipoxygenase-1 and of soybean lipoxygenase-1 by quercetin was studied in detail (Sadik et al., 2003). Quercetin modulates the time course of the lipoxygenase reaction in a complex manner by exerting three distinct effects: (i) prolongation of the kinetic lag period, (ii) instant decrease in the initial rate after the lag phase being overcome, (iii) time-dependent inactivation of the enzyme during reaction, but not in the absence of substrate (Schewe & Sies, 2003). The literature data obviously indicate that quercetin represents one of the most potent inhibitors of different LOXs (Schneider & Bucar[a], 2005; Schneider & Bucar[b], 2005).

Structural analysis reveals that quercetin entrapped within LOX undergoes degradation and the resulting compound has been identified by X-ray analysis as protocatechuic acid (3,4-dihydroxybenzoic acid) positioned near the iron site (Borbulevych et al., 2004).

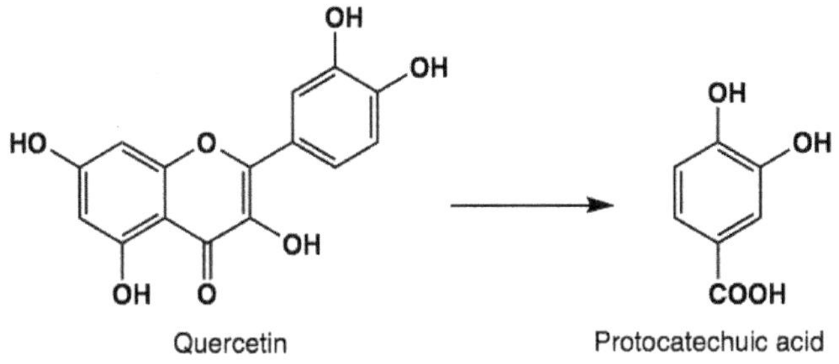

| Quercetin | Protocatechuic acid |

Fig. 4. Product of quercetin degradation by soybean LOX-3 (Borbulevych et al., 2004).

We demonstrated that pH values may influence the molecular interactions between soybean LOX-1 and quercetin, and especially the alcaline pH favours the ionic display of quercetin in order to interact with LOX better (Chedea et al., 2006).

Quercetin inhibited the 12 (S)-hydroxytetraenoic acid production at concentrations below those necessary for growth inhibition in colorectal cancer cells overexpressing the enzyme 12(S)-lipoxygenase with an IC_{50} of 1µM (Bednar et al., 2007). The finding that LOX can turn different compounds into simple catechol derivatives (with one aromatic ring only) might be of importance as an additional small piece of a "jigsaw puzzle" in the much bigger picture of drug metabolism (Borbulevych et al., 2004). Their interactions with LOX can be more complicated than simply blocking the access to the enzyme's active site. The studies on LOX and quercetin contribute to the understanding of biocatalytic properties of this enzyme and its role in the metabolism of this popular (as a medicinal remedy) flavonol and possibly other, similar compounds (Borbulevych et al., 2004). Acting both as a substrate and a source of inhibition, quercetin seems to play an antinomic role (Fiorucci et al., 2008). But this could be explained as quercetin, one of the most representative flavonoids, is a highly functionalized substrate and can thus be activated and degraded following several ways (Fiorucci et al., 2008).

5.6 Inhibition of soybean lipoxygenase by epigallocatechin gallate

Flavanols (or flavan-3-ols or catechins) are a class of flavonoids that include the catechins and the catechin gallates. Catechins are described as colorless, astringent, water-soluble polyphenols found in many fruits and grains, such as coffee, red grapes, prunes and raisins. Their main source however comes from a beverage made from tea leaves of *Camellia sinensis*. (-)-Epigallocatechin gallate (EGCG) together with other galloylated catechins constitute more than 90% of the total catechin content in green tea (Lekli et al., 2010). Laboratory studies strongly indicate that tea inhibits certain cancers, and there is a multitude of evidence confirming the antioncogenic properties of the individual catechins. For instance: EGCG alone shows anticancer effectiveness against carcinogen-induced skin, lung, forestomach, esophagus, duodeum, liver and colon tumors in rodents. It was found to cause apoptosis and/or cell cycle arrest in human carcinoma cells of skin and prostate cancers (Zimeri & Tong, 1999). Catechins have also known inhibitory activity toward dioxygenases with a potential to be utilized in disease prevention and treatment.

The study of Banerjee (2006) shows that green tea polyphenols are very potent inhibitors of mackerel muscle LOX, with EGCG (epigallocatechin gallate) as the most effective inhibitor (IC_{50} 0.13 nM) followed by ECG (epicatechin gallate) (IC_{50} 0.8 nM), EC (epicatechin) (IC_{50} 6.0 nM), EGC (epigallocatechin) (IC_{50} 9.0 nM) and C (catechin) (IC_{50} 22.4 nM). Chocolate and cocoa are also sources of catechins. Epigallocatechin gallate isolated from the seeds of *Theobroma cacao* had the best inhibitory activity on rabit 15-lipoxygenase-1, with an IC_{50}=4M, epicatechin gallate had IC_{50}=5M and epicatechin an IC_{50}=60M (Schewe et al., 2002). A better inhibition of epicatechin (IC_{50} approx. 15M) was registered in the case of recombinant human platelet 12-lipoxygenase.

Obtained from X-ray analysis, the 3D structure of the resulting complex of (-)-epigallocatechin gallate (EGCG) interacting with soybean lipoxygenase-3 reveals the inhibitor depicting (-)-epigallocatechin that lacks the galloyl moiety (Skrzypczak-Jankun et al., 2003). The A-ring is near the iron co-factor, attached by the hydrogen bond to the C-terminus of the enzyme, and the B-ring hydroxyl groups participate in the hydrogen bonds and the van der Waals interactions formed by the surrounding amino acids and water molecules (Skrzypczak-Jancun et al., 2003).

Epigallocatechin gallate   Epigallocatechin

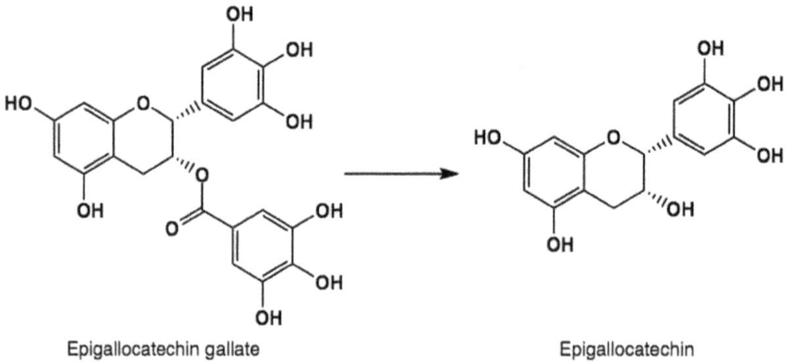

Fig. 5. Lipoxygenase-3 in complex with epigallocatechin gallate as an inhibitor determines the degradation of natural flavonoid to epigallocatechin (Skrzypczak-Jancun et al., 2003).

X-ray analysis of soybean lipoxygenase-3 crystals soaked with EGCG shows the molecular complex of LOX-3 with (-)epigallocatechin molecule.

5.7 Inhibitory effects of soybean isoflavones on lipoxygenase activity

Soybeans are important sources of isoflavone levels (Song et al., 1998), present as 12 derivatives, including free genistin, daidzin, glycitin and their acetyl, malonyl or glycosilated forms. Isoflavones are composed of 2 benzene rings (A and B) linked through a heterocyclic pyrane C ring. The position of the B ring discriminate flavonoid flavones (C2-position) from isoflavones (C3-position).

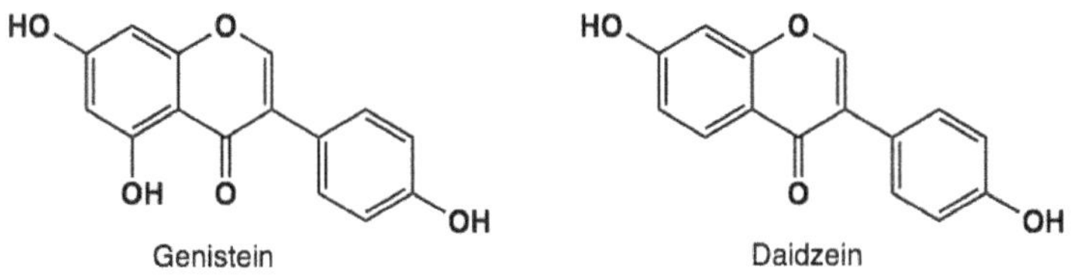

Genistein   Daidzein

Fig. 6. Dietary isoflavones inhibitors of lipoxygenase.

The impact of dietary isoflavones, daidzein and genistein, on the health of adults and infants is well documented, an increasing interest for these compounds being registered due to their biological effects including: estrogen-like activity, prevention of breast (Warri et al., 2008), prostate (Matsumura et al., 2008) and colon cancer (Mac Donald et al., 2005), antioxidant activity (Malencić et al., 2007; Sakthivelu et al., 2008), prevention of menopausal symptoms and osteoporosis (Ma et al., 2008), and heart disease (Xiao, 2008).

In the work of Mahesha et al. (2007) the inhibition of soy lipoxygenase-1 and 5-lipoxygenase from human polymorph nuclear lymphocyte by isoflavones, genistein and daidzein as glycosilated and unglycosilated compounds was studied. Soybean isoflavones inhibit LOX either as aglycons, or as glucosides. Isoflavones exert combined dual actions as inhibitors: they compete with the hydroperoxide formation to prevent the generation of LOX active ferric state (1) and also are capable of reducing the ferric enzyme to its inactive ferrous form (2) (Mahesha et al., 2007).

Vicaş et al. (2011) showed that genistein was almost twice more potent inhibitor than daidzein at similar concentration with concentration that induces 50% soybean lipoxygenase-1 inhibition values of 5.33 mM versus 11.53 mM. Genistein and daidzein proved to be noncompetitive with inhibition constants K_i of 33.65 and 43.45 mM, respectively (Vicaş et al., 2011). The inhibitory efficiency of the genistein and daidzein depended both on their concentration and on the substrate's concentration (Vicaş et al., 2011).

5.8 Inhibition of soybean lipoxygenase by carotenoids

There are over 600 fully characterized, naturally occurring molecular species belonging to the class of carotenoids. In humans, some carotenoids (the provitamin A carotenoids: α-carotene β-carotene, γ-carotene and the xanthophyll, β-cryptoxanthin) are best known for converting enzymatically into vitamin A; diseases resulting from vitamin A deficiency remain among the most significant nutritional challenges worldwide. Also, the role that carotenoids play in protecting those tissues that are the most heavily exposed to light (e.g. photo protection of the skin, protection of the central retina) is perhaps most evident, while other potential roles for carotenoids in the prevention of chronic diseases (cancer, cardiovascular disease) are still being investigated. Because carotenoids are widely consumed and their consumption is a modifiable health behaviour (via diets or supplements), health benefits for chronic disease prevention, if real, could be very significant for public health (Mayne, 2010). Carotenoids are isoprenoid molecules which contain a polyene chains, with or without cyclisation at the ends.

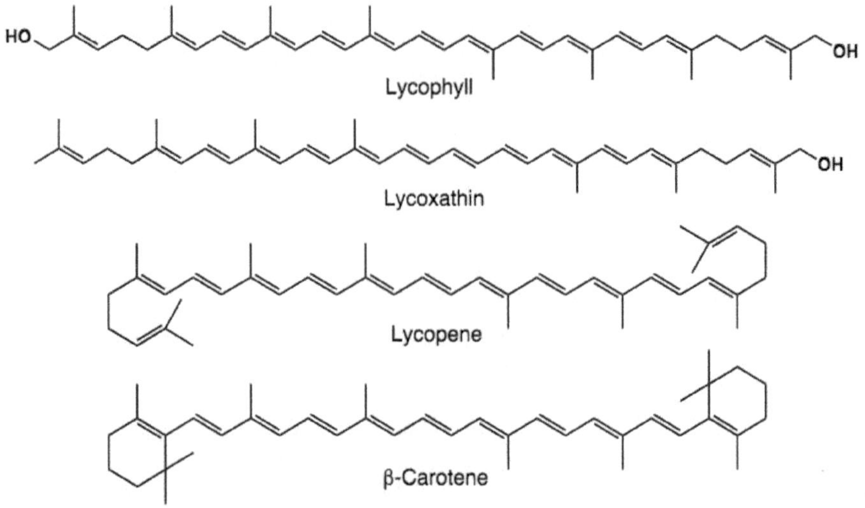

Fig. 7. Carotenoids serving as lipoxygenase co-substrates

The existence of an enzyme "carotene oxidase" in soybeans, which catalyzes the oxidative destruction of carotene was reported by Bohn and Haas in 1928 (Bohn & Haas, 1928). Four years later, Andre and Hou found that soybeans contained an enzyme, lipoxygenase (linoleate oxygen oxidoreductase), which they termed *"lipoxidase"*, catalyzing the peroxidation of certain unsaturated fatty acids (Andre & Hou, 1932).

In 1940 the observation that "lipoxydase" is identical to "carotene oxidase" was published (Sumner & Sumner, 1940). These early findings of lipoxygenase peroxidizing the unsaturated fats and bleaches the carotene were reported as the result of studies on the oxidation of crystalline carotene or carotene dissolved in unsaturated oil. Surprisingly it was found that the carotene oxidase had an almost negligible bleaching action upon the crystalline carotene. On the contrary, when one employs carotene dissolved in a small quantity of fat, the bleaching is extremely rapid. With excessive quantities of fat, the rate of bleaching of the carotene diminishes, and it was concluded that the effect of added fat upon the rate of bleaching of carotene is probably due to a coupled oxidation (Sumner & Sumner, 1940).

Studying the soya-lipoxygenase-catalyzed degradation of carotenoids from tomato Biacs and Daood (2000) found that β-carotene was the most sensitive component, followed by lycoxanthin and lycopene. Their results also implied that β-carotene can actively perform its antioxidant function during the course of lipid oxidation. It seems that oxidative degradation and, accordingly, antioxidant activity of each carotenoid depends on the rate of its interaction with the peroxyl radical produced through the lipoxygenase pathway (Biacs & Daood, 2000) and thus is able to inhibit lipoxygenase. The inhibition of the hydroperoxide formation by carotenoids has been attributed to their lipid peroxyl radical-trapping ability (Burton & Ingold, 1984).

In vitro, lycopene is a substrate of soybean lipoxygenase. The presence of this enzyme also significantly increased the production of lycopene oxidative metabolites (dos Anjos Ferreira et al., 2004; Biacs & Daood, 2000). It was reported that during the co-oxidation of β-carotene by LOX-mediated hydroperoxidation reactions, inhibition of LOX activity takes place also (Lomnitski et al., 1993; Trono et al., 1999; Pastore et al., 2000). The activity of soybean lipoxygenase-1 was inhibited by β-carotene which breaks the chain reaction at the beginning stage of linoleic acid hydroperoxidation (Serpen & Gökmen, 2006). Besides soybean lipoxygenase (Ikedioby & Snyder, 1977; Hildebrand & Hymowitz, 1982) carotene oxidation during lipoxygenase-mediated linoleic acid oxidation has been reported in various studies for the enzymes extracted from potato (Aziz et al., 1999), pea (Yoon & Klein, 1979; Gökmen et al., 2002), wheat (Pastore et al., 2000), olive (Jaren-Galan et al., 1999) and pepper (Jaren-Galan & Minguez-Mosquera, 1997). Soybean lipoxygenase-1 and recombinant pea lipoxygenase-2 and lipoxygenase-3, oxidizing β-carotene, yield apocarotenal, epoxycarotenal, apocarotenone and epoxycarotenone (Wu et al., 1999).

Through molecular modeling Hazai et al. (2006) predicted that lycopene and lycophyll bind with high affinity in the superficial cleft at the interface of the β-barrel and the catalytic domain of 5-LOX (the "cleavage site") suggesting potential direct competitive inhibition of 5-LOX activity by these molecules after in vivo supplementation, particularly in the case of the dial metabolite.

5.9 Quinone and semiquinone formation during the lipoxygenase inhibition reaction

In his excellent review from 2001, Kulkarni presents the studies up to that date indicating the semiquinone and quinone formation in different lipoxygenase catalyzed reactions of xenobiotics oxidation. Diethylstilbestrol (DES) is a human transplacental carcinogen. DES-quinone, one of the metabolites of DES, binds to DNA and is presumed to be the ultimate toxicant. Although DES-quinone formation by human tissue lipoxygenase has yet to be examined, soybean lipoxygenase has been shown to initiate one-electron oxidation to DES semiquinone in the presence of H_2O_2 (Nunez-Delicado et al., 1997). Subsequent dismutation of two molecules of DES semiquinone yields one molecule each of DES-quinone and DES.

Although phenol is oxidized slowly by different lipoxygenase isoenzymes, potato 5-lipoxygenase (Cucurou et al., 1991) and soybean lipoxygenase-1 (Cucurou et al., 1991; Mansuy et al., 1988), substituted phenols and catechols undergo extensive one-electron oxidation and yield the corresponding reactive phenoxyl radicals or semiquinones. These free radicals polymerize to yield a mixture of complex metabolites.

However, it seems that quercetin may act, in most of cases, after being metabolically activated (Metodiewa et al., 1999), and despite a constant increase of knowledge on both positive and negative biological effects of this natural product, it remains often unclear which activated form should play a role in a given process (Fiorucci et al., 2007). Indeed, semiquinone and quinone forms of quercetin, deriving from the abstraction of respectively one or two H•, are involved in many oxidative processes (Metodiewa et al., 1999; Gliszczynska-Swiglo et al., 2003; Hirakawa et al., 2002). For instance, quercetin reduces peroxyl radicals involved in lipid peroxidation, and through this reaction, a semiquinone species is produced, which then undergoes a disproportionation to generate a quinone form (Fiorucci et al., 2007). Three semiquinone forms for quercetin have been considered by Fiorucci et al. (2008) in order to study the quercetin binding to lipoxygenase-3 by molecular modeling simulations. In the case of lipoxygenase–catechol complexes, the formation of the catechol-iron(III) complex of soybean lipoxygenase 1 gradually results in reduction of the cofactor and release of the semiquinone but no evidence of quinone formation in the UV-visible spectra of samples of the native enzyme treated with catechol was obtained (Spaapen et al., 1980; Nelson, 1988; Pham et al., 1998).

Besides their antioxidant properties, catechins have been described to display pro-oxidant activity having the potential to oxidize the quinones or semiquinones resulting in redox cycling and reactive oxygen species production as well as in thiol, DNA and protein alkylation (Galati & O'Brien, 2004; van der Woude et al., 2006).

Our previous study shows that the oxidation products of catechins are formed within the cellular matrix but also in the extracellular medium (Chedea et al., 2010). We have demonstrated by UV-Vis spectroscopy, that the quinones are involved in the modulation of lipoxygenase activity in the presence of catechins within the cells (Chedea et al., 2010). This conclusion is in agreement with that of Sadik et al. (2003) and Banerjee (2006). An irreversible covalent modification of soybean LOX by flavonoids has been suggested by Sadik et al. (2003) whereby during the formation of fatty acid peroxyl radical in the LOX pathway, the flavonoids are co-oxidized to a semi-quinone or quinone, which in turn may bind to sulfhydryl or amino groups of the enzyme causing inhibition (Banerjee, 2006).

6. Conclusions

The knowledge presented in this study addressed the lipoxygenase pathway inhibition by antioxidant polyphenols at two levels: human diet and human health or to a larger extent human disease prevention and treatment. Review articles (Jachak, 2006; Schneider & Bucar[a], 2005; Schneider & Bucar[b], 2005) summarize natural products with inhibitory properties toward LOX (Skrzypczak-Jankun et al., 2007). Natural remedies almost never consist of a single ingredient and usually are a mixture of many in proper proportions, with a synergistic effect of their simultaneous action being absolutely necessary for beneficial medicinal results. Thus, one should proceed with caution, since the action of a selected single compound may not be the same (Skrzypczak-Jankun et al., 2007). The X-ray studies of soybean complexes with quercetin, curcumin, EGCG, EGC indicated conversion of these inhibitors into their metabolites (Skrzypczak-Jankun et al., 2007), which is not surprising considering the co-oxidative activity of LOXs (Kulkarni, 2001). As already presented a question arises concerning the lipoxygenase inhibion: "What is really inhibiting LOX, a given chemical or its LOX metabolite?" (Skrzypczak-Jankun et al., 2007). The results presented so far indicate a complex mode of inhibition involving the inhibitor itself but also its reaction product with lipoxygenase.

The lipoxygenase researcher faces the next antinomy: despite the difference in the number of amino acids between plant and mammalian LOXs, these proteins are amazingly similar in topology with high similarities in the active site of these enzymes. It is believed that all LOXs follow the same catalytic mechanism; however, it is probably the vicinity of the iron site that determines the regio and stereospecificity of the particular enzyme (Skrzypczak-Jankun et al., 2003). In this contradictorily state of facts soybean lipoxygenases stands as a control point in terms of structure, activity and thus inhibition.

7. Aknowledgements

V.S. Chedea is a Japan Society for the Promotion of Science (JSPS) postdoctoral fellow. The authors wish to thank Dr. Ewa Skrzypczak-Jankun for the enlightening discussions during this manuscript writing and Ms. Nana Henmi for helping with the structures drawing.

8. References

André, E. & Hou, K.-W. (1932). La présence d'une oxydase des lipides ou lipoxydase dans graine de soja, glycine soja lieb, *Comptes Rendus*, Vol.194, pp. 645–647.

Aziz, S., Wu, Z. & Robinson, D.S. (1999). Potato lipoxygenase catalyzed co-oxidation of β-carotene. *Food Chemistry* Vol.64, pp. 227–230.

Banerjee, S. (2006). Inhibition of mackerel (*Scomber scombrus*) muscle lipoxygenase by green tea polyphenols, *Food Research International*, Vol.39, pp. 486–491.

Barlow, S. M. (1990). Toxicological aspects of antioxidants used as food additives, In B. J. F. Hudson (Ed.). *Food antioxidants*, London: Elsevier Applied Science, pp. 253–308.

Bednar, W., Holzmann, K. & Marian, B. (2007). Assessing 12(S)-lipoxygenase inhibitory activity using colorectal cancer cells overexpressing the enzyme, *Food and Chemical Toxicology*, Vol.45, pp. 508–514.

Beecher, G.R. (2003). Overview of dietary flavonoids: Nomenclature, occurrence and intake, *Journal of Nutrition*, Vol. 133, pp. 3248S–3254S.

Bergström[a], S., Krabisch, L., Samuelsson, B., & Sjövall, J. (1962). Preparation of prostaglandin F from prostaglandin E, *Acta Chemica Scandinavica*, Vol.16, pp.969-974.

Bergström[b], S., Dressler F., Ryhage R., Samuelsson B. & Sjövall, J. (1962). The isolation of two further prostaglandins from sheep prostate glands, *Arkiv foer Kemi*, Vol.19, pp. 563-567.

Berk, Z. (1992). Technology of production of edible flours and protein products from soybeans, In: *FAO Agricultural Services Bulletin*. Vol. 97. Date of access: 19.01.2010, Available from: http://www.fao.org/docrep/t0532e/t0532e01.htm.

Bhattacharya, S., Mathew, G., Jayne, D.G., Pelengaris, S. & Khan, M. (2009). 15-Lipoxygenase-1 in colorectal cancer: A review, *Tumor Biology*, Vol. 30, pp. 185–199.

Biacs, P.A. & Daood, H.G. (2000). Lipoxygenase-catalysed degradation of carotenoids from tomato in the presence of antioxidant vitamins. *Biochemical Society Transactions*, Vol. 28, pp. 839.

Bohn, R.M. & Haas, L.W. (1928). Chemistry and methods of enzymes, in *Chemistry and methods of enzymes*, ed by Sumner JB and Somers GF, Academic Press, New York.

Borbulevych, O.Y., Jankun, J., Selman, S. & Skrzypezak-Jankun, E. (2004). Lypoxygenase interactions with natural flavonoid, quercetin, reveal a complex with protocatechuic acid in its X-Ray structure at 2.1 Å resolution. *Proteins, Structure, Function and Bioinformatics*, Vol. 54, pp.13-19.

Borrelli[a], G.M., Troccoli, A., Fares, C., Trono, D., De Leonardis, A.M., Padalino, L., Pastore, D., Del Giudice, L. & Di Fonzo, N. (1999). Lipoxygenase in durum wheat: What is the role in pasta colour?, *Options Mediterraneennes*, pp. 497-499, Date of access 12.02.2011, Available from: http://ressources.ciheam.org/om/pdf/a40/00600082.pdf

Borrelli[b], G.M., Troccoli, A., Di Fonzo, N. & Fares, C. (1999). Durum wheat lipoxygenase activity and other quality parameters that affect pasta colour, *Cereal Chemistry*, Vol. 76, pp. 335-340.

Bors, W., Heller, W., Michael, C. & Saran, M. (1990). Flavonoids as antioxidants: Determination of radical-scavenging efficiencies. In L. Packer & A. N. Glazer (Eds.). *Methods in enzymology* (Vol. 186, pp. 343–355). San Diego: Academic Press.

Boyington, J. C., Gaffney, B. J. & Amzel L. M. (1993). Structure of soybean lipoxygenase-1. *Biochemical Society Transactions*, Vol. 21, pp. 744–748.

Burton, G.W. & Ingold, K.U. (1984). β-carotene: An unusual type of lipid antioxidant. *Science*, Vol. 224, pp. 569-573.

Casey, R. (1999). *Lipoxygenases in seed proteins*, pp. 685-708. Shewry, P.R. and Casey, R.eds. Kluwer Academic Publishers.

Chawengsub, Y., Gauthier, K. M. & Campbell, W. B. (2009). Role of arachidonic acid lipoxygenase metabolites in the regulation of vascular tone, *American Journal of Physiology- Heart Circulatory Physiology*, Vol. 297, pp. H495–H507, ISSN: 0363-6135, eISSN: 1522-1539.

Chedea, V. S., Vicaş, S.I., Oprea, C. & Socaciu, C. (2006). Lipoxygense kinetic in relation with lipoxygenase's inhibition by quercetin, *Proceedings of the XXXVI Annual Meeting of the European Society for New Methods in Agricultural Research (ESNA)*, 10 - 14 September 2006, Iasi-Romania, pp. 449-456, ISBN (10) 973-7921-81-X ; ISBN (13) 978-973-7921-81-9.

Chedea, V.S., Braicu, C. & Socaciu, C. (2010). Antioxidant/prooxidant activity of a polyphenolic grape seed extract, *Food Chemistry*, Vol. 121, pp. 132–139.

Coffa, G., Schneider, C. & Brash, A.R. (2005). A comprehensive model of positional and stereo control in lipoxygenases. *Biochemical and Biophysical Research Communications*, Vol. 338, pp. 87–92, ISSN: 0006-291X.

Cucurou, C., Battioni, J. P., Daniel, R. & Mansuy, D. (1991). Peroxidase-like activity of lipoxygenase: different substrate specificity of potato 5-lipoxygenase and soybean 15-lipoxygenase and particular affinity of vitamin E derivatives for the 5-lipoxygenase. *Biochimica et Biophysica Acta*, Vol. 1081, pp. 99–105.

Cumbee, B., Hildebrand, D.F. & Addo, K. (1997). Soybean flour lipoxygenase isozymes effects on wheat flour dough rheological and breadmaking properties. *Journal of Food Science*, Vol. 62, pp. 281–283, ISSN:1750-3841.

dos Anjos Ferreira, A. L., Yeum, K. J., Russell, R. M., Krinsky, N. I. & Tang, G. (2004). Enzymatic and oxidative metabolites of lycopene. *The Journal of Nutritional Biochemistry*, Vol. 15, pp. 493-502, ISSN: 0955-2863.

Duroudier, N.P., Tulah, A.S. & Sayers, I. (2009). Leukotriene pathway genetics and pharmacogenetics in allergy. *Allergy*, Vol. 64, pp. 823–839.

Feussner, I. & Wasternack, C. (2002). The Lipoxygenase Pathway. *Annual Review in Plant Biology*, Vol. 53, pp. 275-297.

Fiorucci, S., Golebiowski, J., Cabrol-Bass, D. & Antonczak, S. (2007). DFT Study of quercetin activated forms involved in antiradical, antioxidant, and prooxidant biological processes, *Journal of Agricultural and Food Chemistry*, Vol. 55, pp. 903-911.

Fiorucci, S., Golebiowski, J., Cabrol-Bass, D. & Antonczak, S. (2008). Molecular simulations enlighten the binding mode of quercetin to lipoxygenase-3. *Proteins*, Vol. 73, pp. 290-298.

Fukushima, D. (1994). Recent progress on biotechnology of soybean proteins and soybean protein food products. *Food Biotechnology*, Vol. 8, pp.83–135.

Galati, G. & O'Brien, P. J. (2004). Potential toxicity of flavonoids and other dietary phenolics: Significance for their chemopreventive and anticancer properties. *Free Radical Biology & Medicine*, Vol. 37, pp. 287-303, ISSN: 0891-5849.

Gardner, H.W. (1991). Recent investigations into the lipoxygenase pathway in plants. *Biochimica Biophysica Acta*, Vol. 1084, No. 3, pp. 221-239.

Gillmor, S. A., Villasenor, A., Fletterick, R., Sigal, E. & Browner, M.F. (1997). The structure of mammalian 15-Lipoxygenase reveals similarity to the lipases and the determinants of substrate specificity. *Nature Structural & Molecular Biology*, Vol. 4, pp. 1003–1009, ISSN: 1545-9993.

Gliszczynska-Swiglo, A., van der Woude, H., de Haan, L., Tyrakowska, B., Aatrs, J. M. M. J. G. & Rietjens, I. M. C. M. (2003). The role of quinone reductase (NQO1) and quinone chemistry in quercetin cytotoxicity. *Toxicology in Vitro*, Vol. 17, pp. 423-431, ISSN: 0887-2333.

Gökmen, V., Bahçeci, S. & Acar, J. (2002). Characterization of crude lipoxygenase extract from green pea using a modified spectrophotometric method. *European Food Research and Technology*, Vol. 215, pp. 42–45, ISSN:1438-2377.

Guthrie, N. & Carroll, K. K. (1999). Specific versus non-specific effects of dietary fat on carcinogenesis. *Progress in Lipid Research*, Vol. 38(3), pp. 261-271, ISSN: 0163-7827.

Hazai, E., Bikádi, Z., Zsila, F. & Lockwood, S. F. (2006). Molecular modeling of the non-covalent binding of the dietary tomato carotenoids lycopene and lycophyll, and selected oxidative metabolites with 5-lipoxygenase. *Bioorganic and Medicinal Chemistry*, Vol. 14, pp. 6859–6867.

Hersberger, M. (2010). Potential role of the lipoxygenase derived lipid mediators in atherosclerosis: leukotrienes, lipoxins and resolvins. *Clinical Chemistry and Laboratory Medicine*, Vol. 48(8), pp.1063–1073, ISSN: 1434-6621.

Hildebrand, D.F. & Hymowitz, T. (1982). Carotene and chlorophyll bleaching by soybeans with and without seed lipoxygenase-1. *Journal of Agricultural and Food Chemistry*, Vol. 30, pp. 705–708, ISSN: 0021-8561.

Hirakawa, K., Oikawa, S., Hiraku, Y., Hirosawa, I. & Kawanishi, S. (2002). Catechol and hydroquinone have different redox properties responsible for their differential DNA-damaging ability. *Chemical Research in Toxicology*, Vol. 15, pp.76-82, ISSN: 0893-228X.

Ho, C. T., Chen, Q., Shi, H. & Rosen, R. T. (1992). Antioxidative effect of polyphenol extract prepared from various Chinese teas. *Preventive Medicine*, Vol. 21, pp. 520-525, ISSN: 0091-7435.

Hong, J.G. & Yang, C.S. (2003). Effects of tea polyphenols on arachidonic acid metabolism in human colon. In F. Shahidi et al. (Eds.), ACS symposium series: Vol. 851. *Food factors in health promotion and disease prevention* (pp. 27–38). Washington, DC: American Chemical Society.

Hu, J., Huang, Y., Xiong, M., Luo, S. & Chen, Y. (2006). The effects of natural flavonoids on lipoxygenase-mediated oxidation of compounds with benzene ring structure-a new possible mechanism of flavonoid anti-chemical carcinogenesis and other toxicities. *International Journal of Toxicology*, Vol. 25, pp. 295-301.

Hussain, T., Gupta, S., Adhami, V. M. & Mukhtar, H. (2005). Green tea constituent epigallocatechin-3-gallate selectively inhibits COX-2 without affecting COX-1 expression in human prostate carcinoma cells, *International Journal of Cancer*, Vol. 113, pp. 660–669.

Ikediobi, C.O. & Snyder, H.E. (1977). Co-oxidation of β-carotene by an isoenzyme of soybean lipoxygenase. *Journal of Agricultural and Food Chemistry*, Vol. 25, pp. 124-127, ISSN: 0021-8561.

Jachak, S. M. (2006). Cyclooxygenase inhibitory natural products: current status. *Current Medicinal Chemistry*, Vol. 13, pp. 659-678, ISSN: 0929-8673.

Jaren-Galan, M. & Minguez-Mosquera, M.I. (1997). β-Carotene and capsanthin co-oxidation by lipoxygenase: kinetic and thermodynamic aspects of reaction. *Journal of Agricultural and Food Chemistry*, Vol. 45, pp. 4814-4820, ISSN: 0021-8561.

Jaren-Galan, M., Carmona-Ramon, C. & Minguez-Mosquera, M.I. (1999). Interaction between chloroplast pigments and lipoxygenase enzymatic extract of olives. *Journal of Agricultural and Food Chemistry*, Vol. 47, pp. 2671-2677, ISSN: 0021-8561.

Kanner, J., German, J. B. & Kinsella, J. E. (1987). Initiation of lipid peroxidation in biological systems. *CRC Critical Reviews in Food Science and Nutrition*, Vol. 25, pp. 317-364.

Ke, P. J. & Ackman, R. G. (1976). Metal catalyzed oxidation in mackerel skin and meat lipids. *Journal of the American Oil Chemists Society*, Vol. 53, pp. 636-640.

Kemal, C., Louis-Flamberg, P., Krupinski-Olsen, R. & Shorter, A.L. (1987). Reductive inactivation of soybean lipoxygenase 1 by catechols: a possible mechanism for regulation of lipoxygenase activity. *Biochemistry*, Vol. 26, No. 22, pp. 7064-7072.

Klurfeld, D.M. & Bull, A.W. (1997). Fatty acids and colon cancer in experimental models, *The American Journal of Clinical Nutrition*, Vol. 66, pp. 1530-1538, ISSN: 0002-9165.

Kobayashi, N., Kaneda, H., Kano, Y. & Koshino, S. (1993). The production of linoleic and linolenic acid hydroperoxides during mashing. *Journal of Fermentation and Bioengineering*. Vol. 76, pp. 371-375.

Kulkarni, A.P. (2001). Lipoxygenase – a versatile biocatalyst for biotransformation of endobiotics and xenobiotics. *Cellular and Molecular Life Sciences*, Vol. 58, pp. 1805-1825.

Larreta-Garde, V. (1995). Lipoxygenase in making breads, cookies and crackers. *Oléagineux Corps Gras Lipides*, Vol. 2, pp. 363-365.

Lekli, I., Ray, D. & Das, D.K. (2010). Longevity nutrients resveratrol, wines and grapes. *Genes and Nutrition*, Vol. 5, pp. 55-60, ISSN: 1555-8932.

Lin, C. C. & Lin, C. S. (2005). Enhancement of the storage quality of frozen bonito fillets by glazing with tea extracts, *Food Control*, Vol. 16, pp. 169-175.

Lipkin, M., Yang, K., Edelmann, W., Xue, L., Fan, K., Risio, M., Newmark, H. & Kucherlapati R. (1999). Preclinical mouse models for cancer chemoprevention studies. *Annals of the New York Academy of Sciences*, Vol. 889, pp. 14-19.

Loiseau, J., Benoît, L. V., Macherel, M.-H. & Le Deunff, Y. (2001). Seed lipoxygenases: occurrence and functions. *Seed Science Research*, Vol. 11, pp. 199-211.

Lomnitski, L., Bar-Natan, R., Sklan, D. & Grossman, S. (1993). The interactions between β-carotene and lipoxygenase in plant and animal systems. *Biochimica Biophysica Acta*, Vol. 1167, pp. 331-338.

Ma, D.F., Qin, L.Q., Wang, P.Y. & Katoh, R. (2008). Soy isoflavone intake inhibits bone resorption and stimulates bone formation in menopausal women: meta-analysis of randomized controlled trials. *European Journal of Clinical Nutrition*, Vol. 62, pp. 155-161, ISSN: 0954-3007.

Maccarrone, M., van Aarie, P.G.M., Veldink, G.A. & Vliegenthart, J.F.G. (1994). In vitro oxygenation of soybean biomembranes by lipoxygenase-2. *Biochimica et Biophysica Acta*, Vol. 1190, pp. 164-169.

Macdonald, R.S., Guo, J., Copeland, J., Browning, J.D. JR., Sleper, D., Rottinghaus, G.E. & Berhow, M.A. (2005). Environmental influences on isoflavones and saponins in soybeans and their role in colon cancer. *Journal of Nutrition*, Vol. 135, pp. 1239-1242.

Mahesha, H.G., Singh, S.A. & Appu Rao, A.G. (2007). Inhibition of lipoxygenase by soy isoflavones: Evidence of isoflavones as redox inhibitors. *Archives of Biochemistry and Biophysics*, Vol. 461, pp. 176-185.

Malencic, D., Popovic, M. & Miladinovic, J. (2007). Phenolic content and antioxidant properties of soybean (*Glycine max* (L.) Merr.) seeds, *Molecules*, Vol. 12, pp. 576–581.

Mansuy, D., Cucurou, C., Biatry, B. & Battioni, J. P. (1988). Soybean lipoxygenase-catalyzed oxidations by linoleic acid hydroperoxide: different reducing substrates and dehydrogenation of phenidone and BW755C. *Biochemical and Biophysical Research Communications*, Vol. 151, pp. 339–346.

Matsumura, K., Tanaka, T. & Kawashima, H. (2008). Involvement of the estrogen receptor beta in genistein induced expression of p21 (waf1/cip1) in PC-3 prostate cancer cells. *Anticancer Research*, Vol. 28, pp. 709-714.

Mayne, S.T. (2010). Carotenoids a colorful and timely research field. In *Carotenoids, physical, chemical and biological functions and properties* edited by Landrum T., CRC Press Taylor and Francisc Group, USA.

Metodiewa, D., Jaiswal, A. K., Cenas, N., Dickancaite´, E. & Segura-Aguilar, J. (1999). Quercetin may act as a cytotoxic prooxidant after its metabolic activation to semiquinone and quinoidal product. *Free Radical Biology and Medicine*, Vol. 26, pp. 107-116.

Minor, W., Steczko, J., Stec, B., Otwinowski, Z., Bolin, J.T., Walter, R. & Axelrod, B. (1996). Crystal structure of soybean lipoxygenase L-1 at 1.4 ANG. resolution. *Biochemistry*, Vol. 35 pp. 10687–10701.

Mochizuki, N. & Kwon, Y.-G. (2008). 15-Lipoxygenase-1 in the vasculature: Expanding roles in angiogenesis, *Circulation Research*, Vol.102, pp. 143-145.

Moreno, J.J. (2009). New aspects of the role of hydroxyeicosatetraenoic acids in cell growth and cancer development. *Biochemical Pharmacology*, Vol. 77, pp. 1–10, ISSN: 0006-2952.

Nelson, M. J. (1988). Catecholate complexes of ferric soybean lipoxygenase 1. *Biochemistry*, Vol. 27, pp. 4273-4278.

Nicolas, J. & Potus, J. (1994). Enzymatic oxidation phenomena and coupled oxidations – Effects of lipoxygenase in bread making and of polyphenol oxidase in fruit technology. *Sciences des Aliments*, Vol. 14, pp. 627–642.

Nunez-Delicado, E., Sanchez-Ferrer, A. & Garcia-Carmona, F. (1997). Hydroperoxidative oxidation of diethylstilbestrol by lipoxygenase. *Archives of Biochemistry and Biophysics*, Vol. 348, pp. 411–414, ISSN: 0003-9861.

Pastore, D., Trono, D., Padalino, L., Simone, S., Valenti, D., Fronzo, N.D. & Passarella, S. (2000). Inhibition by a-tocopherol and L-ascorbate of linoleate hydroperoxidation and β-carotene bleaching activities in durum wheat semolina. *Journal of Cereal Science*, Vol. 31, pp. 41–54, ISSN: 0733-5210.

Pham, C., Jankun, J., Skrzypczak-Jankun, E., Flowers, R.A. & Funk, M.O. (1998). Structural and thermochemical characterization of lipoxygenase-catechol complexes. *Biochemistry*, Vol. 37, pp. 17952-17957.

Pidgeon, G. P., Lysaght, J., Krishnamoorthy, S., Reynolds, J. V., O'Byrne, K., Nie, D. & Honn, K. V. (2007). Lipoxygenase metabolism: roles in tumor progression and survival. *Cancer and Metastasis Reviews*, Vol. 26, pp. 503–524, ISSN: 0167-7659.

Prigge, S.T., Boyington, J.C., Faig M., Doctor, K.S., Gaffney, B.J. & Amzel, L.M. (1997). Structure and mechanism of lipoxygenases. *Biochemie*, Vol. 79, pp. 629-636.

Ramadoss, C.S., Pistorius, E.K. & Axelrod, B. (1978). Coupled oxidation of carotene by lipoxygenase requires two isoenzymes. *Archives of Biochemistry and Biophysics*, Vol. 190, pp. 549–552.

Robinson, D.S., Wu, Z.C., Domoney, C. & Casey, R. (1995). Lipoxygenases and the quality of foods. *Food Chemistry*, Vol. 54, pp. 33–43.

Sadik, C.D., Sies, H. & Schewe, T. (2003). Inhibition of 15-lipoxygenases by flavonoids: structure–activity relations and mode of action. *Biochemical Pharmacology*, Vol. 65, pp. 773-781.

Sakthivelu, G., A. Devi, M.K., Giridhar, P., Rajasekaran, T., Ravishankar, G.A., Nikolova, M.T., Angelov, G.B., Todorova, R.M. & Kosturkova, G.P. (2008). Isoflavone composition, phenol content, and antioxidant activity of soybean seeds from India and Bulgaria. *Journal of Agricultural and Food Chemistry*, Vol. 56, pp. 2090-2095, ISSN: 0021-8561.

Schewe, T. & Sies, H. (2003). Flavonoids as protectants against prooxidant enzymes, *Research monographs*, Institut für Physiologische Chemie I; Heinrich-Heine-Universität Düsseldorf– Date of access: 7.07.2008. Available from: http://www.uni-duesseldorf.de/WWW/MedFak/PhysiolChem/index.html.

Schewe, T., Kühn, H. & Rapoport, S. (1986). Positional specificity of lipoxygenases and their suitability for testing potential drugs. *Prostaglandins, Leukotrienes & Medicine*, Vol. 23, pp. 155-160, ISSN: 0952-3278.

Schewe, T., Kühn, H. & Sies H. (2002). Flavonoids of cocoa inhibit recombinant human 5-Lipoxygenase. *The Journal of Nutrition*, Vol. 56, pp. 1825-1829.

Schewe, T., Sadik, C., Klotz, L.-O., Yoshimoto, T., Kühn, H. & Sies, H. (2002). Polyphenols of cocoa: inhibition of mammalian 15-Lipoxygenase. *Biological Chemistry*, Vol. 382, pp. 1687– 1696.

Schneider, I. & Bucar[a] F. (2005). Lipoxygenase inhibitors from natural plant sources. Part I Medicinal plants with action on arachidonate 5-lipoxygenase. *Phytotherapy Research*, Vol. 19, pp. 81-102.

Schneider, I. and Bucar[b] F. (2005). Lipoxygenase inhibitors from natural plant sources. Part II Medicinal plants with action on arachidonate 12-lipoxygenase, 15-lipoxygenase and leukotriene receptor antagonists. *Phytotherapy Research*, Vol. 19, pp. 263-272.

Serpen, A. & Gökmen, V. (2006). A proposed mechanism for the inhibition of soybean lipoxygenase by β-carotene. *Journal of the Science of Food and Agriculture*, Vol. 86, pp. 401–406.

Seto, Y., Lin, C. C., Endo, Y. & Fujimoto, K. (2005). Retardation of lipid oxidation in blue sprat by hot water tea extracts, *Journal of the Science of Food and Agriculture*, Vol. 85, pp. 1119–1124.

Siedow, J.N. (1991). Plant lipoxygenases. Structure and function. *Annual Review of Plant Physiology and Plant Molecular Biology*, Vol. 42, pp. 145-188.

Skrzypczak-Jankun, E., Amzel, L.M., Kroa, B.A. & Funk, M.O. (1997). Structure of soybean lipoxygenase L3 and a comparison with its L1 isoenzyme. *Proteins*, Vol. 29, pp. 15-31.

Skrzypczak-Jankun, E., Zhou, K. & Jankun J. (2003). Inhibition of lipoxygenase by (-)-epigallocatechin gallate: X-ray analysis at 2.1 A reveals degradation of EGCG and shows soybean LOX-3 complex with EGC instead. *International Journal of Molecular Medicine*, Vol. 12, pp. 415-420.

Skrzypczak-Jankun, E., Chorostowska-Wynimko, J., Selman, S.H., & Jankun, J. (2007). Lipoxygenases – A challenging problem in enzyme inhibition and drug development. *Current Enzyme Inhibition*, Vol. 3, pp. 119-132.

Skrzypczak-Jankun, E., Zhou, K., McCabe, N.P., Kernstock, R., Selman, S.H., Funk, M.O., Jr. & Jankun, J. (2002). In *Natural Polyphenols in Complexes with Lipoxygenase - Structural Studies and their Relevance to Cancer*, pp. 205. 93rd Annual Meeting AACR, San Francisco, 2002; AACR: San Francisco.

Song, T.B., Buseman, K. & Murphy, G. (1998). Soy isoflavone analysis: quality control and a new internal standard. *American Journal of Clinical Nutrition*, Vol. 68, pp. 1474-1479.

Spaapen, L. J., Verhagen, J., Veldink, G. A. & Vliegenthart, J. F. (1980). Properties of a complex of Fe(III)-soybean lipoxygenase-1 and 4-nitrocatechol. *Biochimica and Biophysica Acta*, Vol. 617, No.1, pp. 132-140.

Sumner, J.B. & Sumner, R.J. (1940). The coupled oxidation of carotene and fat by carotene oxidase. *Journal of Biological Chemistry*, Vol. 134, pp. 531–533.

Szefel, J., Piotrowska, M., Kruszewski, W.J., Jankun, J., Lysiak-Szydlowska, W. & Skrzypczak-Jankun, E. (2011). Eicosanoids in prevention and management of diseases, *Current Molecular Medicine*, Vol. 11, pp. 13-25.

Szymanowska, U., Jakubczyk, A., Baraniak, B. & Kur, A. (2009). Characterisation of lipoxygenase from pea seeds (*Pisum sativum* var. Telephone L.). *Food Chemistry*, Vol. 116, pp. 906–910.

Theorell, H., Holman, R. T. & Åkeson, Å. (1947). Cristalline lipoxidase. *Acta Chemica Scandinavica*, Vol. 1, pp. 571-576.

Trono, D., Pastore, D. & Fonzo, N.D. (1999). Carotenoid dependent inhibition of durum wheat lipoxygenase. *Journal of Cereal Science*, Vol. 29, pp. 99–102.

van der Woude, Boersma, H., Alink, M. G., Vervoort, G. M., J. & Rietjens, I. M. C. M. (2006). Consequences of quercetin methylation for its covalent glutathione and DNA adduct formation. *Chemico-Biological Interactions*, Vol. 160, pp. 193-203.

Vicaş, S. I., Chedea, V. S. & Socaciu, C. (2011). Inhibitory effects of isoflavones on soybean lipoxygenase-1 activity, *Journal of Food Biochemistry*, Vol. 35, pp. 613-627.

Walther, M., Holzhutter, H.G., Kuban, R.J., Wiesner, R., Rathmann, J. & Kuhn, H. (1999). The inhibition of mammalian 15-lipoxygenases by the anti-inflammatory drug ebselen: dual-type mechanism involving covalent linkage and alteration of the iron ligand sphere *Molecular Pharmacology*, Vol. 56, pp. 196-203.

Warri, A., Saarinen, N.M. & Makela, S. (2008). The role of early life genistein exposures in modifying breast cancer risk. *British Journal of Cancer*, Vol. 98, pp. 1485-1493, ISSN 0007-0920.

Willett, W.C. (1997). Nutrition and cancer. *Salud Publica Mexican*, Vol. 39, No. 4, pp. 298-309.

Wu, Z., Robinson, D. S., Hughes, R. K., Casey, R., Hardy, D. & West, S. I. (1999). Co-oxidation of β-carotene catalysed by soybean and recombinant pea lipoxygenases. *Journal of Agricultural and Food Chemistry*, Vol. 47, pp. 4899–4906.

Xiao, C. W. (2008). Health effects of soy protein and isoflavones in humans. *Journal of Nutrition*, Vol. 138, No.6, pp. 1244 - 1249.

Yoon, S. & Klein, B.P. (1979). Some properties of pea lipoxygenase isoenzymes. *Journal of Agricultural and Food Chemistry*, Vol. 27, pp. 955–962, ISSN: 0021-8561.

Zimeri, J. & Tong, C.H. (1999). Degradation kinetics of (-)epigallocatechin gallate as a function of pH and dissolved oxygen in a liquid model system. *Journal of Food Science*, Vol. 64, pp. 753-758.

6

Soybean Phytoestrogens – Friends or Foes?

Branka Šošić-Jurjević, Branko Filipović and Milka Sekulić
University of Belgrade, Institute for Biological Research „Siniša Stanković",
Serbia

1. Introduction

Proper and balanced nutrition is very important in prevention and treatment of chronic diseases. Many individuals modify their diet and/or take different nutraceuticals expecting to attain optimum health, extend their lifespan and prevent diseases such as cardiovascular, cancer, osteoporosis, obesity, or diabetes type II.

Based on „Japanese phenomenon" (Adlercreutz, 1998), numerous advertisements suggest that soy-based diet, and its phytoestrogens (PE) in particular, provide protection against many chronic diseases and contribute to the long lifespan often observed in Asia. That is why soy and other phytoestrogen - rich plants became increasingly popular in the U.S. and western countries in the past 30 years. Furthermore, in these countries, PEs are often consumed in its purified form, as nutritional supplements, "designed" for special medical purposes. These supplements are freely available in pharmacies, health food shops, grocery shops and are usually consumed without medical control. There is a lack of awareness that uncontrolled consumption of natural PEs may be potentially harmful to human health. Even more concerning is that some people consume supplements in excess of suggested daily dosage (Wuttke et al., 2007).

The soybean (*Glycine max*), compared to other legumes, is richer in protein levels and quality, based on its digestibility and concentration of essential amino acids (Rand et al., 2003). It is also good source of fiber, certain vitamins and minerals, such as folate and potassium (Rochfort and Panozzo, 2007). It has very high antioxidant content, similar to fruits famous for their antioxidant activity (Galleano et al., 2010). Also, despite their high carbohydrate content, the glycemic load of soybeans is relatively low due to their low glycemic index. In addition, soy-food has high levels of iron in the form of ferritin (Lönnerdal et al., 2006). The concentration of calcium in soymilk is much lower than in cow milk, however, its absorption from soy milk is similar to that from cow milk (Reinwald and Weaver, 2010).

Besides the favorable nutritional attributes, soybean contains a number of biologically active components (saponins and lunasin, phytic acids, phytosterols, trypsin inhibitors, and peptides) including isoflavones genistein (G), daidzein (D) and glycitein (Gy). As soybean phytoestrogens, isoflavones are considered the most important in prevention and treatment of hormone-dependent cancers, cardiovascular diseases, osteoporosis, menopausal symptoms and other age-related diseases. In addition, some studies suggest that soy and its isoflavones affect body weight homeostasis.

Modern world is a controversy with ever-increasing obesity on one side, and a high percent of starving people around the globe, on the other side. Having that in mind, combined with observed beneficial health and weight-lowering effects, high nutritional value makes soy probably one of the most strategically important plants.

However, aside from potential beneficial effects (still under intensive investigation and not fully proven), soybean phytoestrogens may also act as endocrine disruptors, by interfering with the function of reproductive system, as well as with other endocrine systems, namely thyroid and adrenal, and may, under some circumstances, increase cancer risk. This is why scientists are intensively trying to precisely evaluate potential benefits versus adverse effects of soy. Due to the importance, the researches are done both in vitro and in vivo, using different experimental approaches, animal models and various human studies. Results obtained so far are highly inconsistent and depend on experimental conditions, applied doses, animals and humans' age and sex, type of diet, presence of other PE sources in the diet, or other factors. Moreover, it remains unclear whether soy extracts, soy concentrate and purified isoflavones have identical effects. This is why the role of soy food in diet became a somewhat confusing topic in recent years. With approximately 2000 soy-related papers published annually, and half of it related to isoflavones (Messina, 2010), it is becoming extremely difficult to compare all of the available data.

Due to the many differences in the chemical composition of soy products, and the fact that two thirds of human population cannot produce equol (Setchell et al., 2002), the authors decided to primarily focus their attention on effects of purified genistein and daidzein. We will evaluate the latest findings, using clear statements from the literature, as well as our own results, focusing on major potential healthful effects while also considering adverse effects of purified soybean phytoestrogens. More important, the authors will try to analyze the data in order to evaluate whether the net beneficial /adverse effect for each targeted organ system depends on sex and age.

2. Structure, absorption and bioavailability of soybean phytoestrogens

Soybeans and its products are the most abundant source of isoflavones in the human diet. Isoflavones are normally taken up with food, absorbed in the gastrointestinal tract, and eliminated via urine. The absorption and bioavailability of isoflavones has been the subject of frequent debates among scientists. One of the main factor influencing the absorption and bioavailability is the chemical structure of the compound (D'Archivio et al., 2010). Structurally, soy isoflavones G, D and Gy are diphenolic compounds, which are present in soy and non-fermented soyfood isoflavones in its glycosylated forms, as glycones genistin, daidzein and glycitin (Setchell, 1999; Fig. 1).

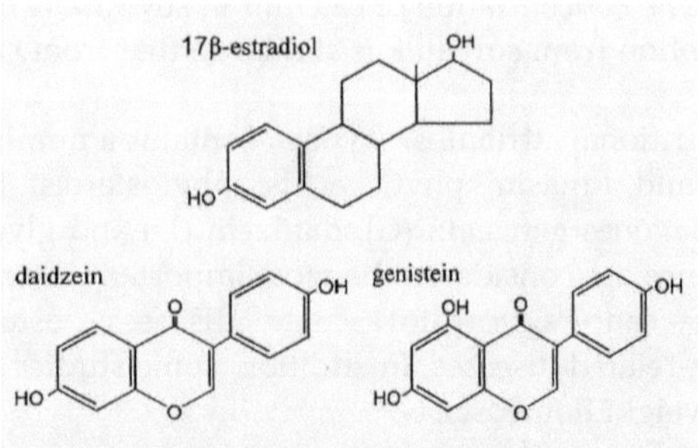

Fig. 1. Structure of soybean phytoestrogens and mammalian 17β – estradiol.

As a prerequisite for absorption, the sugar must be removed from the compound at some point during ingestion (Setchell et al., 2002). Soy isoflavone glycosides are hydrolyzed to their aglycones by lactase phloridizin hydrolase in the apical membrane of the lumen of the small intestine, as well as by bacterial intestinal glucosidases (Wilkinson et el., 2003). Aglycones undergo passive diffusion across the small and large intestinal brush border (Larkin et al., 2008). However, some authors claim that glycosides may be absorbed also through the active sodium–dependent glucose transporter (Gee et al., 2000). Results obtained when we examined effects of soy extract on fluidity of erythrocyte membrane, showed that genistein and isoflavone glucosides intercalate and increase the order and rigidity of the outer layer of cellular membrane. Therefore, isoflavone glucosides may be also transported across the cell membrane directly, via entropy-driven flip-flop (Ajdžanović et al., 2010, 2011). Biological significance of this mechanism is unclear.

The absorption and bioavailability of isoflavones depends to some extent on interaction with other food components (Birt et al., 2001). The assumption that isoflavones are absorbed more efficiently from fermented than from non-fermented soy foods was re-examined and then rejected (Maskarinec et al., 2008). Since intestinal microflora is capable of hydrolyzing the isoflavone glycosides from nonfermented soyfood, recommendations favoring fermented soyfood cannot be justified.

Genistein is stronger than daidzein in its agonistic activity for the ERs, as well as in its antioxidative potential. On the other hand, daidzein can be further metabolized into its bacterial metabolite equol, which has stronger estrogenic and antioxidative properties than both genistein and daidzein, or some other isoflavone metabolites (Mitchell et al., 1998). Although it appears that all animals produce equol following soy ingestion, in humans this is the case in approximately 30% of population (Lampe et al., 1998; Setchell et al., 2002). This is thought to be dependent on inter-individual variability in the presence of specific intestinal bacteria (Rowland et al., 2000). Besides the microflora composition, individual differences in gut transit time and redox potential of colon and genetic polymorphisms are likely to contribute to this great variability (Duffy et al., 2007). When evaluating the effects of age on equol production, it was demonstrated that during the first months of life, equol levels in plasma and urine were significantly lower than in adults, which may be due to the immature intestinal flora (Setchell et al. 2002). Lampe et al. (1998) detected no significant differences in the prevalence of equol production between genders.

3. Biological basis of soybean phytoestrogen actions

The estrogenic effect of the isoflavones was first recognized when examining impaired fertility in grazing animals (Bennetts et al., 1946). Three decades later, Setchell et al. (1987) established that isoflavone-rich soy was a factor in reduced fertility of cheetahs in North American zoos. Isoflavones were classified as phytoestrogens following in vivo and in vitro demonstration of their binding potency of isoflavones for estrogen receptors (ER), as well as for sex-hormone binding globuline (Kuiper et al., 1998).

Testosterone actions in numerous male tissues are mediated through its conversion to estrogen catalyzed by aromatase enzymes. Specific α and β ER are detected in different male and female tissues (Korach, 1994), but the ratio between ERα and ERβ is different (Rosen, 2005). This finding has finally changed the classical view of the estrogens as exclusively female hormones. ERβ is known to modulate ERα transcriptional activity acting as an activator at low concentrations of mammalian estrogen - estradiol 17β (E2) and as an inhibitor at high concentrations of E2. E2 has equal binding affinities for ERα and ERβ, while isoflavones have a higher potency for ERβ.

The molecular structures of genistein, daidzein and E2 are similar in many aspects. The intra-molecular distance between the hydroxyl groups at each end of the molecules is almost identical for both isoflavones and E2. These distances determine hydrogen bond interaction with amino acids of the ligand-binding site of the ER (Vaya and Tamir, 2004). Though molecular binding for ER between isoflavones and E2 are similar, both G and D binding potency for ERs is significantly lesser in comparison to E2. In addition, they bind with higher potency to estrogen receptor (ER) β in comparison to ERα (Kuiper et al., 1998). These features classify them as potential natural selective estrogen receptor modulators (Phyto SERMs). Thus, soybean isoflavones may exert estrogenic, antiestrogenic, or estrogen non-reactive biological actions, depending on their concentration and concentration of endogenous estrogen, tissue, and amount and type of estrogen receptors present in the tissue (Wuttke et al. 2007). Therefore, it is of importance to determine the estrogenic action of isoflavones compared with the effects of E2 (in females) and both testosterone and estradiol (in males) in each individual organ.

Phyto SERMs represent a new and very promising class of potential hormonal therapy agents. Major potential advantage of SERMs over estrogen analogue therapy is that it may demonstrate all of the favorable effects of estrogens. However, in order to declare isoflavones as safe, it needs to be demonstrated that they do not share the risks associated with estrogens used in hormone replacement therapy, osteoporosis treatment or in treatment of prostate carcinoma (Wuttke et al., 2007).

Besides their estrogenic activities, isoflavones also exhibit non-hormonal actions such as antioxidant effects. Antioxidant properties are one of the most important claims for food ingredients, dietary supplements and anticancer products. In addition, the free radical theory of aging continues to be among the most popular theories. Therefore, the antioxidant property of isoflavones offers an additional important mechanism through which they protect against age-related diseases. All soy isoflavones act as antioxidants, playing role in scavenging free radicals that can cause DNA damage and lipid peroxidation (Kruk et al., 2005) and activate antioxidant enzymes such as catalase, superoxide dismutase, glutathione peroxidase and glutathione reductase (Mitchell et al. 1998). The determining factors for isoflavone antioxidant activities are the absence of the 2, 3-double bond and the 4-oxo-group on the isoflavone nucleus and the position of the hydroxyl groups, with hydroxyl substitution being of utmost importance at the 4' position, of moderate importance at the 5 position, and of little significance at the 7 position. That is why G has higher antioxidant capacity than daidzein and the reason why both have stronger antioxidant activity than their glycosides (Cherdshewasart and Sutjit, 2008).

Genistein in high concentration is a potent inhibitor of Tyr kinases (Akiyama et al., 1987), DNA topoisomerases I and II, and ribosomal S6 kinase, resulting in inhibition of cell growth. Tyrosine kinases are responsible for <1% of protein phosphorylation within cells, but they appear to phosphporylate many proteins required for regulation of cell functions. Genistein has been shown to induce cell cycle arrest and apoptosis in numerous cell lines, including ER (+) and ER (-). Many of the reported beneficial effects of isoflavones and particularly those on tumor growth may be attributed to this mechanism. However, some authors critycize this by underlying that the concentrations necessary for such inhibition in the tested cell systems or organs by far exceed the serum concentrations achieved by isoflavone ingestion alone (Jiménez and Montiel 2005; Wuttke et al., 2007).

Growing evidence shows that isoflavones may also modulate the activity/expression of steroidogenic enzymes. These enzymes are present in the adrenal glands and gonads but also in many tissues that have the ability to convert circulating precursors into active hormones (i.e. brain, liver, reproductive tracts, adipose tissue, skin and breast tissue). Genistein and daidzein were reported to inhibit the activity of 3β- hydroxysteroid dehydrogenases (HSD) purified from bovine adrenal microsomes (Wong & Keung, 1999). The same isoflavones were also shown to inhibit 3β-HSD type II in mitochondrial and microsomal preparations of the human adrenocortical H295R cell line, and subsequently a similar inhibition of the conversion of dehydroepiandrosterone (DHEA) to androstenedione by these isoflavones was observed in total membrane fractions of Sf9 insect cells in which human 3β-HSD had been over-expressed (Ohno et al., 2002). However, Mesiano et al. (1999) showed that genistein and daidzein specifically inhibited the activity of 21-hydroxylase (P450c21/CYP 21) in H295 cells but had no effect on other steroidogenic enzymes, including 3β-HSD.

4. Soybean phytoestrogens in prevention and therapy of cancer

The incidence of hormon-dependent cancers, namely breast and prostate, is lower in Asia than in western countries (Messina et al., 2006; Parkin, 2005). Migrants from Asia, who maintained their traditional diet, even when living in the West, had a lower risk of these diseases. However, shifting towards a more of a western diet increased the risk (Ziegler et al., 1993). Once SERM properties of soybean isoflavones were discovered, it was hypothesized that high soy dietary intake might be associated with low incidence of hormone-dependent cancers in Asian population, as well as with other putative health benefits (Setchell, 1999). That is why soyfood and its isoflavones in a form of dietary supplements or concentrated extracts have been increasingly used in the western populations in the recent years.

However, when Patisaul and Jefferson (2010) disscussed potential safety of infant soy formula, they stressed the essential difference between Asians (on a traditional „soy-reach" diet) and Caucasians (on a traditional „Western" diet) in exposure to soy over the lifespan. In Asia, soy consumption is high during entire lifespan, except for a brief breast-feeding period in early infancy. People in the West feed their babies soy infant formula, so the pattern is just the opposite - the highest intake of isoflavones occurs in the first year of life and then drop to near zero, with eventual increase later in advanced adult age. In relation to this, some authors support the opinion that lower incidence of breast cancer in Asian women is due to their continous exposure to soy from early life throuought their whole lifespan (Warri et al., 2008). Maskarinec et al. (2004) concluded that Caucasian women who ate more soy during their lifespan had denser breast tissue (a risk factor for breast cancer) than those who did not.

4.1 Effects on breast cancer

Overexposure to estrogen (early menarche, short duration of breastfeeding and low parity) is a major contributing factor in the development of breast cancer. As soybean isoflavones have a relatively high binding potency for ERs, a concern has been raised that high phytoestrogen intake may promote growth of estrogen-sensitive tumors or put breast cancer survivors at risk of reoccurrence (Helferich et al., 2008; Messina &Loprinzi, 2001).

The data about the role of isoflavones in prevention and therapy of breast cancer are controversial. Some authors proposed that genistein at low, physiologically relevant level, may stimulate ER-positive tumors due to their estrogenic properties, while at higher level, anti-cancer actions of isoflavones may be predominant (Duffy et al., 2007). Shu et al. (2009) also suggested dose-dependent effects of ingested soybean isoflavones: intake of low doses was associated with increased mortality rate and breast cancer recurrence, while intake of more than 40 mg per day appeared to have antiproliferative effects. These results were evident in women with both ER-positive and ER-negative breast cancer. The authors suggested that soy isoflavones protect against breast cancer by competing with estrogens in binding to the estrogen receptor. At the same time soy isoflavones increase the synthesis of sex hormone-binding globulin, lowering the biological availability of sex hormone, inhibit 17β-hydroxysteroid dehydrogenases (thus reducing estrogen synthesis), and increase clearence of steroids from the circulation (Taylor et al., 2009). However, Harris et al. (2004) showed that isoflavones inhibit sulfotransferase (enzymes that catalyze estrogen inactivation in mammary gland) ten times more than sulfatase enzymes, which catalyze local estrogen production. This may lead to increase in free estrogen levels in the tumor tissue, which in turn may stimulate tumor growth.

Recent epidemiological and clinical data were summarized in review article of Messina & Wood (2008) and the authors concluded that isoflavone intake have either a modest protective role or no effect on breast tissue density in pre and postmenopausal women and on breast proliferation in postmenopausal women with or without a history of breast cancer. The results of animal studies are also controversial. Experiments on monkeys that examined effects of soy on mammary gland indicated to the possibility that proliferating effect of estradiol may be antagonized by isoflavone-rich soy protein diet (Jones et al., 2002; Wood et al., 2004).

A number of studies conducted in immunodeficient nu/nu or SCID mice strains demonstrated enhanced proliferation, or no effect of isoflavones on tumor development and progression (Allred et al. 2004; Hsieh et al., 1998). In addition, Heferich and co-workers (2008) implanted estrogen-dependent tumors into ovariectomized mice and found that dietary genistein was able to reduce the inhibitory effect of tamoxifen on tumor growth. However, prevention and inhibition of the progression of experimentally induced mammary tumors by isoflavones was also detected, as well as that post pubertal soy treatment before the induction of tumor had a slightly preventive effect (Pei et al., 2003; Sarkar et al., 2002). Results on Sprag–Dawley rats also proposed that pre-pubertal exposure to soybean isoflavones have highly significant tumor preventive effects (Gallo et al., 2001; Lamartiniere et al., 2002).

More recent review of the animal models used to investigate the health benefits of soy isoflavones concluded that results obtained in different animal models demonstrate minimal effects of isoflavones in breast and prostate cancer prevention (Cooke, 2006).

In vitro genistein inhibited proliferation of ER-positive and ER-negative breast cancer cells at high doses (>10M), but promote tumor growth at lower, more nutritionally relevant doses (Wang et al., 1996). Tamoxifen is the oldest and most-prescribed SERM for breast cancer treatment and it also have mixed effects depending on dose. The SERM-like activities of soy isoflavones makes dietary guidelines particularly difficult to be issued with confidence. Carcinogen-induced mammary cancers predominantly express ERα, and there are some indications that substances that activate mainly ERβ have an antiproliferative effect. In addition, it was reported that genistein may interact with tamoxifen, both synergistically and antagonistically (Shu et al., 2009; Taylor et al., 2009). The inhibition of proliferation in human breast cancer cell line with tamoxifen could be overridden by physiological concentration of genistein (Jones et al., 2002), which indicate that genistein may negate healing effect of tamoxifen on breast cancer patients.

Besides the ER-dependent mechanisms, high doses of genistein may inhibit tumor development and growth by other molecular mechanisms: by antiproliferative actions through inhibition of tyrosine kinase and DNA topoisomerase activities (Akiyama et al., 1987; Markovits et al., 1989), by induction of cell cycle arrest and apoptosis (Bektic et al., 2005), as well as by exerting anti-angiogenic actions (Fotsis et al., 1993).

4.2 Effects on prostate cancer

Lifelong exposure to isoflavones plays a role in the low incidence of prostate cancer observed in Asian males. However, the effects of soy consumption on existing prostate cancer may differ in relation to disease stage. Kurahashi et al. (2007) reported that soy isoflavones in the diet decreased the risk of localized prostate cancer, while soy-containing miso soup increased the risk of advanced prostate cancer. The obtained results may be due to loss of estrogen receptors in advanced tumors, or due to possible errors in food measurement and small sample of men with advanced prostate cancer.

Hamilton-Reeves et al. (2007) reported that soy protein isolate with or without isoflavones affected hormone receptor expression patterns in men at high risk for developing advanced prostate cancer. Intake of soy protein isolate with isoflavones significantly suppressed androgen receptor expression but did not alter estrogen receptor beta expression in prostate, while intake of soy protein isolate without isoflavnoes tended to suppress AR expression (P = 0.09). The authors concluded that soy protein isolate consumption may be beneficial in preventing prostate cancer, and hypothesized that soy isoflavones may attenuate but not prevent progression of latent prostate cancer.

Hussain et al. (2003) found that patients with prostate carcinoma consuming a soy-enriched diet had a statistically significant drop in prostate-specific antigen (PSA) levels, compared to the control group. However, more recent study of deVere White et al. (2010) demonstrated that higher amounts of aglycone isoflavones genistein and daidzein did not lower PSA levels in men with low-volume prostate cancer.

Osterweil (2007) observed a dose-dependent decrease in the risk of localized prostate cancer with isoflavone consumption. Men with higher intake of isoflavones had a decreased risk of prostate cancer compared to those with lower intake of isoflavones.

Few animal studies have been conducted to investigate the role of soy isoflavones on prostate cancer development and progression. Genistein markedly inhibited prostate tumor metastasis in mice (Lakshman et al., 2008). Isoflavone-containing diets retarded the development of prostate cancer in rats (Pollard & Suckow, 2006). In contrast to this, Naik et al. (1994) showed that genistein added to the drinking water or intraperitonealy injected have no effect on the growth of the subcutaneously implanted MAT-LyLu prostate carcinoma in rats.

Zhou et al. (1999) in their in vitro studies found that dietary soy products may inhibit experimental prostate tumor growth through a combination of direct effect on tumor cells and indirect effects on tumor neovasculature. In addition, dietary phytoestrogens down-regulated androgen and estrogen receptor expression in adult male rats prostate (Lund et al., 2004). More recent in vitro studies demonstrated that phytoestrogens at high concentrations exert an anti-androgen effect through the interaction with AR (Mentor-

Marcelet al., 2001). In vitro tests also showed that soy isoflavone genistein induced apoptosis and inhibited growth of both androgen-sensitive and androgen-independent prostate cancer cells (Hussain et al., 2003).

Wuttke et al. (2010) in a recent review provided detailed analysis of both in vitro and animal experimental data and concluded that isoflavones may protect the prostate to make it less prone to develop cancer.

In conclusion, based on inconsistent evidence, it is apparent that the use of phytoestrogens as chemopreventive agents is still in its infancy, justifying a need for further research. Experimental studies based on nutritionally relevant doses are needed to clarify potential health benefits, as well as estrogenic, antiandrogenic and/or nonestrogenic isoflavone activities in the breast and prostate tumors.

5. Soybean phytoestrogens in prevention and therapy of cardiovascular diseases

Soy protein and isoflavones received great attention and provoked heated discussions due to their potential role in reducing risks of cardiovascular diseases. Following is a historical overview of the most relevant results and announcements related to clinical trials, as well as of animal and in vitro research, providing insight into potential mechanisms of isoflavone action.

5.1 Effects on serum lipid levels

Obesity is associated with disruption in lipid and sugar metabolism, and is a principal cause of chronic diseases, namely cardiovascular diseases, hypertension, atherosclerosis and type II diabetes mellitus. This makes obesity a major health problem, which has reached pandemic proportions. The treatment for obesity is lifestyle change, including diet restriction and exercise. However, pharmacological treatment is often necessary. Isoflavones are of particular interest as an alternative to statins or fibrates in potential lowering of serum lipid levels.

Epidemiologic studies demonstrated a reduced rate of mortality due to coronary hearth disease in Japanese postmenopausal women populations consuming a traditional Japanese diet. On the other side expatriate Japanese living in the US had higher blood pressure and cholesterol levels than the Japanese still living in Japan. Some authors proposed that detected differences are not of genetic origin but are due to diet rich in soy products, fish and fiber (Adlercreutz et al., 1998).

Anderson et al. (1995) published a meta-analysis that attracted widespread attention, demonstrating that intake of at least 25g of soy protein per day lowered total and low density lipoprotein (LDL) cholesterol. Lipid lowering potential of soy protein was also demonstrated in various animal studies (Greaves, et al., 1999; Potter, et al., 1995). This led to U.S. Food and Drug Administration (FDA) issuing a health claim for soy protein and coronary hearth disease (1999). FDA also claimed that the evidence did not support significant role of isoflavones in lipid-lowering effects of soy protein. Some more recent reports also demonstrated a significant reduction in plasma concentrations of total and LDL cholesterol in humans exposed to soy proteins (Greany et al., 2004; Teixeira et al., 2000).

Due to their estrogenic activity, isoflavones may be the bioactive component attributed to soy protein. This possibility was examined using different experimental approaches and animal models. Some research studies highlighted a favorable hypolipidemic effect related to isoflavones, at least when consumed in combination with soy proteins. Removal of the isoflavone-containing fraction from soy protein resulted in a loss of its beneficial effect on the serum lipid profile and atherosclerosis progression in mice (Kirk et al., 1998), in golden Syrian hamsters (Lucas, et al., 2001), and in rhesus monkeys (Anthony et al. 1996). High isoflavone, combined with high soy protein intake leads to significantly decreased serum total and LDL cholesterol compared to low isoflavone intake. Some authors reported that ingested purified isoflavones exert lipid-lowering effects (Ae Park et al. 2006; Kojima et al. 2002; Sosić-Jurjević et al., 2007). However, others showed minimal or no effects of isolated isoflavones on blood lipid levels (Greaves et al., 1999; Molsiri et al., 2004).

Clinical trials also show diverse beneficial effects of isoflavone supplements on cardiovascular system. These discrepancies may be a result of different intestinal bacterial flora and hence bioavailability of soy isoflavone metabolites. Other reasons might be differences in dose–response effects (Hooper et al., 2008), sex and length of isoflavone supplementation (Zhan & Ho, 2005), limited number of subjects, or pre-existing metabolic status of subjects included in supplement trials (Villa et al., 2009).

In contrast to previously mentioned data, in 22 random trials, isolated soy protein combined with isoflavones, compared with milk or other proteins, decreased LDL cholesterol by approximately 3%. This reduction was small in comparison to amount of soy protein (average 50g per day) intake (Sacks et al., 2006). There was no detected benefit on level of HDL cholesterol, triglycerides or blood pressure. These authors concluded that soy food may be beneficial to cardiovascular health because of their high content of fiber, vitamins, high content of polyunsaturated fat, rather than and its isoflavone content. Recent review of the animal models used to investigate the health benefits of soy isoflavones also concluded that the efficiency of isoflavones in improving lipid profile is less than earlier research suggested (Cooke, 2006).

For this reason, American Hearth Association issued a discoursing statement, and warned that earlier research indicating clinically important favorable effects of soy products on low density lipoprotein (LDL) is not confirmed by most studies during the past 10 years. U.S. FDA announced its intent to reevaluate the data related to cardio protective effects of soy (2007).

More recent research demonstrated that the combined intervention of genistein and l-carnitine act synergistically in reducing serum lipid and LDL levels, as well as reducing body weight in mice and rats (Che et al., 2011; Yang et al., 2006). In addition, synergy portfolio diet, containing plant sterols, viscous fibers and soy protein reduced serum LDL cholesterol similar to traditional statin drugs (Jenkins et al., 2003). Therefore, soybean isoflavones, either as natural components of food or as nutritional supplements, in combination with other functional food may favorably alter indicators of cardiovascular disease risk.

Though positive effects on metabolism in humans have been widely debated, studies in rodents should help in identifying and evaluating the biologically relevant mechanisms involved in isoflavone actions.

ERs are important mediators of the action of estrogen on lipid metabolism both in males and females. Men with mutations in the aromatase gene (enzyme that converts androgens to estrogens) display truncal obesity, insulin resistance and hyperlipidemia (Carani et al., 1997). Due to structural similarities of isoflavones and E2, G and D might also directly influence the regulation of adipogenesis. However, it must be noticed that genistein preferably binds to ERβ, while ERα is predominantly found in liver. In ovariectomized mice, estradiol and genistein did not increase estrogen-responsive genes in the liver, and the authors suggested that the cholesterol–lowering ability of estrogen requires estrogen receptors (they postulated crosstalk between ERs and NF κβ) but not estrogen receptor-dependent gene transcription (Evans et al., 2001).

Isoflavones may have distinct influences on metabolism in males and females. Males have a different number and distribution of ERs compared to females. It is important to realize the impact of other hormones such as androgens and thyroid hormones on liver and other metabolic tissues. Using ovariectomized Wistar rats Molsiri et al. (2004) obtained no significant difference in serum lipid levels after s.c. genistein injections, while we detected lipid lowering effect of both G and D (similar to this obtained for testosterone-treated groups) in orchidectomized young and middle-aged adults, as well as in testis-intact middle-aged male rats (Sosić-Jurjević et al., 2007 and our unpublished data).

Aside from having estrogenic activity (Potter et al., 1995), both G and D exert „phytofibrate" and or „phytoglitazone"activity, and activate peroxisome proliferator-activated receptors (PPAR) α and γ (Mezei et al., 2006). PPARs bind a wide number of ligands and directly affect lipid metabolism by enhancing transcription of PPAR-regulated genes (Shen et al., 2006). Generally, PPARα controls the transcription of many genes involved in lipid catabolism, whereas PPARγ controls the expression of genes involved in adipocyte differentiation and insulin sensation. PPARα is important for β-oxidation and is mainly expressed in liver, kidney, heart, and muscle, where lipoprotein metabolism is important. PPARγ is mainly expressed in adipose tissues and is considered the master regulator of adipogenesis (Rosen, 2005; Ørgaard & Jensen, 2008).

Isoflavones may also affect lipid metabolism indirectly, via effect on thyroid function and/or thyroid hormone action in liver. T3 and its receptor (TR) play important role in regulation of energy homeostasis, metabolic processes and body weight. Hypothyroidism causes hypercholesterolaemia characterized by increased levels of LDL (Sasaki, et al., 2006).TRβ1 is the major TR in the liver while T3 action is mediated via TRα1 in the heart. TRβ1 agonist KB-141 lower cholesterol, increases metabolic rate and decreases body weight (Grover et al., 2005). Xiao et al. (2007) described that expression of the rat hepatic thyroid hormone receptor β1 is upregulated by isoflavones. In addition, E interplay with TH in regulation of different physiological functions including effects on growth, bone mass, and triglycerides. E can be viewed as a modulator whose response relies on interplay with T3 signaling mechanisms (DiPippo et al., 1995).

Many researchers have tried to link effects of soy intake on lipid metabolism with modulation of thyroid hormone levels. However, it is still difficult to demonstrate clear-cut effects on thyroid (this topic would be analyzed in more details in a subchapter related to endocrine disruptive potential of isoflavones). On the other hand, most researchers who examined lipid-lowering potential of isoflavones did not include in their research examining of the thyroid status, or deiodinase I enzyme activity in liver. When examining the effects of G and D that should mimic exposure to supplements (10mg/kg) in orchidectomized middle-aged male rats our research team obtained that both G and D decreased the serum total cholesterol and LDL levels similar to control testosterone treatment, and brought about an increase in serum triglycerides similar to that observed after control estradiol treatment (Sosić-Jurjević et al., 2007). Within the same animal model we detected significant decrease of serum thyroid hormones (Sosić-Jurjević et al., 2010). However, when we examined deiodinase I enzyme activity in liver of G and D treated rats, it was significantly increased (our unpublished data) in comparison to the control values. Therefore, the local production of T3 in liver was increased and the local increase of T3 might contribute to the detected decrease in total cholesterol and LDL levels.

5.2 Effects on atherosclerosis progression

Atherosclerosis is part of the normal aging process but its progression depends on a wide range of environmental and genetic factors (Davies et al., 2004). Generally, atherosclerosis refers to the formation and hardening of fatty plaques (atheromas) on the inner surface of the arteries. The arteries not only harden, they become narrow. Such narrowed vessels can be easily blocked by constriction or objects in the bloodstream. Atherosclerosis begins with injury to endothelial cells, exposing portions of the artery surface below the endothelium. Free radicals or other irritants could start the process, as well as high blood pressure. Platelets cluster around the injured endothelial cells and release prostaglandins, which cause the endothelial cells to proliferate. LDL-cholesterol particles release their fat into the areas made porous by prostaglandins. Macrophages swell themselves on oxidized LDL-cholesterol until they become "foam cells" that invade atheromas. The atheromas are hardened by fibrin, which forms scar tissue, and finally calcium patches.

The atheroprotective effects of soy-based diets have been partly attributed to the associated reduction in cholesterol levels in human studies (Jenkins et al., 2002). Similar findings have also been reported in nonhuman primates fed soy-based diets (Anthony et al., 1997; Register et al., 2005). Animal studies with rabbits and hamsters, which are considered a good non-primate model for studies of atherosclerosis, demonstrated that soy isoflavones reduce atherosclerotic lesion areas in the aortic arch by means of LDL reduction (Alexandersen et al., 2001; Lucas et al., 2003). In addition, atherosclerotic changes induced by a cholesterol rich diet were prevented by isoflavones in rabbits, hamsters and premenopausal monkeys (Adams, et al. 2005; Lucas et al., 2001). The intake of genistein and daidzein decreases LDL oxidation (Tikkanen, et al., 1998). Both genistein and daidzein have also been shown to protect human umbilical cord endothelial cells and bovine aortic endothelial cells from the atherogenic effect of oxidized LDL (Kapiotis et al., 1997).

However, numerous animal studies suggest that dietary soy inhibits atherosclerotic lesion development by mechanisms other than lowering serum cholesterol.

Isoflavones are reported to prevent lipid peroxidation by scavenging lipid-derived peroxyl radicals (Patel et al., 2001) and inhibit copper-dependent LDL oxidation (Kerry & Abbey, 1998). It is well known that oxidized LDL is more prone to induce atherosclerosis than unoxidized form. In addition, proteome analyses revealed protein targets that in response to soy isoflavones increase the anti-inflammatory response in blood mononuclear cells thereby contributing to the atherosclerosis-preventive activities of a soy-rich diet (Wenzel et al., 2008). Studies in apolipoprotein E knock-out mice showed that atherosclerotic lesions are reduced when fed a soy-containing diet despite unchanged serum lipid levels (Adams et al., 2002). Findings from a recent study in aged lipoprotein receptor knock-out mice has underscored the importance of oxidative stress coupled with a failure to up-regulate There is now compelling evidence that isoflavone supplementation have anti-inflammatory functions and hence can represent an effective therapeutic strategy to enhance Nrf2 activity to protect the aging vasculature (Adams et al., 2002; Mulvihill & Huff, 2010).

Vasodilatatory effects of isoflavones may be also related to their estrogenic actions. Both estrogen receptors α and β are expressed in the arteries (Christian et al., 2006). Estrogens have been shown to stimulate inducible NO synthase in endothelial cells and the increased

NO production causes relaxation of artherial myocites (Mahn et al., 2005). Research on the effect of genistein on plasma nitric oxide concentrations, endothelin-1 levels and endothelium-dependent vasodilatation in postmenopausal women revealed that genistein therapy improved flow-mediated endothelium-dependent vasodilatation in healthy postmenopausal women. This improvement is probably mediated by a direct effect of genistein on vascular function and could be the result of an increased ratio of nitric oxide to endothelin (Squadrito et al., 2002).

In conclusion, despite the fact that dietary soy products and isoflavones are heavily advertised for their hypolipidemic effect, their therapeutic potential is lesser than was previously hoped and depend on many factors related to inter-individual differences.

6. Soybean phytoestrogens in bone protection

Bone remodeling is a continuous process of bone resorption and bone formation for the purpose of maintaining normal bone mass. As a skeletal disease characterized by low bone mass and microarchitectural deterioration of bone tissues, osteoporosis is usually caused by a chronic imbalance in the bone remodeling cycle. This skeletal disorder occurs as part of the natural aging process and is associated with the rapid decline in ovarian function and subsequent reduction of circulating estrogen in women after the menopause and declining testosterone in middle-aged and older men. However, in contrast to postmenopausal osteoporosis in women, the age related bone loss in men is less well-defined. Observational studies have indicated that estrogen administration is important in bone remodeling. Thus, hormone replacement therapy (HRT) administered in a dose-dependent manner, not only significantly reduces bone loss, but also lowers the incidence of hip and vertebral fractures (Lindsay et al., 1976, 1984; Michaelsson et al., 1998). In the other hand, although HRT has a protective effect on bone tissue, it can increase the risk of breast, endometrial ovarian or prostate cancer developing (Davison & Davis, 2003; Loughlin & Richie, 1997; Nelson et al., 2002). For this reason, much attention has been paid to the examination of alternative therapeutic compounds that may have protective effects on bone, without adverse effects on other tissues. Epidemiological studies have demonstrated a low incidence of postmenopausal fractures and high bone mineral density (BMD) in Asian populations with a particularly soy-rich diet (Cooper et al., 1992; Lauderdale et al., 1997; Somekawa et al., 2002). Thus, phytoestrogens have been proposed as an alternative to conventional hormone therapy for preventing osteoporosis and have shown beneficial effects on bone health (Barnes, 2003; Morin, 2004).

Bone remodeling is regulated by the activity of two different cell lines. Osteoblasts stimulate bone formation and calcification, while osteoclasts promote bone resorption. It has been shown that isoflavones affect osteoblastic bone formation and osteoclastic bone resorption in vitro. The anabolic effects of genistein and daidzein on bone metabolism have been investigated in culture using femoral trabecular and cortical bone tissues obtained from elderly female rats (Gao & Yamaguchi, 1999; Yamaguchi & Gao, 1997, 1998). Genistein induced a significant increase in calcium content, alkaline phosphatase activity as a marker of osteoblasts, as well as DNA content, which is an index of bone cell numbers in bone tissues (Yamaguchi & Gao, 1997). In bone tissue culture medium daidzein significantly elevated bone components (Gao & Yamaguchi, 1999). Both genistein and daidzein increased newly synthesized protein content, alkaline phosphatase activity and DNA content in cultures of osteoblastic MC3T3-E1 cells (Sugumoto & Yamaguchi, 2000, 2000a; Yamaguchi & Sugumoto, 2000).

In addition to effects on osteoblasts, many authors have reported that isoflavones are efficacious in suppressing osteoclast activity in vitro. Genistein completely inhibited bone resorption and osteoclast-like multinucleated cells in culture with bone-resorbing factors (Gao & Yamaguchi, 1999a; Yamaguchi & Gao, 1998a). Also, daidzein inhibited the development of osteoclasts from cultures of porcine bone marrow and reduced bone resorption (Rassi et al., 2002).

While in vitro studies reveal possible actions of isoflavones on individual bone cells, in vivo studies provide insight into the effects of isoflavones on the intact system and coupling effects between osteoblasts and osteoclasts. Most of the animal bone studies investigating isoflavone action have been performed in rodents. Aged ovariectomized female and orchidectomized male rats represent a suitable model for simulating osteoporosis due to estrogen or androgen deficiency (Comelekoglu et al., 2007; Filipović et al., 2007; Pantelić et al., 2010; Turner, 2001; Vanderschueren et al., 1992). Using this animal model, supplementation with isoflavones has been shown to prevent bone loss (Fig. 2) induced by gonadal hormone deficiency (Filipović et al., 2010; Khalil et al., 2005; Lee et al., 2004; Om & Shim, 2007; Ren et al., 2007; Soung et al., 2006). In a randomized placebo controlled trial with estrogen and phytoestrogen on ovariectomized nonhuman primates, Ham et al. (2004) failed to show any efficacy of soy phytoestrogens in decreasing all indices of bone turnover as estrogen does, but soy phytoestrogens were able to increase bone volume, trabecular number and decrease trabecular separation, stressing the importance of phytoestrogens in postmenopausal osteoporosis prevention.

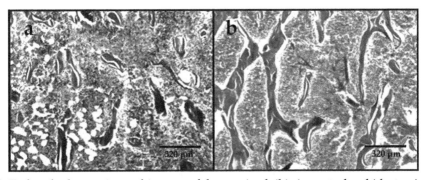

Fig. 2. Trabecular bone microarchitecture of the proximal tibia in control orchidectomized (a) and daidzein-treated orchidectomized (b) rat; azan staining method; unpublished image of Filipovic et al.

Phytoestrogens may elicit a bone sparing effect by both genomic and nongenomic mechanisms. They are able to interact with enzymes and receptors and, their stable structure and low molecular weight enables them to pass through cell membranes (Adlercreutz et al., 1998). The structural similarity of phytoestrogens to mammalian estrogens and their ability to bind to estrogen receptors (Setchell et al., 1999) suggests that the actions of phytoestrogens are mediated via estrogen receptors. ERα and ERβ have been detected in bone (Arts et al., 1997; Onoe et al., 1997). The relative binding affinity of phytoestrogens for ERβ is greater than that for ERα, and the protective effect of phytoestrogens on bone is probably produced through binding to estrogen receptors, particularly ERβ (Kuiper et al., 1998). In addition, phytoestrogens such as coumestrol, genistein and daidzein increase alkaline phosphatase activity in osteoblast-like cells (Kanno et al., 2004). Daidzein stimulates osteoblast differentiation, induces changes in the action of the cytoskeleton responsible for cell adhesion and motility and activates transcription factors associated with cell proliferation and differentiation (de Wilde et al., 2004, Ge et al., 2006; Jia et al., 2003). Also, isoflavones promote insulin-like growth factor-I (IGF-I) production which enhances osteoblastic activity (Ajrmandi et al., 2000)

Isoflavones inhibit bone resorption, via direct targeting of osteoclasts. They can decrease differentiation and increase apoptosis of osteoclasts or interfere with signaling pathways such as intracellular calcium, cAMP or protein kinase and protein tyrosine phosphatase (Gao & Yamaguchi, 2000; Sliwinski et al, 2005). Furthermore, osteoblasts are essential for in vitro osteoclastogenesis through cell-to-cell interactions of cytokines. Isoflavones regulate the expression and osteoblastic production of osteoclastogenesis-regulatory cytokines, such as interleukin-6 (IL-6), which stimulates osteoclast formation, and osteoprotegerin (OPG), which is identical to osteo-clastogenesis inhibitory factor, and the receptor activator of NF-κB ligand (Chen et al., 2002).

In addition to ERs, it has been shown that PPAR are new targets of phytoestrogens. PPAR directly influences osteogenesis and adipogenesis in a divergent way (Dang & Lowik, 2005). These authors suggested that biphasic dose-dependent effects of phytoestrogens are the result of concurrent activation of ERs and PPARs. Dominant ER-mediated effects that increase osteogenesis and decrease adipogenesis can only be seen at low concentrations of phytoestrogens, whereas dominant PPAR-mediated effects that decrease osteogenesis and increase adipogenesis are only evident at high concentrations.

Calcitonin (CT), a hormone secreted from thyroid C cells is known to inhibit osteoclast activity directly through its receptors (Nicholson et al., 1986). It was shown that synthesis and release of CT from thyroid C cells decreased after ovariectomy in rats, due to lack of estrogens (Filipović et al., 2002; Sakai et al., 2000). On the other hand, estrogen treatment had a stimulatory effect on CT secretion in ovariectomized rats (Filipović et al., 2003; Grauer et al., 1993). However, chronic Ca treatment of ovariectomized rats positively affected CT release without any significant changes in morphometric parameters of the C cells, suggesting an important role for estrogen in the regulation of CT synthesis (Filipović et al., 2005). Exogenous CT administration was reported to inhibit CT secretion in rats and therefore CT treatment probably suppresses C cell function due to a negative feedback (Sekulić et al., 2005). Recently, daidzein was found to stimulate CT secreting thyroid C cell activity in addition to increasing trabecular bone mass and decreasing bone turnover (Filipović et al., 2010). These results suggest that, besides direct action, daidzein may affect bone structure indirectly through enhancement of thyroid C cell activity.

Although animal studies demonstrate a clear skeletal benefit of phytoestrogens, clinical trials have given different results. Soy isoflavones were observed to retard bone loss in some (Huang et al., 2006; Newton et al., 2006), but not in other studies (Arjmandi et al., 2005; Brink et al., 2008). To date, only one study indicated that supplementation of intact soy protein providing 83 mg isoflavones daily might increase both hip and spine BMD in men (Newton et al., 2006). Also, the results of meta-analyses of soy foods and isoflavones extracted from soy protein have given conflicting results concerning the prevention of bone loss (Liu et al., 2009; Ma et al., 2008). The large heterogeneity in these conclusions might have arisen because many results were pooled from different individual studies, involving different treatment durations, different doses of soy isoflavone and study quality (Liu et al., 2009). These authors suggested that, because changes in bone mineral density (BMD) occur slowly over time, in short-term intervention studies this change may represent a transient remodeling rather than a long-term steady-state. In addition, a favorable effect on the spine BMD was achieved with large doses of isoflavones ($\geq$ 80 mg/day, median 99 mg/day), but not with lower doses (< 80 mg/day, median 60 mg/day). Thus, the potential of soy isoflavones to prevent bone loss can be achieved by a dosage of 80 mg/day (Huang et al., 2006).

Finally, a wealth of supporting data from many in vitro mechanistic studies on bone cell lines and in vivo investigations using models of osteoporosis shows bone-sparing effects from phytoestrogens. These studies indicate that positive effects of phytoestrogens, as a SERM, may be achieved through estrogen receptors or other mechanisms. However, the results of clinical studies are more inconsistent. The different efficacy of phytoestrogen treatments, in studies involving either animal or human subjects, depended on dose, route and duration of administration. The data are, however, rather tantalizing because it is possible that soy isoflavones may offer the maximum benefit for prevention of osteoporosis. Therefore, it is necessary to perform large-scale clinical dietary intervention studies with phytoestrogens to determine their effects on bone tissue in humans.

7. Soybean phytoestrogens as potential endocrine disruptors

Endocrine systems of vertebrates have essential role in regulation of growth (including bone growth/remodeling), reproduction, stress, lactation, metabolism, energy balance, osmoregulation, and all other processes involved in maintaining homeostasis. Disruption in function of any endocrine system, involving either increased or decreased hormone secretion, result inevitably in disease, the effects of which may extend to many different organs and functions, and may even be life-threatening.

An endocrine-disrupting compound (EDC) is defined by the U.S. Environmental Protection Agency (EPA) as "an exogenous agent that interferes with synthesis, secretion, transport, metabolism, binding action, or elimination of natural blood-borne hormones that are present in the body and are responsible for homeostasis, reproduction, and developmental process." All hormone-sensitive physiological systems are vulnerable to EDCs, including brain and hypothalamic neuroendocrine systems; pituitary; thyroid; adrenal gland; cardiovascular system; mammary gland; adipose tissue; pancreas; ovary and uterus in females; and testes and prostate in males.

The exposure to such chemicals does not necessarily mean that disturbance of the relevant endocrine system will occur, as much depends on the level, duration and timing of exposure. However, even subtle changes, however small, in combination and/or under different conditions and/or in later generations might reduce the ability of humans (animals) to adapt. It may also happen that the magnitude of the disruption becomes evident only in presence of an additional stress factor.

It is beyond the scope of this chapter to discuss the potential interference of soy isoflavones with all endocrine organs; instead, the focus will be on three major endocrine axes that are affected by soybean phytoestrogens: pituitary – gonadal, -thyroid and - adrenocortical systems.

7.1 Effects on female reproductive system

Soybean isoflavones are ligands for both ERα and ERβ, despite the fact that their estrogenic potency is much lower than that of E2. Therefore, they can mimic and/or antagonize the mechanisms of E2 action and thus interfere with both endocrine and reproductive functions of the pituitary-gonadal axis. The rat uterotrophic assay is a widely used screening test for the detection of estrogenic, endocrine-disrupting chemicals. Genistein administration to ovariectomized rats induced a dose-dependent uterine growth and altered expression of estrogen-regulated genes (Diel et al., 2004). As E2 does not stimulate these uterine parameters in ERα KO mice (Couse & Korach, 2001), this test is considered as the proof for estrogenic action of phytoestrogens via ERα.

The first recognized health benefit of isoflavones was their potential to alleviate climacteric complaints, namely hot flushes and night sweats in perimenopausal women (Adlercreutz, 1998). Within the short period of time, numerous isoflavone and soy products became available in a form of food supplements and remedies. They were advertised as natural alternative to hormone replacement therapy, useful in prevention of climacteric symptoms. However, majority of recent placebo-controlled clinical trials support the opinion that isoflavone preparations are not superior to placebo, as placebo effect is 30% to 50% when dealing with psychosomatic climacteric complaints (Patisaul & Jefferson, 2010). Animal studies also demonstrated that only high doses of isoflavones were able to suppress overactivation of hypothalamic gonadotropin –release hormone pulse generator induced by estrogen deprivation (the major cause of hot flushes and other climacteric symptoms (Wuttke et al., 2007). It is important to stress that exposure to high doses of soy isoflavones (150mg/kg) is similar in biological effects to classical hormone replacement therapy. Therefore, their consumption bears a risk of increased proliferation of endometrial and mammary gland tissue with so far unpredictable risk of cancer development.

Multiple human studies demonstrated that exposure of premenopausal women to soybean isoflavones have a suppressive effect on pituitary-gonadal axis; consumption of isoflavones-rich soy food suppresses serum estrogen and progesterone levels and attenuate the preovulatory surge of follicle-stimulating hormone (FSH) and luteinizing hormone (LH) (Hooper et al., 2009; Nagata et al., 1998; Schmidt et al., 2006;). However, some researchers found no impact of isoflavones on female hormone levels (Maskarinec et al., 2002). Soybean phytoestrogens may also affect the women menstrual cycle, but findings are inconsistent. It was shown that a diet with soy protein delays menstruation and prolongs the follicular phase of the menstrual cycle (Cassidy et al., 1994). Other studies demonstrated increased or unchanged follicular phase length, decreased or unchanged midcycle LH and FSH, increased, decreased or unchanged estradiol, decreased dehidroepiandrosterone sulfate, and decreased or unhanged luteal phase progesterone in relation to isoflavone ingestion (Cassidy et al., 1994; Duncan et al., 1999). Therefore, women who try to become pregnant or have menstrual cycle irregularities should be cautious with consumption of isoflavone-enriched soy products or supplements.

Animal studies in rodents produced clear evidence of adverse effects of G on the female reproductive system following treatment during development (Chen et al., 2007; Kouki et al., 2003; National Toxicilogy Program, 2008). Studies that demonstrated clear evidence of developmental toxicity for G involved treatment during the period of lactation in rodents, as well as multigenerational studies that included exposure during gestation, lactation, and post-weaning. In adulthood, the effects of neonatal exposure to 50 mg G/kg bw/day were manifested as a lower number of live pups per litter (Padilla-Banks et al., 2006), a lower number of implantation sites and corpora lutea (Jefferson et al., 2005), and a higher incidence of histomorphological changes of the reproductive tract (i.e., cystic ovaries, progressive proliferative lesions of the oviduct, cystic endometrial hyperplasia, and uterine carcinoma) relative to control females (Newbold et al., 2001).

In addition, the reproductive performance of the neonatally-treated mice was tested during adulthood and there was a significant negative trend for the number of dams with litters. Because the effects were more pronounced in animals at 6 months of age than at 2 or 4 months of age, the authors suggested that reproductive senescence may occur earlier in these animals as a result of the neonatal G treatments (Jefferson et al., 2005). These authors explained that, although G-treated mice ovulate under exogenous hormonal influence, the ovulation rate was changed. The lower doses of G treatment enhanced ovulation rate, while the higher doses decreased this parameter. Ovulation of too many oocytes early in life may reduce the number of oocytes available for fertilization and lead to lower fertility rates later in life (McLachlan et al., 1982). The development of the ovary and ovarian follicles was altered following neonatal G treatment (Jefferson et al., 2002). Ovaries of G-treated mice contained multioocyte follicles (MOFs) at 19th postnatal day. This phenotype is a marker for altered development of the ovary, which lead to oocytes of poor quality (Jefferson et al., 2005). These oocytes are less potent, since the oocytes derived from single oocyte follicles were far more likely to be fertilized in vitro than oocytes derived from MOFs (Iguchi et al., 1990). In our laboratory, results obtained on the ovaries of immature rats treated with 50mg G/kg for three days (from 19th till 21th postnatal day) showed that G disturbed the follicular parenchyma-ovarian stroma ratio (Fig. 3), induced increase of total ovary volume (Medigović et al., 2009).

Data from experiments using DNA microarray analysis for examining the effects of genistein in the developing rat uterus indicate that genistein alters the expression of 6-8 times as many genes as does E2, most of which were down-regulated (Barnes, 2004).

Data are not consistent about onset of puberty and sexual maturation in rats and mice following exposure during gestation and lactation or continuous exposure to soy diet or supplements. An earlier onset of vaginal opening was observed in mice exposed directly to G during the period of lactation (Nikaido et al., 2004.) and in rats treated by sc injection as neonates with 10 mg G/kg bw/day (Bateman & Patisaul, 2008). However, other authors reported delay in vaginal opening (Anzalone et al., 1998).

Only a very small number of studies have been published on D and its estrogenic metabolite equol, and no studies have evaluated the effects of developmental exposure to glycitein. Detection of typical estrogenic effects in these studies are controversial. Kouki et al. (2003) reported no effect on estrous cyclicity in rats treated by sc injection with ~19 mg D/kg bw/day on PND1-5. In contrast, treatment with the same dose levels of G caused the predicted estrogenic effect in all of these studies. Similar to these authors, in our laboratory (unpublished data) no uterotrophic response was detected after subcutaneous injection of immature female rats with 50mg D/kg/day (treatment lasted from 19th postnatal till 21th postnatal day), though the same treatment with G caused predicted estrogenic response.

Isoflavones can pass from mother to fetus through placenta. However, this exposure is considerably lower than in infants fed with soy formula. Initially developed as an alternative to bovine milk formulas for babies with a milk allergy, use of soy infant formula became more popular among environmentally oriented population with vegetarian life style. A recent prospective study in human infants observed that female infants fed soy-based formulas exhibit estrogenized vaginal epithelium at times when their breast fed or cow milk- based formula fed peers did not (Bernbaum et al., 2008). Patisaul and Jefferson

(2010) concluded that further determination if soy infant formula have long-term reproductive health effects should be a public health imperative.

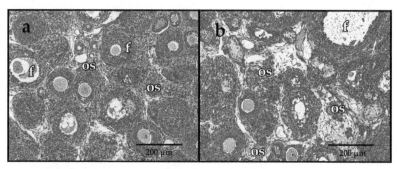

Fig. 3. Ovaries of 21 day old control (a) and genistein – treated rat; hematoxylin - eosin staining method; OS, ovarian stroma; f, follicular parenchyma; hematoxylin – eosin staining method; unpublished image of Medigović et al.

7.2 Effects on male reproductive system

Soy phytoestrogens, alone or in combination with some other EDC, may adversely affect androgen hormone production, spermatogenesis, sperm capacitation and fertility. Results of recent meta-analysis suggest that neither soy foods nor isoflavone supplements alter bioavailable T concentration in adult men (Hamilton-Reeves et al., 2010). However, Tanaka et al. (2009) reported that short-term administration of soy isoflavones decreased testosterone and dihydrotestosterone (DHT) and increased sex hormone-binding globulin levels.

Male reproductive system is particularly sensitive in prenatal stage and during early infancy, when disruption of the hormonal balance in favor of estrogens can lead to irreversible abnormalities in sex specific physiology and behavior in the adulthood (Patisaul & Jafferson, 2010).

Only few animal studies reported results on the developmental effects of exposure to soy infant formula. Their study designs were based on the same group of male marmosets treated during infancy, and assessed either as juveniles (Sharpe et al., 2002) or adults (Tan et al., 2006). The soy infant formula-fed male marmosets had significantly lower plasma testosterone levels than their cow milk formula-fed co-twins. Histopathological analysis on the testes of a subset of the co-twins revealed an increase in Leydig cell abundance per testes in the soy infant formula-fed marmosets compared to their cow milk formula–fed co-twin, in the absence of a significant change in testicular weight. A follow up study was conducted on the remaining animals when they were sexually mature (80 weeks of age or older). The males fed with soy infant formula as infants had significantly heavier testes and increased number of both Leydig and Sertoli cells per testicle compared to cow milk formula-fed controls. In addition, there was no significant onset of puberty, level of adult plasma testosterone, or fertility. The authors suggest that the increase in testes weight was likely due to an increase in testicular cell populations. Therefore, these results demonstrated permanent effects on testicular cell populations, but no obvious effects on reproductive function, namely fertility or permanent changes in testosterone levels of experimental animals.

Some studies on rats and mice demonstrated increased testicular weight when animals were treated with soy diet or isoflavone supplements during gestation and lactation or continuous exposure, similar to the effect described above in marmosets treated with soy infant formula during infancy (Akingbemi et al., 2007; McVey et al., 2004; Piotrowska et al., 2011; Ruhlen et al., 2008; Wisniewski et al., 2005). Other authors reported a decrease (Atanassova et al., 1999; Wisniewski et al., 2003) or no effect on testicular weight (Fielden et al., 2003; Kang et al., 2002).

Controversial results are found as to the effects of lifelong exposure of rodents to phytoestrogens on reproductive function, namely fertility or changes in testosterone levels. The litter size was not affected when male rats were exposed to dietary soy throughout life (Atasnassova et al., 1999). Also, chronic dietary exposure to G did not adversely affect spermatogenesis or seminal vesicle weight in rats (Delclos et al., 2001; Roberts et al., 2000). On the other hand, a few studies indicate negative effects of phytoestrogens on male reproductive success. Thus, a continuous exposure to low combined doses of G and vinclozolin affects male rats' reproductive health by inducing reproductive developmental anomalies, alterations in sperm production and quality, and fertility disorders (Eustache, 2009).

Exposure to G was found to induce hyperplasia of Leydig cells in mice (Lee et al., 2004b). The exposure to isoflavones during 5 weeks decreased the level of circulating testosterone, depending on the dose used (Weber et al., 2001). No significant differences in serum testosterone concentration was detected in rats receiving high doses of G and D from intrauterine life through sexual maturity (Piotrowska et al., 2011). In vitro investigation showed that G can promote the testosterone production of rat Leydig cells at a low concentration, but both D and G can inhibit it at a higher concentration (Zhu et al., 2009).

Effect of phytoestrogens on male reproduction system is a complex process that depends on developmental stage and time of exposure, applied dosage, and other factors. Together, these factors determine the potential risk for adverse consequences with long-lasting effects on male reproductive function. At present, the evidence is insufficient to determine whether soy products cause or do not cause adverse developmental effect on male reproductive system, due to the small number of studies, limitations in their experimental designs, and failure to detect adverse functional effects.

7.3 Effects on pituitary-thyroid axis

Goitrogenic effects of a soybean diet in animals were reported in 1933 (McCarrison, 1933). Similar to animals, goiter and hypothyroidism were reported in infants fed with adapted soy formula without adequate iodine supply (Van Wik et al., 1959). This effect was eliminated by supplementing commercial soy infant formulas with iodine, or by switching to cow milk (Chorazy et al., 1995). However, infants with congenital hypothyroidism that were fed with iodine supplemented diet still needed higher doses of L-thyroxine (Jabbar et al., 1997). In addition, the incidence rate of autoimmune thyroid disease was doubled in teenage children who consumed soy formula as infants (Fort et al., 1990). However, results of clinical studies with adults are not consistent: some authors suggest that isoflavones have a mild or no effect on thyroid function (Dillingham et al., 2007; Duncan et al., 1999), while others indicate that isoflavones suppress the thyroid function (Haselkorn et al., 2003; Ralli, 2003; Sathyapalan et al., 2011).

Rats provide a useful risk assessment model for various thyroid toxins (Choksi et al., 2003). However, compared to the human, rodent thyroid gland is more sensitive to adverse chemicals (Capen, 1997). Several investigators have reported induction of goiter in iodine-deficient rats maintained on a soybean diet (Ikeda et al., 2000; Kajiya et al., 2005; Kimura et al., 1976), although only in cases of iodine deficiency or presence of some other goitrogenic factor. Rats receiving low iodine diet that included 20% of defatted soybeans developed severe hypothyroidism, characterized by a reduction in serum thyroxin and an increase in serum TSH (Ikeda et al., 2000). In addition, a diet containing higher percentage of soy (40% of defatted soybeans) in combination with iodine deficiency induced the development of thyroid carcinoma in rats (Kimura et al., 1976).

Doerge and his associates demonstrated that genistein and daidzein inhibit the activity of thyroid peroxidase (TPO), the key enzyme in the synthesis of thyroid hormones (TH), both in vitro and in vivo (Divi et al., 1997; Chang & Doerge, 2000; Doerge et al., 2002). However, despite significant inactivation of this enzyme, serum thyroid hormone levels were unaffected by isoflavone treatments in young adult rats of both sexes. Most other authors, who performed their studies on young adult animals of both sexes, also reported that soy or isoflavones alone, in the absence of other goitrogenic stimulus, did not affect thyroid weights, histopathology and the serum levels of TSH and thyroid hormones (Chang & Doerge, 2000, Schmutzler et al., 2004). The authors suggested that soy could cause goiter, but only in animals or humans consuming diets marginally adequate in iodine, or who were predisposed to develop goiter, or exposed to additional goitrogenic compounds such as perchlorate, a potent inhibitor of the sodium-iodide-symporter (NIS) of thyrocytes.

Increasing evidence is available that set points of the HPT axis change during various life phases and tend to be less sensitive to negative feedback by thyroid hormones in aging individuals. However, the results on isoflavone effects in aged humans and rodents are scarce. In rodent models, we are the first who demonstrated that both genistein and daidzein induce micro-follicular changes in the thyroid tissue, including hypertrophy of Tg-immunopositive follicular epithelium and colloid depletion (Fig.4), and reduce the level of serum thyroid hormones in orchidectomized (Orx) middle-aged male rats, a model of andropause (Šošić-Jurjević et al., 2010). The concentration of total T4 in serum decreased more prominently than concentration of total T3 in serum in comparison to the corresponding control values. This reduction consequently led to a feedback stimulation of pituitary TSH cells, detected by the increase in cell volume and relative volume density of TSHβ-immunopositive cells per pituitary unit volume, as well as by the increased concentration of TSH in serum. Besides the TPO, there might be other molecular targets for isoflavone interference with the pituitary-thyroid axis.

Soy isoflavones may interfere with thyroid hormones at binding sites of serum distribution proteins such as transthyretin (TTR). In vitro analysis demonstrated that soy isoflavones are potent competitors for T4 binding to TTR in serum and cerebrospinal fluid (Radović et al., 2006). As an outcome of this interference, isoflavones may alter free thyroid hormone concentrations, resulting in altered availability and metabolism of thyroid hormones in target tissues (Köhrle, 2008; Radović et al., 2006). The role of serum binding proteins for thyroid hormone in thyroid homeostasis is not well understood. No single serum T4 - binding protein is essential for good health or for the maintenance of euthyroid state in humans (Robbins, 2000). There are a number of clinical situations in which serum binding proteins are elevated or reduced (even completely absent) and the thyroid state remain normal (Refetoff, 1989). In contrast, there is evidence that the role of serum binding proteins is to allow the equal distribution of hormone delivery to tissues (Mendel et al., 1987). In rats, TTR is a major serum transport protein of thyroid hormones. In humans TTR is produced in the choroid plexus and appears to be important for thyroid hormone action in the brain (Richardson et al., 2007). Thus, TTR may mediate transport of environmental chemicals into various compartments such as placenta (Meerts et al., 2002). Chemical binding to the TTR may not only decrease the availability of thyroid hormone to various tissues, it may also selectively target these chemicals for transport and uptake.

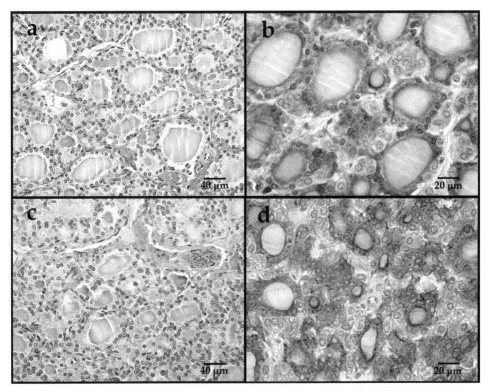

Fig. 4. Thyroid gland tissue of control orchidectomized (a and b) and daidzein-treated orchidectomized (c and d) rat; hematoxylin - eosin and immuno-staining for thyroglobulin; unpublished image of Šošić-Jurjević et al.

In order to accurately asses thyroid function it must be understood that deiodinase enzymes are essential control points of cellular thyroid activity that determine intracellular activation and deactivation of thyroid hormones. Apart from the hormone synthesis by the thyroid gland, deiodination pathways in liver and kidney are the main contributors to thyroid hormone metabolism, turnover and homeostasis. Enzyme 5'-deiodinase type I (5'DI) is the key enzyme in thyroid hormone activation and inactivation in extra thyroidal tissues. This enzyme catalyzes deiodination of the thyroid hormone precursor thyroxine (T4) to the biologically active triiodo-thyronine (T3), as well as the inactivation of T4 and T3 to „reverse" T3 and T2. It is expressed in different tissues, with highest expression rate found in rat liver, kidney, thyroid gland and pituitary (Bianco et al., 2002). It is regulated in a TH-dependent manner (Köhrle, 2002). In response to iodine deficiency or hypothyroidism, plasma TH values are reduced, TSH is increased and the organism tries to restore normal T3 levels by down-regulation of 5'DI in brain and liver, respectively. In addition, activity of 5'DI in different tissues seems to be sex- (Köhrle et al., 1995; Lisboa et al., 2001) and age- dependent (Corrêa da Costa et al., 2001). According to Corrêa da Costa et al. (2001) decreased serum T3 was detected only in old males, which was explained by a two-times-higher hepatic deiodination of T4 to T3, detected in aged females in comparison to males. Genistein was shown to increase hepatic 5'DI activity of about 33% in young adult female rats, but the detected increase was not statistically significant (Schmutzler et al., 2004). In our model – system (orchidectomized middle-aged rats) G significantly increased (p<0.05) 5'DI activity by 33% (unpublished data). However, neither 5'DI in thyroid nor pituitary 5'DII activity were affected by G or D treatment (unpublished data). These data indicate that although pituitary-thyroid axis in male rats is more vulnerable compared to the one in young adults, it still has great ability to compensate the adverse effects of isoflavones.

Isoflavones may also affect the thyroid function indirectly, via its estrogenic action. Estrogen receptors were located both in pituitary thyrotrophs and in thyroid follicular cells (González et al., 2008; Hampl et al., 1985). Donda et al. (1990) found that pituitary TSH cells in adult female rats have a higher density of T_3 and TRH receptors than in male rats, probably due to a modulatory effect of estradiol. Males are more prone to develop goitrogenesis in response to goitrogenic stimuli, probably due to higher TSH levels in comparison to females (Capen, 1997). It seems that estradiol make the TSH cells more sensitive to the negative feedback regulation with thyroid hormones (Ahlquist et al., 1987). In orchidectomized middle-aged rats we demonstrated that pharmacological doses of testosterone and estradiol disturbed the endocrine homeostasis of pituitary-thyroid axis, but in different directions. Testosterone acted stimulatory, probably through central stimulation of pituitary TSH cells, since both serum TSH and T4 levels were increased. Estradiol acted inhibitory and, though detected structural changes corresponded to centrally induced hypothyroidism, the level of TSH in serum was not significantly altered, suggesting that estradiol may interfere with TSH action within the thyrocytes (Sekulic et al., 2010).

Estrogen was also demonstrated to inhibit activity of thyroid follicular cells in the absence of TSH both in vitro and in vivo (Furlanetto et al., 2001; Vidal et al., 2001). Our previous research of a young adult and middle-aged rat menopause models indicated that chronic estradiol treatment modulated pituitary TSH cells and thyroid structure and decreased serum levels of thyroid hormones, with no significant changes in serum TSH level (Šošić-Jurjević et al., 2005, 2006). Genistein acted as estrogen agonist in an estrogen-responsive pituitary cell line (Stahl et al., 1998).

In conclusion, though there are multiple molecular targets for interference of isoflavones with pituitary-thyroid-peripheral network, this system has considerable capacity to compensate disturbances of its feedback mechanism. If thyroid function is impaired, the risk of developing hypothyroidism increases. Elderly population and individuals with thyroid dysfunction should be aware of potential risk when use isoflavone supplements.

7.4 Effects on pituitary-adrenocortical axis

Results concerning the potential effects of the soy phytoestrogens on pituitary-adrenocortical axis in humans are very limited. The animal studies and in vitro experiments demonstrated remarkable influence of isoflavones on morphology and function of adrenal cortex. The continuous administration of genistein (40mg/kg) to weanling rats resulted in greater total protein content in zona fasciculata (ZF) and zona reticularis (ZR) of adrenal cortex, and low serum corticosterone concentration (corticosterone is a major glucocorticoid hormone in rats; Ohno et al., 2003). Genistein administration to orchidectomized middle-aged rats, as a model of andropause, increased zona glomerulosa (ZG), ZF (Fig. 5) and ZR cell volumes, and decreased serum aldosterone and corticosterone concentrations (p<0.05), whereas serum DHEA concentration significantly increased (Ajdžanović et al., 2009a). Genistein and daidzein increased androgen and decreased glucocorticoid production (Mesiano et al., 1999) in human adrenocortical cells in a culture. Recent study on human adrenocortical H295R cell line demonstrated that daidzein and genistein strongly inhibited secretion of cortisol with IC50 values below 1 μM (Ohlsson et al., 2010).

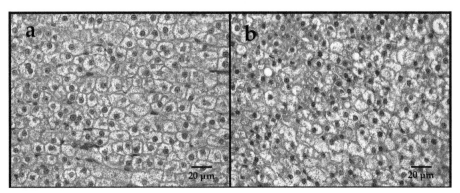

Fig. 5. Zona fasciculata of adrenal cortex in control orchidectomized (a) and genistein-treated orchidectomized (b) rat; azan staining method; unpublished image of Ajdžanović et al.

The isoflavones possess structural features similar to estradiol, which enables them to act via ERs (Lephart et al., 2004). Production of steroids in human fetal adrenocortical cells is modulated by estrogens (Fujieda et al., 1982, Mesiano & Jaffe, 1993; Voutilainen et al., 1979). It was shown that 17β-estradiol in high concentrations increased ACTH-stimulated androgen production and inhibited glucocorticoid synthesis in cultured human fetal adrenal cortical cells (Mesiano & Jaffe, 1993). Although these results indicate the influence of estrogens on the adrenocortical cells in vitro, their physiological significance is still unclear. Under physiological conditions the endogenous estrogen concentration does not reach 1 μM/L in nonpregnant adults. However, it is possible that dietary phytoestrogens, as estrogen-related compounds, could reach circulating levels high enough to exert estrogenic actions. Consuming the large amounts of soy-derived foods, for example in Japanese diet, circulating concentrations of phytoestrogens can reach higher levels (1-5 micromole/L) (Adlercreutz & Mazur, 1997).

Isoflavones may also affect activity or expression of steroidogenic enzymes, which seems to be the case for its action on rat adrenal cortex (Malendowicz et al., 2006; Mesiano et al., 1999; Ohno et al., 2003). Within the adrenals, steroids are produced through the action of five forms of cytochrome P450 and 3β-hydroxysteroid dehydrogenase (3βHSD) (Simpson & Waterman, 1992). Differential expression of these enzymes in the three adrenocortical zones leads to the production of specific steroids within each zone (Suzuki et al., 2000). As a precursor of steroidogenesis, the glomerulosa cells use pregnenolone which can be metabolized by either 3βHSD or 17α-hydroxylase, 17, 20-lyase. The relative expression of these enzymes influences the synthesis of aldosterone and cortisol/corticosterone in ZG and ZF, as well as adrenal androgens in ZR (Conley & Bird, 1997). The major physiological regulators of adrenal aldosterone production are angiotenzin II (Ang II) and potassium. Ang II stimulates aldosteron production through the activation of multiple intracellular signaling pathways including a number of tyrosine kinases (Berk & Corson, 1997; Ishida et al., 1995). It was showed that genistein, as a potent inhibitor of various tyrosine kinases may inhibit aldosteron production (Akiyama et al., 1987; Dhar et al., 1990). Genistein and daidzein are also potent competitive inhibitors of human adrenocortical 3βHSD and cytochrome P450 21-hydroxylase, suppressing cortisol and stimulating DHEA production in vitro (Mesiano et al., 1999). Part of the inhibition of aldosteron production may result from an increase in 17α-hydroxylase, 17, 20-lyase activity, which removes the substrate from the pathway leading to aldosterone and directs it towards the synthesis of adrenal androgens (Sirianni et al., 2001). Isoflavones could also affect adrenal function indirectly, by affecting pituitary ACTH cells. It was previously reported that estrogen replacement lowered the *proopiomelanocortin* (POMC) gene mRNA level and the ACTH response to repeated stressful stimuli in ovariectomized rats (Redei et al., 1994). A certain synergism between CRH (*corticotrophin releasing hormone*) and the various cytokines, namely IL-1, IL-2 and IL-6, has been shown to exist in stimulation of the pituitary ACTH secretion (Bateman et al., 1989; Besedowsky & del Ray, 1996). Genistein may interrupt the stimulatory effects of CRH and cytokines on POMC gene transcription and reduce the level of ACTH, through inhibition of tyrosine kinase phosphorylation cascades (Katahira et al., 1998), but the biological significance of this mechanism is still unclear. We treated orchidectomized middle-aged rats with different doses of genistein or daidzein (10 and 30mg/kg body weight); (Ajdžanović et al., 2009; Ajdžanović et al., 2010, Milošević et al., 2009), and detected similar decrease in pituitary ACTH cellular volume and plasma ACTH levels. Corticosterone levels were also decreased, supporting that some other mechanism, aside from feedback regulation, is involved in effect of isoflavones on pituitary ACTH cell regulation. Keeping in mind that aging is associated with augmented activity of the pituitary-adrenal axis and higher incidence of stress-related psychiatric disorders (Hatzinger et al. 2000), this decline might be considered beneficial at some point.

On the other hand, chronic treatment of weanling rats with genistein (40mg/kg body weight) elevated ACTH level, most probably due to decreased serum corticosterone level and thus release from a negative feedback regulation (Ohno et al., 2003). This finding is of importance since glucocorticoids have important "programming" effects during development. This means that alternations in the circulating levels of glucocorticoid hormones may affect the timing and set points of other endocrine axes (Manojlovic-Stojanoski et al., 2010), as well as brain development, memory and learning capabilities in adults (de Kloet et al., 1988).

Based on animal studies and in vitro research it may be concluded that soy isoflavones interfere with the function of pituitary-adrenocortical axis. This hormonal axis plays a major role in control of stress response and regulation of numerous body processes (digestion, metabolism of carbohydrates, protein and fat, attenuation of the inflammatory response, mood, emotions and sexuality). Therefore, the biological impact of this interference is high. Potential health risks for various age groups should be further assessed.

8. Conclusion

So, are soy isoflavones friends or foes? The answer is complex and may ultimately depend on age, sex, health status, quantity of intake, and even the composition of an individual's intestinal micro flora. In vitro and animal research, as well as human research including both clinical and epidemiologic data, suggests that isoflavone-containing products pose a risk to estrogen-sensitive breast cancer patients and in women at high risk of developing this disease. Results of animal and human studies suggest a modest benefit in prevention of prostate cancer. Exposure to isoflavones by feeding soy infant formula bears a risk of adverse effects on the long–term development of infants. Women who tend to get pregnant or have irregularities in menstrual cycle, as well as persons who are at risk of thyroid dysfunction, should avoid soy isoflavone supplements. The usage of soy protein (with or without isoflavones) seems to have a modest beneficial effect on cardiovascular system and protective role in prevention and treatment of osteoporosis. Research of potential synergy of isoflavones and drugs, and/or other functional food could be a new promising strategy in reducing risk of age-related diseases, improving life quality and expanding life span.

9. Acknowledgment

This work was supported by the Ministry of Education and Science of the Republic of Serbia, Grant No. 173009. Publishing of this chapter was in part financially supported by SOJAPROTEIN, member of Victoria Group, Soybean Processing Company – PLC, Bečej, Serbia. The authors express their sincere thanks to late Dr. Dana Brunner for her guidance and contribution, and to Mr. Kristijan Jurjevic for his valued assistance with English manuscript preparation.

10. References

Adams, MR., Golden, DL., Register, TC., Anthony, MS., Hodgin, JB, Maeda, N. & Williams, JK. (2002). The atheroprotective effect of dietary soy isoflavones in apolipoprotein E-/- mice requires the presence of estrogen receptor-alpha. *Arterioscler Thromb Vasc Biol*, Vol. 22, pp. 1859-1864

Adams, MR., Golden, DL., Williams, JK., Franke, AA., Register, TC. & Kaplan, JR. (2005). Soy protein containing isoflavones reduces the size of atherosclerotic plaques without affecting coronary artery reactivity in adult male monkeys. *J Nutr*, Vol. 135, pp. 2852-2856

Adlercreutz, H. & Mazur W. (1997). Phyto-oestrogens and Western diseases. *Ann Med*, Vol. 29, pp. 95-120

Adlercreutz, H. (1998), Evolution, nutrition, intestinal microflora, and prevention of cancer: a hypothesis. *Proc Soc Exp Biol Med*, Vol. 217, pp. 241–246

Adlercreutz, H., Höckerstedt, K., Bannwart, C., Bloigu, S., Hämäläinen, E., Fotsis, T. & Ollus, A. (1987). *J Steroid Biochem*, Vol.27, pp. 1135-1144

Ae Park, S., Choi, MS., Cho, SY., Seo, JS., Jung, UJ., Kim, MJ., Sung, MK., Park, YB. & Lee, MK.(2006). Genistein and daidzein modulate hepatic glucose and lipid regulating enzyme activities in C57BL/KsJ-db/db mice. *Life Sci*, Vol. 79, pp. 1207-1213

Ahlquist, JA., Franklyn, JA., Wood, DF., Balfour, NJ., Docherty, K., Sheppard, MC. & Ramsden, DB. (1987) Hormonal regulation of thyrotropin synthesis and secretion. *Horm Metab Res Suppl*, Vol. 17, pp. 86-89

Ajdžanović, V., Sosić-Jurjević, B., Filipović, B., Trifunović, S., Brkić, D., Sekulić, M. & Milošević, V. (2009). Genistein affects the morphology of pituitary ACTH cells and decreases circulating levels of ACTH and corticosterone in middle-aged rats. *Biol Res*, 42, pp. 13-23

Ajdžanović, V., Sosić-Jurjević, B., Filipović, B., Trifunović, S., Manojlović-Stojanoski, M., Sekulić M. & Milošević V. (2009) a. Genistein-induced histomorphometric and hormone secreting changes in the adrenal cortex in middle-aged rats. *Exp Biol Med*, 234, pp. 148-156

Ajdžanović, V., Spasojević, I., Filipović, B., Sosić-Jurjević, B., Sekulić, M. & Milosević V. (2010). Effects of genistein and daidzein on erythrocyte membrane fluidity: an electron paramagnetic resonance study. *Can J Physiol Pharmacol*, Vol. 88, pp. 497-500

Ajdžanović, V., Spasojević, I., Sosić-Jurjević, B., Filipović, B., Trifunović, S., Sekulić, M. & Milošević, V. (2011). The negative effect of soy extract on erithrocyte membrane fluidity: An electron paramagnetic resonance study. *J Membrane Biol*, Vol. 239, pp. 131-135

Akingbemi, BT., Braden, TD., Kemppainen, BW., Hancock, KD., Sherrill, JD., Cook, SJ., He, X. & Supko, JG. (2007). Exposure to phytoestrogens in the perinatal period affects androgen secretion by testicular Leydig cells in the adult rat. *Endocrinology*, Vol. 148, pp. 4475-4488

Akiyama, T., Ishida, J., Nakagawa, S., Ogawara, H., Watanabe, S., Itoh, N., Shibuya, M. & Fukami, Y. (1987). Genistein a specific inhibitor of tyrosine-specific protein kinases. *J Biol Chem*, 262, pp. 5592-5595.

Alexandersen, P., Haarbo, J., Breinholt, V. & Christiansen, C. (2001). Dietary phytoestrogens and estrogen inhibit experimental atherosclerosis. *Climacteric*, Vol. 4, pp. 151-159

Allred, CD., Allred, KF., Ju, YH., Goeppinger, TS., Doerge, DR. & Helferich, WG. (2004). Soy processing influences growth of estrogen-dependent breast cancer tumors. *Carcinogenesis*, Vol. 25. pp. 1649-1657

Anderson, JW., Johnstone, BM. & Cook-Newell, ME. (1995). Meta-analysis of the effects of soy protein intake on serum lipids. *N Engl J Med*, Vol. 333, pp. 276-282

Anthony, MS., Clarkson, TB., Bullock, BC. & Wagner, JD. (1997). Soy protein versus soy phytoestrogens in the prevention of diet-induced coronary artery atherosclerosis of male cynomolgus monkeys. *Arterioscler Thromb Vasc Biol*, Vol. 17, pp. 2524-2531

Anthony, MS., Clarkson, TB., Hughes, CL. Jr Morgan, TM. & Burke, GL. (1996). Soybean isoflavones improve cardiovascular risk factors without affecting the reproductive system of peripubertal rhesus monkeys. *J Nutr*, Vol. 126, pp. 43-50

Anzalone, CR., Lu, JK. & LaPolt, PS. (1998). Influences of age and reproductive status on ovarian ovulatory responsiveness to gonadotropin stimulation. *Proc Soc Exp Biol Med*, Vol. 217, pp. 455-460

Arjmandi, BH., Lucas, EA., Khalil, DA., Devareddy, L., Smith, BJ., McDonald, J., Arquitt, AB., Payton, ME. & Mason, C. (2005). One year soy protein supplementation has positive effects on bone formation markers but not bone density in postmenopausal women. *Nutr J*, Vol. 4, pp. 8

Arts, J., Kuiper, GG., Janssen, JM., Gustafsson, JA., Lowik, CW., Pols, HA. & van Leeuwen, JP. (1997). Differential expression of estrogen receptors alpha and beta mRNA during differentiation of human osteoblast SV-HFO cells. *Endocrinology*, Vol. 138, pp. 5067- 5070

Atanassova, N., McKinnell, C., Walker, M., Turner, KJ., Fisher, JS., Morley, M., Millar, MR., Groome, NP. & Sharpe, RM. (1999). Permanent effects of neonatal estrogen exposure in rats on reproductive hormone levels, Sertoli cell number, and the efficiency of spermatogenesis in adulthood. *Endocrinology*, Vol. 140, pp. 5364-5373

Barnes, S. (2003). Phyto-oestrogens and osteoporosis: what is a safe dose? *Br J Nutr*, Vol. 89, pp S101-S108

Barnes, S. (2004). Soy isoflavones--phytoestrogens and what else? *J Nutr*, Vol. 134, pp. 1225S-1228S

Bateman, A., Singh, A., Kral, T. & Solomon S. (1989). The immune hypothalamic-pituitary-adrenal axis. *Endocr Rew*, 10, pp. 92-112

Bateman, HL. & Patisaul, HB. (2008). Disrupted female reproductive physiology following neonatal exposure to phytoestrogens or estrogen specific ligands is associated with decreased GnRH activation and kisspeptin fiber density in the hypothalamus. *Neurotoxicology*, Vol. 29, pp. 988-997

Bektic, J., Guggenberger, R., Eder, IE., Pelzer, AE., Berger, AP., Bartsch, G. & Klocker, H. (2005). Molecular effects of the isoflavonoid genistein in prostate cancer. *Clin Prostate Cancer*, Vol. 4, pp. 124-129

Bennetts, HW., Underwood, EJ. & Shier, FL. (1946). A specific breeding problem of sheep on subterranean clover pastures in Western Australia. *Br Vet J*, Vol. 102, pp. 348-352

Berk BC & Corson MA. (1997). Angiotensin II signal transduction in vascular smooth muscle. *Circulation Research*, 80, pp. 607–616

Bernbaum, JC., Umbach, DM., Ragan, NB., Ballard, JL., Archer, JI., Schmidt-Davis, H. & Rogan, WJ. (2008). Pilot studies of estrogen-related physical findings in infants. *Environ Health Perspect*, Vol. 116, pp.416-20

Besedovsky, HO., & del Rey, A. (1996). Immune-neuro-endocrine interactions: facts and hypotheses. *Endocr Rev*, 17, pp. 64–102

Bianco, AC., Salvatore, D., Gereben, B., Berry, MJ. & Larsen, PR. (2002). Biochemistry, cellular and molecular biology, and physiological roles of the iodothyronine selenodeiodinases. *Endocr Rev*, Vol. 23, pp. 38-89

Birt, DF., Hendrich, S. & Wang, W. (2001). Dietary agents in cancer prevention: flavonoids and isoflavonoids. *Pharmacol Ther*, Vol. 90, pp.157-177

Brink, E., Coxam, V., Robins, S., Wahala, K., Cassidy, A. & Branca, F. (2008). Long-term consumption of isoflavone-enriched foods does not affect bone mineral density, bone metabolism, or hormonal status in early postmenopausal women: a randomized, double-blind, placebo controlled study. *Am J Clin Nutr*, Vol. 87, pp. 761-770

Capen C. (1997). Mehanicistic data and risk assesment of selected toxic end points of the thyroid gland. *Toxicologic Pathology*, Vol. 25, pp. 39-48

Carani, C., Qin, K., Simoni, M., Faustini-Fustini, M., Serpente, S., Boyd, J., Korach, KS. & Simpson, ER. (1997). Effect of testosterone and estradiol in a man with aromatase deficiency. *N Engl J Med*, Vol. 337, pp. 91-95

Cassidy, A., Bingham, S. & Setchell KDR. (1994). Biological effects of a diet of soy protein rich in isoflavones on the menstrual cycle of premenopausal women. *Am J Clin Nutr*, Vol. 60, pp. 333–340

Chang, HC. & Doerge, DR. (2000). Dietary genistein inactivates rat thyroid peroxidase in vivo without an apparent hypothyroid effect. *Toxicol Appl Pharmacol,* Vol. 168, pp. 244-252

Che, JH., Kwon, E., Kim, SH., You, JR., Kim, BH., Lee, SJ., Chung, JH. & Kang, BC. (2011). Acute and subchronic toxicity of FCD, a soybean extract combined with l-carnitine, in Sprague-Dawley rats. *Regul Toxicol Pharmacol,* Vol. 59, pp. 285-292

Chen, XW., Garner, SC. & Anderson, JJ. (2002), Isoflavones regulate interleukin-6 and osteoprotegerin synthesis during osteoblast cell differentiation via an estrogen-receptordependent pathway. *Biochem Biophys Res Commun,* Vol. 295, pp. 417–422

Chen, Y., Jefferson, WN., Newbold, RR., Padilla-Banks, E. & Pepling, ME. (2007). Estradiol, progesterone, and genistein inhibit oocyte nest breakdown and primordial follicle assembly in the neonatal mouse ovary in vitro and in vivo. *Endocrinology* Vol. 148, pp. 3580–3590

Cherdshewasart, W. & Sutjit, W. (2008). Correlation of antioxidant activity and major isoflavonoid contents of the phytoestrogen-rich Pueraria mirifica and Pueraria lobata tubers. *Phytomedicine,* Vol. 15, pp. 38-43

Choksi, NY., Jahnke, GD., St Hilaire, C. & Shelby, M. (2003) Role of thyroid hormones in human and laboratory animal reproductive health. *Birth Defects Res B Dev Reprod Toxicol,* Vol. 68, pp. 479-491

Chorazy, PA., Himelhoch, S., Hopwood, NJ., Greger, NG. & Postellon, DC. (1995). Persistent hypothyroidism in an infant receiving a soy formula: case report and review of the literature. *Pediatrics,* Vol. 96, pp. 148-50

Christian, RC., Liu, PY., Harrington, S., Ruan, M., Miller, VM. & Fitzpatrick, LA. (2006). Intimal estrogen receptor (ER)beta, but not ERalpha expression, is correlated with coronary calcification and atherosclerosis in pre- and postmenopausal women. *J Clin Endocrinol Metab,* Vol. 91, pp. 2713-2720

Comelekoglu, U., Bagis, S., Yalin, S., Ogenler, O., Yildiz, A., Sahin, NO., Oguz, I. & Hatungil, R. (2007). Biomechanical evaluation in osteoporosis: ovariectomized rat model. *Clin Rheumatol,* Vol. 26, pp. 380-384

Conley, AJ., & Bird, IM. (1997) The role of cytochrome P450 17α-hydroxylase and 3β-hydroxysteroid dehydrogenase in the integration of gonadal and adrenal steroidogenesis via the delta 5 and delta 4 pathways of steroidogenesis in mammals. *Biology of Reproduction,* Vol. 56, pp. 789–799

Cooke, GM. (2006). A review of the animal models used to investigate the health benefits of soy isoflavones. *J AOAC Int,* Vol. 89, pp. 1215-1227

Cooper, C., Campion, G. & Melton, LJD. (1992). Hip fractures in the elderly, a world-wide projection. *Osteoporosis International,* Vol. 2, pp. 285-289

Corrêa da Costa, V.M., Moreira, D.G. & Rosenthal, D. (2001) Thyroid function and aging: gender-related differences. *J Endocrinol,* Vol. 171, pp. 193-198

Couse, JF. & Korach, KS. (2001). Contrasting phenotypes in reproductive tissues of female estrogen receptor null mice. *Ann N Y Acad Sci,* Vol. 948, pp.1-8

Dang, ZC. & Lowik, C. (2005), Dose-dependent effects of phytoestrogens on bone. *Trends Endocrinol Metab,* Vol. 16, pp. 207– 213

D'Archivio, M., Filesi, C., Varì, R., Scazzocchio, B. & Masella, R. (2010). Bioavailability of the polyphenols: status and controversies. *Int J Mol Sci,* Vol. 11, pp. 1321-1342

Davies, JR., Rudd, JH. & Weissberg, PL. (2004). Molecular and metabolic imaging of atherosclerosis. *J Nucl Med,* Vol. 45, pp. 1898-1907

Davison, S. & Davis, SR. (2003). Hormone replacement therapy: current controversies. *Clin Endocrinology,* Vol. 58, pp. 249–261

De Kloet, ER., Rosenfeld, P., Van Eekelen, JA., Sutanto, W. & Levine, S. (1988). Stress, glucocorticoids and development. *Prog Brain Res,* Vol. 73, pp. 101-120

de Wilde, A., Heberden, C., Chaumaz, G., Bordat, C. & Lieberherr, M. (2006), Signaling networks from Gbeta1 subunit to transcription factors and actin remodeling via a membrane- located ERbeta-related protein in the rapid action of daidzein in osteoblasts. *J Cell Physiol,* Vol. 209, pp. 786–801

Delclos, KB., Bucci, TJ., Lomax, LG., Latendresse, JR., Warbritton, A., Weis, CC. & Newbold, RR. (2001). Effects of dietary genistein exposure during development on male and female CD (Sprague-Dawley) rats. *Reproductive Toxicology,* Vol. 15, pp. 647–663

deVere White, RW., Tsodikov, A., Stapp, EC., Soares, SE., Fujii, H. & Hackman, RM. (2010). Effects of a high dose, aglycone-rich soy extract on prostate-specific antigen and serum isoflavone concentrations in men with localized prostate cancer. *Nutr Cancer,* Vol. 62, pp. 1036-1043

Dhar, A., Paul, AK. & Shukla SD. (1990). Platelet-activating factor stimulation of tyrosine kinase and its relationship to phospholipase C in rabbit platelets: studies with genistein and monoclonal antibody to phosphotyrosine. *Mol Pharmacol,* Vol. 37, pp. 519-525

Diel, P., Schmidt, S., Vollmer, G., Janning, P., Upmeier, A., Michna, H., Bolt, HM. & Degen, GH. (2004). Comparative responses of three rat strains (DA/Han, Sprague-Dawley and Wistar) to treatment with environmental estrogens. *Arch Toxicol,* Vol. 78, pp. 183-193

Dillingham, BL., McVeigh, BL., Lampe, JW. & Duncan, AM. (2007). Soy protein isolates of varied isoflavone content do not influence serum thyroid hormones in healthy young men.*Thyroid,* Vol. 17, pp. 131-137

DiPippo, VA., Lindsay, R. & Powers, CA. (1995). Estradiol and tamoxifen interactions with thyroid hormone in the ovariectomized-thyroidectomized rat. *Endocrinology,* Vol. 136, pp. 1020-1033

Divi, RL., Chang, HC. & Doerge, DR. (1997). Anti-thyroid isoflavones from soybean: isolation, characterization, and mechanisms of action. *Biochem Pharmacol,* Vol. 54, pp. 1087-1096

Doerge, DR. & Chang, HC. (2002). Inactivation of thyroid peroxidase by soy isoflavones, in vitro and in vivo. *J Chromatogr B Analyt Technol Biomed Life Sc,* Vol. 777, pp. 269-279

Donda, A., Reymond, F., Rey, F. & Lemarchand-Beraud, T. (1990) Sex steroids modulate the pituitary parameters involved in the regulation of TSH secretion in the rat. *Acta Endocrinol (Copenh),* Vol. 122, pp. 577-584

Duffy, C., Perez, K. & Partridge, A. (2007). Implications of phytoestrogen intake for breast cancer. *CA Cancer J Clin,* Vol. 57, pp. 260-277

Duffy, C., Perez, K. & Partridge, A. (2007). Implications of phytoestrogen intake for breast cancer. *CA Cancer J Clin,* Vol. 57, pp. 260-277

Duncan, AM., Underhill. KE., Xu. X., Lavalleur. J., Phipps. WR. & Kurzer. MS. (1999). Modest hormonal effects of soy isoflavones in postmenopausal women. *J Clin Endocrinol Metab,* Vol. 84: 3479-3484

Eustache, F., Mondon, F., Canivenc-Lavier, MC., Lesaffre, C., Fulla, Y., Berges, R., Cravedi, J. P., Vaiman, D. & Auger, J. (2009). Chronic dietary exposure to a low-dose mixture of genistein and vinclozolin modifies the reproductive axis, testis transcriptome and fertility. *Environmental Health Perspectives*, Vol. 117, pp. 1272–1279

Evans, MJ., Eckert, A., Lai, K., Adelman, SJ. & Harnish, DC. (2001). Reciprocal antagonism between estrogen receptor and NF-kappaB activity in vivo. *Circ Res*, Vol. 89, pp. 823-830

Fielden, MR., Samy, SM., Chou, KC. & Zacharewski, TR. (2003). Effect of human dietary exposure levels of genistein during gestation and lactation on long-term reproductive development and sperm quality in mice. *Food and Chemical Toxicology*, Vol. 41, pp. 447–454

Filipović, B., Sosić-Jurjević, B., Ajdzanović, V., Brkić, D., Manojlović-Stojanoski, M., Milosević, V. & Sekulić M. (2010). Daidzein administration positively affects thyroid C cells and bone structure in orchidectomized middle-aged rats. *Osteoporos Int*, Vol. 21, pp. 1609-1616

Filipović, B., Šošić-Jurjević, B., Ajdžanović, V., Trifunović, S., Manojlović Stojanoski, M., Ristić, N., Nestorović, N., Milošević, V. & Sekulić M. (2007) The effect of orchidectomy on thyroid C cells and bone histomorphometry in middle-aged rats. *Histochem Cell Biol*, Vol. 128, pp. 153–159

Filipović, B., Šošić-Jurjević, B., Manojlović Stojanoski, M., Nestorović, N., Milošević V. & Sekulić M. (2005). The effect of chronic calcium treatment on thyroid C cells in ovariectomized rats. *Life Sci*, Vol. 77, pp. 121-129

Filipović, B., Šošić-Jurjević, B., Manojlović-Stojanoski, M., Nestorović, N., Milošević, V. & Sekulić, M. (2002). The effect of ovariectomy on thyroid C cells of adult rats. *Yugoslov Med Biohem*, Vol. 21, pp. 345-350

Filipović, B., Šošić-Jurjević, B., Nestorović, N., Manojlović Stojanoski, M., Kostić N., Milošević, V. & Sekulić M. (2003). The thyroid C cells of ovariectomized rats treated with estradiol. *Histochem Cell Biol*, Vol. 120, pp. 409-414

Food and Drug Administration, Health claims and qualified health claims; dietary lipids and cancer, soy protein and coronary heart disease, antioxidant vitamins and certain cancers, and selenium and certain cancers. Reevaluation. (2007). Available from http://www.fda.gov/OHRMS/DOCKETS/98fr/E7-24813.pdf

Fort, P., Moses, N., Fasano, M., Goldberg, T. & Lifshitz, F. (1990). Breast and soy-formula feedings in early infancy and the prevalence of autoimmune thyroid disease in children. *J Am Coll Nutr*, Vol. 9, pp. 164–167

Fotsis, T., Pepper, M., Adlercreutz, H., Fleischmann, G., Hase, T., Montesano, R. & Schweigerer, L. (1993). Genistein, a dietary-derived inhibitor of in vitro angiogenesis. *Proc Natl Acad Sci USA*, Vol. 90, pp. 2690-2694

Fujieda, K., Faiman, C., Reyes, FI. & Winter JSD. (1982) The control of steroidogenesis by human fetal adrenal cells in tissue culture. IV. The effects of exposure to placental steroids. *J Clin Endocrinol Metab*, Vol. 54, pp. 89 –94

Furlanetto, T.W., Nunes, R.B., Sopelsa, A.M.I., Maciel, R.M.B. (2001) Estradiol decreases iodide uptake by rat thyroid follicular FTRL-5 cells. Brazil J Biol Res 34: 259-263

Galleano, M., Pechanova, O. & Fraga, CG. (2010). Hypertension, nitric oxide, oxidants, and dietary plant polyphenols. *Curr Pharm Biotechnol*, Vol. 11, pp. 837-848

Gallo, D., Giacomelli, S., Cantelmo, F., Zannoni, GF., Ferrandina, G., Fruscella, E., Riva, A., Morazzoni, P., Bombardelli, E., Mancuso, S. & Scambia, G. (2001). Chemoprevention of DMBA-induced mammary cancer in rats by dietary soy. Breast *Cancer Res Treat,* Vol. 69, pp. 153-64

Gao YH. & Yamaguchi, M. (1999) a. Inhibitory effect of genistein on osteoclast-like cell formation in mouse marrow cultures. *Biochem Pharmacol,* Vol. 58, pp. 767-772

Gao, YH. & Yamaguchi M. (1999). Anabolic effect of daidzein on cortical bone in tissue culture: comparison with genistein effect. *Mol Cell Biochem,* Vol. 194, pp. 93-98

Gao, YH. & Yamaguchi, M. (2000). Suppressive effect of genistein on rat bone osteoclasts: involvement of protein kinase inhibition and protein tyrosine phosphatase activation. *Int J Mol Med,* Vol. 5, pp. 261-267

Ge, Y., Chen, D., Xie, L. & Zhang, R. (2006). Enhancing effect of daidzein on the differentiation and mineralization in mouse osteoblast-like MC3T3-E1 cells. *Yakugaku Zasshi,* 126, 651-656

Gee, JM., DuPont, MS., Day, AJ., Plumb, GW., Williamson, G. & Johnson, IT. (2000). Intestinal transport of quercetin glycosides in rats involves both deglycosylation and interaction with the hexose transport pathway. *J Nutr,* Vol. 130, pp. 2765-2771

González, M., Reyes, R., Damas, C., Alonso & R., Bello, AR. (2008). Oestrogen receptor alpha and beta in female rat pituitary cells: an immunochemical study. *Gen Comp Endocrinol,* Vol. 155, pp. 857-868

Greany, KA., Nettleton, JA., Wangen, KE., Thomas, W. & Kurzer, MS. (2004). Probiotic consumption does not enhance the cholesterol-lowering effect of soy in postmenopausal women. *J Nutr,* Vol. 134, pp. 3277-3283

Greaves, KA., Parks, JS., Williams, JK. & Wagner, JD. (1999). Intact dietary soy protein, but not adding an isoflavone-rich soy extract to casein, improves plasma lipids in ovariectomized cynomolgus monkeys. *J Nutr,* Vol. 129, pp. 1585-1592

Grover, GJ., Mellstrom, K. & Malm, J. (2005). Development of the thyroid hormone receptor beta-subtype agonist KB-141: a strategy for body weight reduction and lipid lowering with minimal cardiac side effects. *Cardiovasc Drug Rev,* Vol. 23, pp.133-148

Ham, KD., Carlson, C.S. (2004). Effects of estrogen replacement therapy on bone turnover in subchondral bone and epiphyseal metaphyseal cancellous bone of ovariectomized cynomolgus monkeys.*J Bone Miner Res, Vol.*19(5)pp. 823-829.

Hamilton-Reeves, JM., Rebello, SA., Thomas, W., Slaton, JW. & Kurzer, MS. (2007). Isoflavone-rich soy protein isolate suppresses androgen receptor expression without altering estrogen receptor-beta expression or serum hormonal profiles in men at high risk of prostate cancer. *J Nutr,* Vol. 137, pp. 1769-1775

Hamilton-Reeves, JM., Vazquez, G., Duval, SJ., Phipps, WR., Kurzer, MS. (2009). Messina, MJ. Clinical studies show no effects of soy protein or isoflavones on reproductive hormones in men: results of a meta-analysis. *Fertil Steril,* Vol.94, pp. 997-1007

Hampl, R., Nemec, J., Jeresova, J., Kimlova, I. & Starka, L. (1985) Estrogen receptors in human goitrous and neoplastic thyroid. *Endocrinol Exp,* Vol. 19, pp.227-230

Harris, RM., Wood, DM., Bottomley, L., Blagg, S., Owen, K., Hughes, PJ., Waring, RH.. & Kirk, CJ. (2004). Phytoestrogens are potent inhibitors of estrogen sulfation: implications for breast cancer risk and treatment. *J Clin Endocrinol Metab,* Vol. 89, pp. 1779-1787

Haselkorn, T., Stewartm SL. & Horn-Rossm, PL. (2003). Why are thyroid cancer rates so high in southeast asian women living in the United States? The bay area thyroid cancer study. *Cancer Epidemiol Biomarkers Prev*, Vol. 12, pp. 144-150

Hatzinger, M., Wotjak, CT., Naruo, T., Simchen, R., Keck, ME., Landgraf, R., Holsboer, F. & Neumann, ID. (2000). Endogenous vasopressin contributes to hypothalamic-pituitary-adrenocortical alterations in aged rats. *J Endocrinol*, Vol. 164, pp. 197-205

Hooper, L., Kroon, PA., Rimm, EB., Cohn, JS., Harvey, I., Le Cornu, KA., Ryder, JJ. & Hall, WL. (2008). Cassidy A.Flavonoids, flavonoid-rich foods, and cardiovascular risk: a meta-analysis of randomized controlled trials. *Am J Clin Nutr*, Vol. 88, pp. 38-50

Hooper, L., Ryder, JJ., Kurzer, MS., Lampe, JW., Messina, MJ., Phipps, WR. & Cassidy, A. (2009). Effects of soy protein and isoflavones on circulating hormone concentrations in pre- and post-menopausal women: a systematic review and meta-analysis. *Hum Reprod Update*, Vol. 15, pp. 423-440

Hsieh, CY., Santell, RC., Haslam, SZ. & Helferich, WG. (1998). Estrogenic effects of genistein on the growth of estrogen receptor-positive human breast cancer (MCF-7) cells in vitro and in vivo. *Cancer Res*, Vol. 58, pp. 3833-3838

Huang, HY., Yang, HP., Yang, HT., Yang, TC., Shieh, MJ. & Huang, SY. (2006). One-year soy isoflavone supplementation prevents early postmenopausal bone loss but without a dosedependent effect. *J Nutr Biochem*, Vol. 17, pp. 509–517

Hussain, M., Banerjee, M., Sarkar, FH., Djuric, Z., Pollak, MN., Doerge, D., Fontana, J., Chinni, S., Davis, J., Forman, J., Wood, DP. & Kucuk, O. (2003). Soy isoflavones in the treatment of prostate cancer. *Nutr Cancer*, Vol. 47, pp. 111-117

Iguchi, T., Fukazawa, Y., Uesugi, Y. & Takasugi, N. (1990). Polyovular follicles in mouse ovaries exposed neonatally to diethylstilbestrol in vivo and in vitro. *Biol Reprod*, Vol. 43, pp. 478–484

Ikeda, T., Nishikawa, A., Imazawa, T., Kimura, S. & Hirose, M. (2000). Dramatic synergism between excess soybean intake and iodine deficiency on the development of rat thyroid hyperplasia. *Carcinogenesis*, Vol. 21, pp. 707-713

Ishida, M., Marrero, MB., Schieffer, B., Ishida, T., Bernstein, KE. & Berk BC. (1995). Angiotensin II activates pp60c-src in vascular smooth muscle cells. *Circulation Research*, Vol. 77, pp. 1053–1059

Jabbar, MA., Larrea, J. & Shaw, RA. (1997). Abnormal thyroid function tests in infants with congenital hypothyroidism: the influence of soy-based formula. *J Am Coll Nutr*, Vol. 16, pp. 280-282

Jefferson, WN., Couse, JF., Padilla-Banks, E., Korach, KS. & Newbold, RR. (2002). Neonatal exposure to genistein induces estrogen receptor (ER)alpha expression and multiocyte follicles in the maturing mouse ovary: evidence for ERbeta-mediated and nonestrogenic actions. *Biol Reprod*, Vol. 67, pp.1285–1296

Jefferson, WN., Padilla-Banks, E. & Newbold, RR. (2005). Adverse effects on female development and reproduction in CD-1 mice following neonatal exposure to the phytoestrogen genistein at environmentally relevant doses. *Biol Reprod*, Vol. 73, pp. 798–806

Jenkins, DJ., Kendall, CW., Jackson, CJ., Connelly, PW., Parker, T., Faulkner, D., Vidgen, E., Cunnane, SC., Leiter, LA. & Josse, RG. (2002). Effects of high- and low-isoflavone soyfoods on blood lipids, oxidized LDL, homocysteine, and blood pressure in hyperlipidemic men and women. *Am J Clin Nutr*, Vol. 76, pp. 365-372

Jenkins, DJ., Kendall, CW., Marchie, A., Faulkner, D., Vidgen, E., Lapsley, KG., Trautwein, EA., Parker, TL., Josse, RG., Leiter, LA. & Connelly, PW. (2003). The effect of combining plant sterols, soy protein, viscous fibers, and almonds in treating hypercholesterolemia. Metabolism, Vol. 52, pp. 1478-1483

Jia, TL., Wang, HZ., Xie, LP., Wang, XY. & Zhang, RQ. (2003), Daidzein enhances osteoblast growth that may be mediated by increased bone morphogenetic protein (BMP) production. Biochem Pharmacol, Vol. 65, pp. 709–715

Jiménez, E. & Montiel, M. (2005). Activation of MAP kinase by muscarinic cholinergic receptors induces cell proliferation and protein synthesis in human breast cancer cells. J Cell Physiol, Vol. 204, pp. 678-686

Jones, JL., Daley, BJ., Enderson, BL., Zhou, JR. & Karlstad, MD. (2002). Genistein inhibits tamoxifen effects on cell proliferation and cell cycle arrest in T47D breast cancer cells. Am. Surg, Vol. 68, pp. 575-577; discussion 577-578

Kajiya, H., Takekoshi, S., Miyai, S., Ikeda, T., Kimura, S. & Osamura, RY. (2005). Dietary soybean enhances Pit-1 dependent pituitary hormone production in iodine deficient rats. J Mol Histol, Vol. 36, pp. 265-274

Kanemori, M. & Prygrocki, M. (2005). Results of breast conservation therapy from a single-institution community hospital in Hawaii with a predominantly Japanese population. Int J Radiat Oncol Biol Phys, Vol. 62, pp. 193-197

Kang, KS., Che, JH. & Lee, YS. (2002). Lack of adverse effects in the f1 offspring maternally exposed to genistein at human intake dose level. Food and Chemical Toxicology, Vol. 40, pp. 43–51

Kanno, S., Hirano, S. & Kayama, F. (2004), Effects of the phytoestrogen coumestrol on RANK-ligand-induced differentiation of osteoclasts. Toxicology, Vol. 203, pp. 211–220

Kapiotis, S., Hermann, M., Held, I., Seelos, C., Ehringer, H. & Gmeiner, BM. (1997). Genistein, the dietary-derived angiogenesis inhibitor, prevents LDL oxidation and protects endothelial cells from damage by atherogenic LDL. Arterioscler Thromb Vasc Biol, Vol. 17, pp. 2868-2874

Katahira, M., Iwasaki, Y., Aoki, Y., Oiso, Y. & Saito H. (1998). Cytokine regulation of the rat proopiomelanocortin gene expression in AtT-20 cells. Endocrinology, Vol.139, pp. 2414-2422

Kerry, N. & Abbey, M. (1998). The isoflavone genistein inhibits copper and peroxyl radical mediated low density lipoprotein oxidation in vitro. Atherosclerosis, Vol. 140, pp. 341-347

Khalil, DA., Lucas, EA., Smith, BJ., Soung, DY., Devareddy, L., Juma, S., Akhter, MP., Recker, R. & Arjmandi, BH. (2005). Soy isoflavones may protect against orchidectomy-induced bone loss in aged male rats. Calcif Tissue Int. Vol. 76, pp. 56–62

Kimura, S., Suwa, J., Ito, M. (1976). Sato, H. Development of malignant goiter by defatted soybean with iodine-free diet in rats. Gann, Vol. 67, pp. 763-765

Kirk, EA., Sutherland, P., Wang, SA., Chait, A. & LeBoeuf, RC. (1998). Dietary isoflavones reduce plasma cholesterol and atherosclerosis in C57BL/6 mice but not LDL receptor-deficient mice. J Nutr, Vol. 128, pp. 954-959

Köhrle J. (2002). Iodothyronine deiodinases. Methods Enzymol, Vol. 347, pp. 125-167

Köhrle, J. (2008). Environment and endocrinology: the case of thyroidology. *Ann Endocrinol (Paris)*, Vol. 69, pp. 116-122

Köhrle, J., Schomburg, L., Drescher, S., Fekete, E. & Bauer, K. (1995). Rapid stimulation of type I 5'-deiodinase in rat pituitaries by 3,3',5-triiodo-L-thyronine. *Mol Cell Endocrinol*, Vol. 108, pp. 17-21

Kojima, T., Uesugi, T., Toda, T., Miura, Y. & Yagasaki, K. (2002). Hypolipidemic action of the soybean isoflavones genistein and genistin in glomerulonephritic rats. *Lipids*, Vol. 37, pp. 261-265

Korach, KS. (1994). Insights from the study of animals lacking functional estrogen receptor. *Science*, Vol. 266, pp. 1524-1527

Kouki, T., Kishitake, M., Okamoto, M., Oosuka, I., Takebe, M. & Yamanouchi, K. (2003). Effects of neonatal treatment with phytoestrogens, genistein and daidzein, on sex difference in female rat brain function: estrous cycle and lordosis. *Horm Behav*, Vol. 44, pp.140–145

Kruk, I., Aboul-Enein, HY., Michalska, T., Lichszteld, K. & Kładna, A. (2005). Scavenging of reactive oxygen species by the plant phenols genistein and oleuropein. *Luminescence*, Vol. 20, pp. 81-89

Kuiper, GG., Lemmen, JG., Carlsson, B., Corton, JC., Safe, SH., van der Saag, PT., van der Burg, B. & Gustafsson, JA. (1998). Interaction of estrogenic chemicals and phytoestrogens with estrogen receptor beta. *Endocrinology*, Vol. 139, pp. 4252–4263

Kurahashi, N., Iwasaki, M., Sasazuki, S., Otani, T., Inoue, M. & Tsugane, S. (2007). Japan Public Health Center-Based Prospective Study Group. Soy product and isoflavone consumption in relation to prostate cancer in Japanese men. *Cancer Epidemiol Biomarkers Prev*, Vol. 16, pp. 538-545

Lakshman, M., Xu, L., Ananthanarayanan, V., Cooper, J., Takimoto, CH., Helenowski, I., Pelling, JC. & Bergan, RC. (2008). Dietary genistein inhibits metastasis of human prostate cancer in mice. *Cancer Res*, Vol. 68, pp. 2024-2032

Lamartiniere, CA., Cotroneo, MS., Fritz, WA., Wang, J., Mentor-Marcel, R. & Elgavish, A. (2002). Genistein chemoprevention: timing and mechanisms of action in murine mammary and prostate. *J Nutr*, Vol. 132, pp. 552S-558S

Lampe, JW., Karr, SC., Hutchins, AM. & Slavin, JL. (1998). Urinary equol excretion with a soy challenge: influence of habitual diet. *Proc Soc Exp Biol Med*, Vol. 217, pp. 335-339

Larkin, T., Price, WE. & Astheimer, L. (2008). The key importance of soy isoflavone bioavailability to understanding health benefits. *Crit Rev Food Sci Nutr*, Vol. 48, pp. 538-552

Lauderdale, DS., Jacobsen, SJ., Furner, SE., Levy, PS., Brody, JA. & Goldberg, J. (1997). Hip fractures incidence among elderly Asian-American populations. *American Journal of Epidemiology*, Vol. 146, pp. 502-509

Lee, BJ., Jung, EY., Yun, YW., Kang, JK., Baek, IJ., Yon, JM., Lee, YB., Sohn, HS., Lee, JY., Kim, KS. & Nam, SY. (2004) a. Effects of exposure to genistein during pubertal development on the reproductive system of male mice. *J Reprod Dev*, Vol. 50, pp. 399–409

Lee, YB., Lee, HJ., Kim, KS., Lee, JY., Nam, SY., Cheon, SH. & Sohn HS. (2004). Evaluation of the preventive effect of isoflavone extract on bone loss in ovariectomized rats. *Bioscience, Biotechnology and Biochemistry*, Vol. 68, pp. 1040-1045

Lephart, ED., Setchell, KD., Handa, RJ.& Lund TD. (2004). Behavioral effects of endocrine-disrupting substances: phytoestrogens. *ILAR J*, Vol. 45, pp. 443-454

Lindsay, R., Hart, DM. & Clark, DM. (1984). The minimum effective dose of estrogen for prevention of postmenopausal bone loss. *Obstet Gynecol*, Vol. 63, pp. 759–763

Lindsay, R., Hart, DM., Aitken, JM., MacDonald, ED., Anderson, JB. & Clarke, AC. (1976). Long-term prevention of postmenopausal osteoporosis by oestrogen. *Lancet*, Vol. 1, pp. 1038–1041

Lisbôa, PC., Curty, FH., Moreira, RM., Oliveira, KJ. & Pazos-Moura, CC. (2001). Sex steroids modulate rat anterior pituitary and liver iodothyronine deiodinase activities. *Horm Metab Res*, Vol. 33, pp. 532-535

Liu, J., Ho, SC., Su, YX., Chen, WQ., Zhang, CX. & Chen, YM. (2009). Effect of long-term intervention of soy isoflavones on bone mineral density in women: a meta-analysis of randomized controlled trials. *Bone*, Vol. 44, pp. 948–953

Lönnerdal, B., Bryant, A., Liu, X. & Theil, EC. (2006). Iron absorption from soybean ferritin in nonanemic women. *Am J Clin Nutr*, Vol. 83, pp.103-107

Loughlin, K. & Richie, J. (1997). Prostate cancer after exogenous testosterone treatment for impotence. *J Urology*, Vol. 157, pp.1845

Lucas, EA., Khalil, DA., Daggy, BP. & Arjmandi, BH. (2001). Ethanol-extracted soy protein isolate does not modulate serum cholesterol in golden Syrian hamsters: a model of postmenopausal hypercholesterolemia. *J Nutr*, Vol. 131, pp. 211-214

Lucas, EA., Lightfoot, SA., Hammond, LJ., Devareddy, L., Khalil, DA., Daggy, BP., Soung do, Y. & Arjmandi, BH. (2003). Soy isoflavones prevent ovariectomy-induced atherosclerotic lesions in Golden Syrian hamster model of postmenopausal hyperlipidemia. *Menopause*, Vol. 10, pp. 314-321

Lund, TD., Munson, DJ., Adlercreutz, H., Handa, RJ. & Lephart, ED. (2004). Androgen receptor expression in the rat prostate is down-regulated by dietary phytoestrogens. *Reprod Biol Endocrinol*, Vol. 2, pp. 5

Ma, DF., Qin, LQ., Wang, PY. & Katoh, R. (2008). Soy isoflavone intake increases bone mineral density in the spine of menopausal women: meta-analysis of randomized controlled trials. *Clin Nutr*, Vol. 27, 57–64

Mackey, R. & Eden, J. (1998). Phytoestrogens and the menopause. *Climateric*, Vol. 1, pp. 302-308

Mahn, K., Borrás, C., Knock, GA., Taylor, P., Khan, IY., Sugden, D., Poston, L., Ward, JP., Sharpe, RM., Viña, J., Aaronson, PI. & Mann, GE. (2005). Dietary soy isoflavone induced increases in antioxidant and eNOS gene expression lead to improved endothelial function and reduced blood pressure in vivo. *FASEB J*, Vol. 19, pp. 1755-1757

Malendowicz, LK., Trejter, M., Rebuffat, P., Ziolkowska, A., Nussdorfer, GG. & Majchrzak, M. (2006). Effects of some endocrine disruptors on the secretory and proliferative activity of the regenerating rat adrenal cortex. *Int J Mol Med*, 18, pp. 197-200

Manojlović-Stojanoski, M., Nestorović, N., Ristić, N., Trifunović, S., Filipović, B., Sosić-Jurjević, B. & Sekulić, M. (2010). Unbiased stereological estimation of the rat fetal pituitary volume and of the total number and volume of TSH cells after maternal dexamethasone application. *Microsc Res Tech*, Vol. 73, pp. 1077-1085

Markovits, J., Linassier, C., Fossé, P., Couprie, J., Pierre, J., Jacquemin-Sablon, A., Saucier, JM., Le Pecq, JB. & Larsen, AK. (1989). Inhibitory effects of the tyrosine kinase inhibitor genistein on mammalian DNA topoisomerase II. *Cancer Res*, Vol. 49, pp. 5111-5117

Maskarinec, G., Franke, AA., Williams, AE., Hebshi, S., Oshiro, C., Murphy, S. & Stanczyk, FZ. (2004). Effects of a 2-year randomized soy intervention on sex hormone levels in premenopausal women. *Cancer Epidemiol Biomarkers Prev*, Vol. 13, pp. 1736-1744

Maskarinec, G., Watts, K., Kagihara, J., Hebshi, SM. & Franke, AA. (2008). Urinary isoflavonoid excretion is similar after consuming soya milk and miso soup in Japanese-American women. *Br J Nutr*, Vol. 100, pp. 424-429

Maskarinec, G., Williams, AE., Inouye, JS., Stanczyk, FZ. & Franke, AA. (2002). A randomized isoflavone intervention among premenopausal women. *Cancer Epidemiol Biomarkers Prev*, Vol. 11, pp. 195-201

McCarrison, R. (1993). A Paper on FOOD AND GOITRE. *Br Med J*, Vol. 14, pp. 671-675

McLachlan, JA., Newbold, RR., Shah, HC., Hogan, MD. & Dixon, RL. (1982). Reduced fertility in female mice exposed transplacentally to diethylstilbestrol (DES). *Fertil Steril*, Vol. 38, pp. 364–371

McVey, MJ., Cooke, GM., & Curran, IH. (2004). Increased serum and testicular androgen levels in F1 rats with lifetime exposure to soy isoflavones. *Reproductive toxicology (Elmsford, N.Y)*, Vol. 18, pp. 677-685

Medigović, I., Ristić, N., Manojlović-Stojanoski, M., Šošić-Jurjević, B., Pantelić, J., Milošević, V., Nestorović, N. (2009). The effects of genistein and estradiol on the ovaries of immature rats: Stereological study. *Second Congress of Physiological Sciences of Serbia with International Participation*, p. 108, Kragujevac, Serbia, September 17-20, 2009

Meerts, IA., Assink, Y., Cenijn, PH., Van Den Berg, JH., Weijers, BM., Bergman, A., Koeman. JH. & Brouwer, A. (2002). Placental transfer of a hydroxylated polychlorinated biphenyl and effects on fetal and maternal thyroid hormone homeostasis in the rat. *Toxicol Sci*, Vol. 68, pp. 361–371

Mendel, CM., Weisiger, RA., Jones, AL. & Cavalieri, RR. (1987). Thyroid hormone-binding proteins in plasma facilitate uniform distribution of thyroxine within tissues: a perfused rat liver study. *Endocrinology*, Vol. 120, pp. 1742–1749

Mentor-Marcel, R., Lamartiniere, CA., Eltoum, IE., Greenberg, NM. & Elgavish, A. (2001). Genistein in the diet reduces the incidence of poorly differentiated prostatic adenocarcinoma in transgenic mice (TRAMP). *Cancer Res*, Vol. 61, pp. 6777-6782

Mesiano, S. & Jaffe, RB. (1993). Interaction of insulin-like growth factor-II and estradiol directs steroidogenesis in the human fetal adrenal toward dehydroepiandrosterone sulfate production. *J Clin Endocrinol Metab*, Vol. 77, pp.754 –758

Mesiano, S., Katz, SL., Lee, JY. & Jaffe RB. (1999). Phytoestrogens alter adrenocortical function: genistein and daidzein suppress glucocorticoid and stimulate androgen production by cultured adrenal cortical cells. *J Clin Endocrinol Metab*, Vol. 84, pp. 2443-2448

Messina, M. (2010). Insights gained from 20 years of soy research. *J Nutr*, Vol. 140, pp. 2289S-2295S

Messina, M., Nagata, C. & Wu, AH. (2006). Estimated Asian adult soy protein and isoflavone intakes. *Nutr Cancer*, Vol. 55, pp. 1-12

Messina, MJ. & Loprinzi, CL. (2001). Soy for breast cancer survivors: a critical review of the literature. *J Nutr*, Vol. 131, pp. 3095S-3108S

Messina, MJ. & Wood, CE. (2008). Soy isoflavones, estrogen therapy, and breast cancer risk: analysis and commentary. *Nutr J*, Vol. 7, pp. 17

Mezei, O., Li, Y., Mullen, E., Ross-Viola, JS. & Shay, NF. Dietary isoflavone supplementation modulates lipid metabolism via PPARalpha-dependent and -independent mechanisms. *Physiol Genomics*, Vol. 26, pp. 8-14

Michaelsson, K., Baron, JA., Farahmand, BY., Johnell, O., Magnusson, C., Persson, PG., Persson, I. & Ljunghall S. (1998). Hormone replacement therapy and risk of hip fracture: population based case-control study. The Swedish Hip Fracture Study Group. *BMJ*, Vol. 316, pp. 1858–1863

Milošević, V., Ajdžanović, V., Sosić-Jurjević, B., Filipović, B., Brkić, M., Nestorović, N. & Sekulić, M. (2009). Morphofunctional characteristics of ACTH cells in middle-aged male rats after treatment with genistein. *Gen Physiol Biophys*, 28, pp. 94-97

Mitchell, JH., Gardner, PT., McPhail, DB., Morrice, PC., Collins, AR. & Duthie, GG. (1998). Antioxidant efficacy of phytoestrogens in chemical and biological model systems. *Arch Biochem Biophys*, Vol. 360, pp. 142-148

Molsiri, K., Khemapech, S., Patumraj, S. & Siriviriyakul, P. (2004). Preventive mechanism of genistein on coronary endothelial dysfunction in ovariectomized rats: an isolated arrested heart model. *Clin Hemorheol Microcirc*, Vol. 31, pp. 59-66

Morin, S. (2004). Isoflavones and bone health. *Menopause*; Vol. 11, pp. 239- 241

Mulvihill, EE. & Huff, MW. (2010). Antiatherogenic properties of flavonoids: implications for cardiovascular health. *Can J Cardiol*, Vol. 26, pp. 17A-21A

Nagata, C., Takatsuka, N., Inaba, S., Kawakami, N. & Shimizu, H. (1998). Effect of soymilk consumption on serum estrogen concentrations in premenopausal Japanese women. *J Natl Cancer Inst*, Vol. 90, pp. 1830-1835

Naik, HR., Lehr, JE. & Pienta, KJ. (1994). An in vitro and in vivo study of antitumor effects of genistein on hormone refractory prostate cancer. *Anticancer Res*, Vol. 14, pp. 2617-2619

National Toxicology Program. (2008). NTP Multigenerational Reproductive Study of Genistein (CAS no. 446-72-0) in Sprague-Dawley Rats (Feed Study). Technical Report 539. Research Triangle Park, NC:National Toxicology Program

Nelson, HD., Humphrey, LL., Nygren, P., Teutsch, SM. & Allan, JD. (2002). Postmenopausal hormone replacement therapy: scientific review. *JAMA*. Vol. 288, pp. 872–881

Newbold, RR., Banks, EP., Bullock, B. & Jefferson, WN. (2001). Uterine adenocarcinoma in mice treated neonatally with genistein. *Cancer Res*, Vol. 61, pp. 4325–4328

Newton, KM., LaCroix, AZ., Levy, L., Li, SS., Qu, P., Potter, JD. & Lampe, JW. (2006) Soy protein and bone mineral density in older men and women: a randomized trial. *Maturitas*, Vol. 55, pp. 270–277

Nicholson, CG., Mosley, JM., Sexton, PM., Mendelsohn, FAO. & Martin, TJ. (1986). Abundant calcitonin receptors in isolated rat osteoclasts. *J Clin Invest*, Vol. 78, pp.355-360

Nikaido, Y., Yoshizawa, K., Danbara, N., Tsujita-Kyutoku, M., Yuri, T., Uehara, N. & Tsubura, A. (2004). Effects of maternal xenoestrogen exposure on development of the reproductive tract and mammary gland in female CD-1 mouse offspring. *Reprod Toxicol*, Vol. 18, pp. 803-811

Ohlsson, A., Ullerås, E., Cedergreen, N. & Oskarsson, A. (2010). Mixture effects of dietary flavonoids on steroid hormone synthesis in the human adrenocortical H295R cell line. *Food Chem Toxicol,* Vol. 48, pp. 3194-3200

Ohno, S., Nakajima, Y., Inoue, K., Nakazawa, H. & Nakajin, S. (2003). Genistein administration decreases serum corticosterone and testosterone levels in rats. *Life Sci,* Vol. 74, pp. 733-742

Ohno, S., Shinoda, S., Toyoshima, S., Nakazawa, H., Makino, T. & Nakajin, S. (2002). Effects of favonoid phytochemicals on cortisol production and on activities of steroidogenic enzymes in human adrenocortical H295R cells. *J Steroid Biochem Mol Biol,* Vol. 80, pp. 355-363

Om, AS. & Shim, JY. (2007). Effect of daidzein, a soy isoflavone, on bone metabolism in Cd-treated ovariectomized rats. *Acta Biochim Pol,* Vol. 54, pp. 641-646

Onoe, Y., Miyaura, C., Ohta, H., Nozawa, S. & Suda, T. (1997). Expression of estrogen receptor in rat bone. *Endocrinology,* Vol. 138, pp. 4509-4512

Ørgaard, A. & Jensen, L. (2008). The effects of soy isoflavones on obesity. *Exp Biol Med (Maywood),* Vol. 233, pp. 1066-1080

Osterweil, N. (2007). Soy as prostate cancer protection yields paradoxical results. Medpage Today March 16, 2007

Padilla-Banks, E., Jefferson, WN. & Newbold, RR. (2006). Neonatal exposure to the phytoestrogen genistein alters mammary gland growth and developmental programming of hormone receptor levels. *Endocrinology,* Vol. 147, pp. 4871-4882

Pantelić, J., Filipović, B., Sosić-Jurjević, B., Medigović, I. & Sekulić, M. (2010). Effects of testosterone and estradiol treatment on bone histomorphometry in orchidectomized middle-aged rats. *Proceedings of 4th Serbian Congress for Microscopy,* pp. 143-144, Belgrade, Serbia, October 11-12, 2010

Parkin, DM., Whelan, SL., Ferlay, J. & Storm, H. (2005). Cancer incidence in five continents, IARC Cancer Base, 2005; No. 7. Lyon:IARC Press

Patel, RP., Boersma, BJ., Crawford, JH., Hogg, N., Kirk, M., Kalyanaraman, B., Parks, DA., Barnes, S. & Darley-Usmar, V. 82001). Antioxidant mechanisms of isoflavones in lipid systems: paradoxical effects of peroxyl radical scavenging. *Free Radic Biol Med,* Vol. 31, pp. 1570-1581

Patisaul, HB. & Jefferson, W. (2010). The pros and cons of phytoestrogens. *Front Neuroendocrinol,* Vol. 31, pp. 400-419

Pei, RJ., Sato, M., Yuri, T., Danbara, N., Nikaido, Y. & Tsubura, A. (2003). Effect of prenatal and prepubertal genistein exposure on N-methyl-N-nitrosourea-induced mammary tumorigenesis in female Sprague-Dawley rats. *In Vivo,* Vol. 17, pp. 349-357

Piotrowska, K., Baranowska-Bosiacka, I., Marchlewicz, M., Gutowska, I., Noceń, I., Zawiślak, M., Chlubek, D. & Wiszniewska, B. (2011). Changes in male reproductive system and mineral metabolism induced by soy isoflavones administered to rats from prenatal life until sexual maturity. *Nutrition,* Vol 27, pp. 372-379

Pollard, M. & Suckow, M. (2006). Dietary prevention of hormone refractory prostate cancer in Lobund-Wistar rats: a review of studies in a relevant animal model. *Comp Med,* Vol. 56, pp. 461-467

Potter, SM. (1995). Overview of proposed mechanisms for the hypocholesterolemic effect of soy. *J Nutr,* Vol. 125, pp. 606S-611S

Radović, B., Mentrup, B., Köhrle, J. (2006). Genistein and other soya isoflavones are potent ligands for transthyretin in serum and cerebrospinal fluid. *Br J Nutr,* Vol. 95, pp. 1171-1176

Ralli, M. (2003). Soy and the Thyroid: Can This Miracle Food Be Unsafe? *Nutrition Noteworthy,* Vol. 6, pp. 6

Rand, WM., Pellett, PL. & Young, VR. (2003). Meta-analysis of nitrogen balance studies for estimating protein requirements in healthy adults. *Am J Clin Nutr,* Vol. 77, pp. 109-127

Rassi, CM., Lieberherr, M., Chaumaz, G., Pointillart, A. & Cournot, G. (2002). Down-regulation of osteoclast differentiation by daidzein via caspase 3. *J Bone Miner Res,* Vol. 17, pp. 630-638

Redei, E., Li, L., Halasz, I., McGivern, RF. & Aird, F. (1994). Fast glucocorticoid feedback inhibition of ACTH secretion in the ovariectomized rat: effect of chronic estrogen and progesterone. *Neuroendocrinology,* Vol. 60, pp. 113-123

Refetoff, S. (1989). Inherited thyroxine-binding globulin abnormalities in man. *Endocr Rev,* Vol. 10, pp. 275–293

Register, TC., Cann, JA., Kaplan, JR., Williams, JK., Adams, MR., Morgan, TM., Anthony, MS., Blair, RM., Wagner, JD. & Clarkson, TB. (2005). Effects of soy isoflavones and conjugated equine estrogens on inflammatory markers in atherosclerotic, ovariectomized monkeys. *J Clin Endocrinol Metab,* Vol. 90, pp. 1734-1740

Reinwald, S. & Weaver, CM. (2010). Soy components vs. whole soy: are we betting our bones on a long shot? *J Nutr,* Vol. 140, pp. 2312S-2317S

Ren, P., Ji, H., Shao, Q., Chen, X., Han, J. & Sun, Y. (2007).Protective effects of sodium daidzein sulfonate on trabecular bone in ovariectomized rats. *Pharmacology,* Vol. 79, pp. 129-136

Richardson, SJ., Lemkine, GF., Alfama, G., Hassani, Z. & Demeneix, BA. (2007). Cell division and apoptosis in the adult neural stem cell niche are differentially affected in transthyretin null mice. *Neurosci Lett,* Vol. 421, pp. 234–238

Robbins, J. (2000) Thyroid hormone transport proteins and the physiology of hormone binding. In: Braverman L, Utiger R, eds. Werner & Ingbar's The Thyroid: a fundamental and clinical text. 8th ed. Philadelphia: Lippincott, Williams and Wilkins, pp. 105–120

Roberts, D., Veeramachaneni, DN., Schlaff, WD. & Awoniyi, CA. (2000). Effects of chronic dietary exposure to genistein, a phytoestrogen, during various stages of development on reproductive hormones and spermatogenesis in rats. *Endocrine,* Vol. 13, pp. 281–286

Rochfort, S. & Panozzo, J. (2007). Phytochemicals for health, the role of pulses. *J Agric Food Chem,* Vol. 55, pp. 7981-7894

Rosen, ED. (2005). The transcriptional basis of adipocyte development. *Prostaglandins Leukot Essent Fatty Acids,* Vol. 73, pp. 31-34

Rowland, IR., Wiseman, H., Sanders, TA., Adlercreutz, H. & Bowey, EA. (2000). Interindividual variation in metabolism of soy isoflavones and lignans: influence of habitual diet on equol production by the gut microflora. *Nutr Cancer,* Vol. 36, pp. 27-32

Ruhlen, RL., Howdeshell, KL., Mao, J., Taylor, JA., Bronson, FH., Newbold, RR., Welshons, WV. & vom Saal, FS. (2008). Low phytoestrogen levels in feed increase fetal serum estradiol resulting in the "fetal estrogenization syndrome" and obesity in CD-1 mice. *Environ Health Perspect*, Vol. 116, pp. 322-328

Sacks, FM., Lichtenstein, A., Van Horn, L., Harris, W., Kris-Etherton, P. & Winston, M. (2006). Soy protein, isoflavones, and cardiovascular health: a summary of a statement for professionals from the american heart association nutrition committee. *Arterioscler Thromb Vasc Biol*, Vol. 26, pp. 1689-1992

Sarkar, FH. & Li, Y. (2002). Mechanisms of cancer chemoprevention by soy isoflavone genistein. *Cancer Metastasis Rev*, Vol. 21, pp. 265-280

Sasaki, S., Kawai, K., Honjo, Y. & Nakamura, H. (2006). Thyroid hormones and lipid metabolism. *Nippon Rinsho*, Vol. 64, pp. 2323-2329

Sathyapalan, T., Manuchehri, AM., Thatcher, NJ., Rigby, AS., Chapman, T., Kilpatrick, ES. & Atkin, SL. (2011). The Effect of Soy Phytoestrogen Supplementation on Thyroid Status and Cardiovascular Risk Markers in Patients with Subclinical Hypothyroidism: A Randomized, Double-Blind, Crossover Study. *J Clin Endocrinol Metab*, [Epub ahead of print]

Schmidt, S., Degen, GH., Seibel, J., Hertrampf, T., Vollmer, G. & Diel, P. (2006). Hormonal activity of combinations of genistein, bisphenol A and 17beta-estradiol in the female Wistar rat. *Arch Toxicol*, Vol. 80, pp. 839-845

Schmutzler, C., Hamann, I., Hofmann, PJ., Kovacs, G., Stemmler, L., Mentrup, B., Schomburg, L., Ambrugger, P., Grüters, A., Seidlova-Wuttke, D., Jarry, H., Wuttke, W. & Köhrle, J. (2004). Endocrine active compounds affect thyrotropin and thyroid hormone levels in serum as well as endpoints of thyroid hormone action in liver, heart and kidney. *Toxicology*, Vol. 205, pp. 95-102

Sekulić, M., Šošić-Jurjević, B., Filipović, B., Milošević, V., Nestorović, N. & Manojlović-Stojanoski, M. (2005). The effects of synthetic salmon calcitonin on thyroid C and follicular cells in adult female rats. *Folia Histochem Cytobiol*. Vol. 43, pp. 103-108

Sekulić, M., Sosić-Jurjević, B., Filipović, B., Pantelić, J., Nestorović, N., Manojlović-Stojanoski, M. & Milošević, V. (2010). Testosterone and estradiol differently affect thyroid structure and function in orchidectomizrd middle-aged rats. *Proceedings of 14th International Thyroid Congress*, p. 302 Paris, France, September 11-16, 2010

Setchell, KD., Brown, NM. & Lydeking-Olsen, E. (2002). The clinical importance of the metabolite equol-a clue to the effectiveness of soy and its isoflavones. *J Nutr*, Vol. 132, pp. 3577-3584

Setchell, KD., Gosselin, SJ., Welsh, MB., Johnston, JO., Balistreri, WF., Kramer, LW., Dresser, BL. & Tarr, MJ. (1987). Dietary estrogens--a probable cause of infertility and liver disease in captive cheetahs. *Gastroenterology*, Vol. 93, pp. 225-233

Sharpe, RM., Martin, B., Morris, K., Greig, I., McKinnell, C., McNeilly, A. S., Walker, M., Setchell, KD., Cassidy, A., Ingram, D., Sanders, K., Kolybaba, M., Lopez, D., Fitzpatrick, M., Fort, P., Moses, N., Fasano, M., Goldberg, T., Lifshitz, F., Chang, H. C., Doerge, DR., Bingham, S., Setchell, K., Herman-Giddens, ME., Slora, EJ., Wasserman, RC., Bourdony, CJ., Bhapkar, MV., Koch, GG., & Hasemeier, CM. (2002). Infant feeding with soy formula milk: effects on the testis and on blood testosterone levels in marmoset monkeys during the period of neonatal testicular activity. *Hum Reprod*, Vol. 17, pp. 1692-1703

Shen, P., Liu, MH., Ng, TY., Chan, YH. & Yong EL. (2006). Differential effects of isoflavones, from Astragalus membranaceus and Pueraria thomsonii, on the activation of

PPARalpha, PPARgamma, and adipocyte differentiation in vitro. *J Nutr*, Vol. 136, pp. 899-905

Shu, XO., Zheng, Y., Cai, H., Gu, K., Chen, Z., Zheng, W. & Lu, W. (2009). Soy food intake and breast cancer survival. *JAMA*, Vol. 302, pp. 2437-2443

Simpson, ER. & Waterman, MR. (1992). Regulation of expression of adrenocortical enzymes. In *The Adrenal Gland*, edn 2. Ed VHT James. New York: Raven Press

Sirianni, R., Sirianni, R., Carr, BR., Pezzi, V. & Rainey WE. (2001). A role for src tyrosine kinase in regulating adrenal aldosterone production. *J Mol Endocrinol*, 26, pp. 207-215

Sliwinski, L., Folwarczna, J., Janiec, W., Grynkiewicz, G. & Kuzyk, K. (2005). Differential effects of genistein, estradiol and raloxifene on rat osteoclasts in vitro. *Pharmacol Rep*, Vol. 57, pp. 352–359

Somekawa, Y., Chiguchi, M., Ishibashi, T. & Aso, T. (2001). Soy intake related to menopausal synptoms, serum lipids, and bone mineral density in postmenopausal Japanese women. *Obstetrics and Gynecology*, Vol. 97, pp.109-115

Sosić-Jurjević, B., Filipović, B., Ajdzanović, V., Brkić, D., Ristić, N., Stojanoski, MM., Nestorović, N., Trifunović, S. & Sekulić, M. (2007). Subcutaneously administrated genistein and daidzein decrease serum cholesterol and increase triglyceride levels in male middle-aged rats. *Exp Biol Med*, Vol. 232, pp. 1222-1227

Sosić-Jurjević, B., Filipović, B., Ajdžanović, V., Savin, S., Nestorović, N., Milošević, V., Sekulić, M. (2010). Suppressive effects of genistein and daidzein on pituitary-thyroid axis in orchidectomized middle-aged rats. *Exp Biol Med*, Vol. 255, pp. 590-598

Sosić-Jurjević, B., Filipović, B., Milosević, V., Nestorović, N., Negić, N. & Sekulić, M. (2006). Effects of ovariectomy and chronic estradiol administration on pituitary-thyroid axis in adult rats. *Life Sci*, Vol. 79, pp. 890-897

Sosić-Jurjević, B., Filipović, B., Milošević, V., Nestorović, N., Manojlović Stojanoski, M. & Sekulić, M. (2005) Chronic estradiol exposure modulates thyroid structure and decreases T4 and T3 serum levels in middle-aged female rats. *Horm Res*, Vol. 63, pp. 48-54

Soung, DY., Devareddy, L., Khalil, DA., Hooshmand, S., Patade, A., Lucas, EA. & Arjmandi, BH. (2006). Soy affects trabecular microarchitecture and favorably alters select bone-specific gene expressions in a male rat model of osteoporosis. *Calcif Tissue Int*, Vol. 78, pp. 385–391

Squadrito, F., Altavilla, D., Morabito, N., Crisafulli, A., D'Anna, R., Corrado, F., Ruggeri, P., Campo, GM., Calapai, G., Caputi, AP. & Squadrito, G. (2002). The effect of the phytoestrogen genistein on plasma nitric oxide concentrations, endothelin-1 levels and endothelium dependent vasodilation in postmenopausal women. *Atherosclerosis*, Vol. 163, pp. 339-347

Stahl, S., Chun, TY. & Gray, WG. (1998). Phytoestrogens act as estrogen agonists in an estrogen-responsive pituitary cell line. *Toxicol Appl Pharmacol*, Vol. 152, pp. 41-48

Sugimoto, E. & Yamaguchi, M. (2000) a. Stimulatory effect of Daidzein in osteoblastic MC3T3-E1 cells. *Biochem Pharmacol*, Vol. 59, pp. 471-475

Sugimoto, E. & Yamaguchi, M. (2000). Anabolic effect of genistein in osteoblastic MC3T3-E1 cells. *Int J Mol Med*, Vol. 5, pp. 515-520

Suzuki, T., Sasano, H., Tadeyama, J., Kaneko, C., Freije, WA., Carr, BR. & Rainey, WE. (2000). Developmental changes of steroidogenic enzymes in human postnatal adrenal cortex: immunohistochemical studies. *Clinical Endocrinology*, 53, pp. 739–747

Tan, KA., Walker, M., Morris, K., Greig, I., Mason, JI. & Sharpe, RM. (2006). Infant feeding with soy formula milk: effects on puberty progression, reproductive function and testicular cell numbers in marmoset monkeys in adulthood. *Hum Reprod*, Vol. 21, pp. 896-904

Tanaka, M., Fujimoto, K., Chihara, Y., Torimoto, K., Yoneda, T., Tanaka, N., Hirayama, A., Miyanaga, N., Akaza, H. & Hirao, Y. (2009). Isoflavone supplements stimulated the production of serum equol and decreased the serum dihydrotestosterone levels in healthy male volunteers. *Prostate Cancer Prostatic Dis*, Vol. 12, pp. 247–252

Taylor, CK., Levy, RM., Elliott, JC. & Burnett, BP. (2009). The effect of genistein aglycone on cancer and cancer risk: a review of in vitro, preclinical, and clinical studies. *Nutr Rev*, Vol. 67, pp. 398-415

Teixeira, SR., Potter, SM., Weigel, R., Hannum, S., Erdman, JW Jr. & Hasler, CM. (2000). Effects of feeding 4 levels of soy protein for 3 and 6 wk on blood lipids and apolipoproteins in moderately hypercholesterolemic men. *Am J Clin Nutr*, Vol. 71, pp. 1077-1084

Tikkanen, MJ., Wähälä, K., Ojala, S., Vihma, V. & Adlercreutz, H. (1998). Effect of soybean phytoestrogen intake on low density lipoprotein oxidation resistance. *Proc Natl Acad Sci U S A*, Vol. 95, pp. 3106-3110

Turner, AS. (2001). Animal models of osteoporosis--necessity and limitations. *Eur Cell Mater*, Vol. 1, pp. 66-81

Van Wyk, JJ., Arnold, MB., Wynn, J. & Pepper, F. (1959). The effects of a soybean product on thyroid function in humans. *Pediatrics*, Vol. 24, pp. 752-760

Vanderschueren, D., Van Herck, E., Suiker, AMH., Visser, WJ., Schot, LPC. & Bouillon, R. (1992). Bone and mineral metabolism in aged male rats: short- and long-term effects of androgen deficiency. *Endocrinology*, Vol. 130, pp. 2906-2916

Vaya, J. & Tamir, S. (2004). The relation between the chemical structure of flavonoids and their estrogen-like activities. *Curr Med Chem*, Vol. 11, pp. 1333-1343

Vidal, S., Cameselle-Teijeiro, J., Horvath, E., Kovacs, K. & Bartke, A. (2001) Effects of protracted estrogen administration on the thyroid of Ames dwarf mice. *Cell Tissue Res*, Vol. 304, pp. 51-58

Villa, P., Costantini. B., Suriano. R., Perri, C., Macrì, F., Ricciardi, L., Panunzi, S. & Lanzone, A. (2009). The differential effect of the phytoestrogen genistein on cardiovascular risk factors in postmenopausal women: relationship with the metabolic status. *J Clin Endocrinol Metab*, Vol. 94, pp. 552-558

Voutilainen, R., Kahri, AI. & Salmenpera, M. (1979) The effects of progesterone, pregnenolone, estriol, ACTH and hCG on steroid secretion of cultured human fetal adrenals. *J Steroid Biochem*, Vol. 10, pp. 695–700

Wang, TT., Sathyamoorthy, N. & Phang, JM. (1996). Molecular effects of genistein on estrogen receptor mediated pathways. *Carcinogenesis*, Vol. 17, pp. 271-275

Warri, A., Saarinen, NM., Makela, S. & Hilakivi-Clarke, L. (2008). The role of early life genistein exposures in modifying breast cancer risk. *Br J Cancer*, Vol. 98, pp. 1485-1493

Weber, KS., Setchell, KDR., Stocco, DM. & Lephard, ED. (2001). Dietary soydphytoestrogens decrease testosterone levels and prostate weight without altering LH, prostate 5a-reductase or testicular steroidogenic acute regulatory peptide levels in adult male Sprague-Downey rats. *J Endocrinol*, Vol. 170, pp. 591–599

Wenzel, U., Fuchs, D. & Daniel, H. (2008). Protective effects of soy-isoflavones in cardiovascular disease. Identification of molecular targets. *Hamostaseologie*, Vol. 28, pp. 85-88

Wilkinson, AP., Gee, JM., Dupont, MS., Needs, PW., Mellon, FA., Williamson, G. & Johnson, IT. (2003). Hydrolysis by lactase phlorizin hydrolase is the first step in the uptake of daidzein glucosides by rat small intestine in vitro. *Xenobiotica*, Vol. 33, pp. 255-264

Wisniewski, AB., Cernetich, A., Gearhart, JP. & Klein, SL. (2005). Perinatal exposure to genistein alters reproductive development and aggressive behavior in male mice. *Physiol Behav*, Vol. 84, pp. 327–334

Wisniewski, AB., Klein, SL., Lakshmanan, Y. & Gearhart, JP. (2003). Exposure to genistein during gestation and lactation demasculinizes the reproductive system in rats. *Journal of Urology*, Vol. 169, pp. 1582–1586

Wong, CK. & Keung, WM. (1999). Bovine adrenal 3β-hydrogenase E.C.1.1.1.145)/5-ene-4ene isomerase (E>C>5.3.3.5.1); characterization and its inhibition by iosoflavones. *J Steroid Biochem Mol Biol*, Vol. 71, pp. 191–202

Wood, CE., Register, TC., Anthony, MS., Kock, ND. & Cline, JM. (2004). Breast and uterine effects of soy isoflavones and conjugated equine estrogens in postmenopausal female monkeys. *J Clin Endocrinol Metab*, Vol. 89, pp. 3462-3468

Wuttke, W., Jarry, H. & Seidlova-Wuttke, D. (2007). Isoflavones – safe food additives or dangerous drugs? *Ageing Res Rev*, Vol. 6, pp. 150–188

Wuttke, W., Jarry, H. & Seidlova-Wuttke, D. (2010). Plant-derived alternative treatments for the aging male: facts and myths. *Aging Male*, Vol. 13, pp. 75-81

Xiao, CW., Mei, J., Huang, W., Wood, C., L'abbé, MR., Gilani, GS., Cooke, GM. & Curran, IH. Dietary soy protein isolate modifies hepatic retinoic acid receptor-beta proteins and inhibits their DNA binding activity in rats. *J Nutr*, Vol. 137, pp. 1-6

Yamaguchi, M. & Gao YH. (1998). Anabolic effect of genistein and genistin on bone metabolism in the femoral-metaphyseal tissues of elderly rats: the genistein effect is enhanced by zinc. *Mol Cell Biochem*, Vol. 178, pp. 377-382

Yamaguchi, M. & Gao, YH. (1997). Anabolic effect of genistein on bone metabolism in the femoral-metaphyseal tissues of elderly rats is inhibited by the anti-estrogen tamoxifen. *Res Exp Med (Berl)*, Vol. 197, pp. 101-107

Yamaguchi, M. & Sugimoto, E. (2000). Stimulatory effect of genistein and daidzein on protein synthesis in osteoblastic MC3T3-E1 cells: activation of aminoacyl-tRNA synthetase. *Mol Cell Biochem*, Vol. 214, pp. 97-102

Yamaguchi, M. (2006). Regulatory mechanism of food factors in bone metabolism and prevention of osteoporosis. *Yakugaku Zasshi*, Vol. 126, pp. 1117-1

Yamaguchi, M.. & Gao YH. (1998) a. Inhibitory effect of genistein on bone resorption in tissue culture. *Biochem Pharmacol*, Vol. 55, pp. 71-76

Yang, JY., Lee, SJ., Park, HW. & Cha, YS. (2006). Effect of genistein with carnitine administration on lipid parameters and obesity in C57Bl/6J mice fed a high-fat diet. *J Med Food*, Vol. 9, pp. 459-467

Zhan, S. & Ho, SC. (2005). Meta-analysis of the effects of soy protein containing isoflavones on the lipid profile. *Am J Clin Nutr,* Vol. 81, 397-408

Zhou, JR., Gugger, ET., Tanaka, T., Guo, Y., Blackburn, GL. & Clinton, SK. (1999). Soybean phytochemicals inhibit the growth of transplantable human prostate carcinoma and tumor angiogenesis in mice. *J Nutr,* Vol. 129, pp. 1628-1635

Zhu, FR., Wang, YG., Chen, J., Hu, YX., Han, FS. & Wang HY. (2009). Effects of phytoestrogens on testosterone production of rat Leydig cells. *Zhonghua Nan Ke Xue,* Vol. 15, pp. 207–211

Ziegler, RG., Hoover, RN., Pike, MC., Hildesheim, A., Nomura, AM., West, DW., Wu-Williams, AH., Kolonel, LN., Horn-Ross, PL., Rosenthal, JF. & Hyer, MB. (1993). Migration patterns and breast cancer risk in Asian-American women. *J Natl Cancer Inst,* Vol. 85, pp. 1819-1827

7

The Main Components Content, Rheology Properties and Lipid Profile of Wheat-Soybean Flour

Nada Nikolić and Miodrag Lazić
University of Niš, Faculty of Technology, Leskovac
Serbia

1. Introduction

Soybean (*Glycine max* (L.) Merr.) is a species of legume, native to a Eastern Asia and an important global crop, today. Soybean is rich in high quality proteins, contains essential amino acids, similar to those found in meat, minerals such as Fe, Zn, Cu, Mn, Ca and Mg as well as phytic acid. The bulk of seed soybean proteins contains albumins and globulins as major components, but there are minor, undesirable components such as inhibitors of trypsyn and chymotrypsin, and sugar-binding lecitins. The inhibitors and lecitins are generally inactivated by heat treatment. A new immunochemical methods can be used for quantitative detection of soybean proteins and production of healthful foods (Brandon & Friedman, 2002). The soybean lipid contains a significant amount of unsaturated acids: α-linolenic acid, known as omega-3 acid, linoleic, γ-linolenic and arachidonic acid, known as omega-6, and oleic acids known as omega-9 acid and are very important in human nutrition (Liu, 1997). The soybean lipid also contains saturated acids: palmitic and stearic acid (Bressani, 1972; Olguin et al., 2003; Bond et al., 2005), as well as tocopherols (Ortega-Garcia et.al., 2004, Yoshida et al., 2006). These soybean components make the products with soybean have higher nutritional value.

1.1 Soybean in bread making industry

In the bread making industry, the soybean is used with the aim to increase the bread protein value and decrease carbohydrate value. In ordinary white bread protein content ranges from 8 to 9% and by including soybean, the protein content can be made up to 16% (Ribotta et al. 2010) and at the same time, the dough and bread are richer with lipoxygenase enzymes preparation which can make the dough physical properties better. The fact is that bread made with soybean costs less and it is especially important in countries where wheat is not a major domestic crop. Bread and products with higher protein content and lower carbohydrate content are more suitable for use in some diets than bread and products formulations currently used (Mohamed et al., 2006). As the main protein component in white bread is gluten, a component which causes celiac disease, the usage of soy is useful for decreasing the gluten content in bread. A portion of 0.3 to 5% of soybean flour portion is usually added (Auerman, 1979), but rational addition for increasing bread protein value is 20-30% of soybean flour. Besides whole soybean flour, different soybean products can also be used: defatted soy flour (Mashayekh et al., 2008), physically modified soy flour (Maforimbo et al., 2008), soy flour and durum wheat flour mixture (Sabanis & Tzia, 2009), commercial soy protein isolate (Roccia et al., 2009), diffrerent kinds of soy protein powder (Qian et al., 2006) and part of soy seeds such as a hulls (Anjum et a., 2006). Based on these investigations results, different bread formulations are defined, and soy is used in portion up to 20%. When soy flour in wheat flour was to a level of 10% and in durum wheat flour up to 20%, the produced bread was without any negative effects in quality attributes such as colour, hardness and flavour, promising nutritious and healthy alternative to consumers (Sabanis & Tzia, 2009). By investigating the effect of defatted soy flour on sensory and rheological properties of wheat bread, Mashayekh et al. 2008, concluded that adding 3 and 7% defatted soy flour gives as good a loaf of bread as the 100% wheat bread and acceptable consumer attribute with rheological and sensory characteristics. Adding small quantity of soy protein powder of 3% to wheat flour did not change the sensory properties of bread and a large quantity of soy flour adding, exceeding 7%, can lead to stickiness and leguminous flavour (Roccia et al., 2009). The results of investigation (Anjum et al., 2006) of soy hulls usage showed the content of 4.5% soy hulls combined with wheat flour is acceptable and suitable level by the consumers.

1.2 Wheat-soybean bread manufacturing

The manufacture of bread from flour without gluten represents considerable technological difficulties (Jong et al., 1968; Schober et al., 2003) because gluten is the most important structure forming protein for making bread (Gujural et al., 2003; Moore et al,. 2004) and by using appropriate soy-wheat flour mixture these difficulties can be avoided. When soy was added to wheat, the soy globulins interact with wheat gluten proteins forming aggregates of high molecular weight. As reduction-reoxidation treatment facilitated the interaction of glutenin subunits and soy proteins (11S subunits), interaction probably occurs through the oxidation of SH groups (Maforimbo, 2008). Investigations of the changes in glutenin macro polymer content, protein composition and free sulfhydryl content showed that active soy flour decreased glutenin macro polymer content due to gluten depolymerization and glutenin macro polymer content increased by inactive soy flour because soy proteins became insoluble and precipitated together. Soy proteins were associated to wheat protein through physical interaction and covalent and non-covalent bonds during mixing and resting and these interactions produced large and medium-size polymers. This increased solubility of insoluble gluten proteins, producing a weakening of the gluten network (Perez et al., 2008) and decreasing availability of water to build up in gluten network (Roccia et all. 2009). Physicochemical status of soy protein in the product had a great influence on how wheat-soy proteins will interact (Perez et al., 2008). Incorporation of soy proteins changes the rheolocical and bread properties. The investigations showed that adding soy protein powder depresses loaf volume, gives poor crumb characteristics and decreases acceptability by consumers (Qian et al., 2006.) Ribotta et al. 2010. tested different additive combinations for improved bread quality obtained from soy-wheat flour in ratio of 90:10 w/w and found that the combination with transglutaminase showed a major improving effect on dough rheological properties and crumb uniformity.

1.3 Aims of investigations

Dough rheological properties have great relevance in predicting the mixing behaviour, sheeting and baking performance (Dobraszczyk & Morgenstain, 2003) and supplementation wheat flour with other changed these properties. Ribotta et al. 2005, presented data about effect of soybean addition on farinographic properties such as water absorption (WA), dough development time (DT), dough stability (Dst) and dough degree of softening (DSf). The present work has been undertaken with the objective to investigate the effect of whole soybean seed flour (»full-fat soy flour«) portion of 3 to 30% on dough farinographic, but extensographic and amylographic properties, too. Based on the composition of wheat and soybean flour lipids, the aim is to obtain lipid composition in wheat-soybean flour mixtures, and compare it to wheat flour only, with an emphasis on content of total saturated fatty acids (TS), total monounsaturated fatty acids (TMUS), total polyunsaturated fatty acids (TPUS) and total unsaturated fatty acids (TU). In order to value wheat-soybean flour mixtures, they were classified into groups by using statistical analysis and Euclidean distances and the correlations between content of some lipid components and rheological properties were found.

2. Experimental

2.1 Soybean seed
The whole soybean seeds (*Glycine max* L.) cultivars ZP Lana, grown in Serbia in summer of 2006 were used. The seeds were purchased in local store „Green Apple" (Leskovac, Serbia) and milled. The particle size was determined by method of insemination via riddles with gaps size from 0.315 to 3.15 mm. The overall particle size was determined using equations:

$$100 / d_{sr} = \sum \Delta d_i / d_i \tag{1}$$

where Δd_i is weight of fraction with appropriate particle size in %, and

$$d_i = (d_{i-1} + d_{i+1}) / 2 \tag{2}$$

where d_{i-1} is bottom riddle gap size and d_{i+1} is upper riddle gap size.

2.2 Chemicals
Chemicals used for oil extraction were high quality chemicals (Centrohem, Serbia) and for HPLC and GC analysis they were analytical grade (Riedel-de Haën, Honeywell Specialty Chemicals, Germany).

2.3 Wheat and soybean flour and flour mixtures
The wheat flour, Kikinda Mill, Serbia (WF) was bought from the local market. The soybean flour (SF) as »full-fat soy flour« was obtained by soybean seeds milling (IKA Model M120), to an overall particle size (d_{sr}) of 0.4 mm. Quantities of 291, 285, 270, 240 and 210 g of wheat flour and 9, 15, 30, 60 and 90 g of soybean flour, respectively, were used to make flour mixture with 3, 5, 10, 20 and 30 % (w/w) soybean flour portion, without adding additives.

2.4 Flour analyses
Flour protein content was determined by the Kjeldahl method (Nx5.95). The moisture content was determined by Scaltec SMO 01 (Scaltec instruments, Germany) instruments: flour (5 g) was put into the disk plate analyzer, dried at 110°C to a constant weight, and the

moisture content was read out on the display. The ash content was determined by staking at 800°C during 5 h. For gluten content determination, the dough was prepared by adding a sodium chloride solution first; wet gluten was isolated by dough washing and weighed. The starch content was determined by polarimetry according to Grossfeld's method and the total carbohydrates content according to Luff-Schoorl's method (Trajković et al., 1983). The values for samples are from triplicate analysis and followed by standard deviation.

2.5 The wheat and soybean flour lipid content

The wheat or soybean flour (50g) were put into Erlenmeyer flask, 500 ml of trichloroethylene was added and extracted for 30 minutes, under reflux and by mixing (200 min^{-1}) at approximately 88°C. The extract was separated by using Buchner funnel under weak vacuum. The plant material was extracted three more times by the same method; the extracts were mixed together and eluted by water in the separation funnel (3 x 10 ml). The eluted extracts' volume was recorded and an aliquot (3 ml) was taken for the dry residue determination test. This test was performed by drying at 110°C to a constant weight and the dry residue content was read out on the analyzer display (Scaltec SMO 01, Scaltec instruments, Germany). The lipid content in wheat and soybean flour was calculated based on average value of three measurements. The remnant of lipid extracts, after dry residue determination test, is evaporated under vacuum and obtained residue was used for HPLC and GC analysis.

2.6 Rheology measurements

The Brabender farinograph (Brabender Model 8 10 101, Duisburg, Germany) according to ISO 5530-1 test procedure, was used for water absorption values (WA value in ml/100g), development time (DT in minutes), dough stability (DSt in minutes), degree of softening (DSf in BU) and farinograph quality number (QN) determination. For extensograph measurement, the Brabender extensograph (Brabender, Model 8600-01, Duisburg, Germany) and test procedure ISO 5530-2 were used. The samples were prepared from wheat flour and wheat-soybean flour mixtures, distilled water and salt, and data for energy (E in cm^2), resistance (R in EU), extensibility (Ex in mm) and ration number (R/Ex) were recorded on extensograph curve. To obtain amylograph data, such as gelatinization temperature (T_{max} in °C) and gelatization maximum (η_{max} in AU), the amylograph (Brabender Model PT 100, Duisburg, Germany) and ISO 7973 test procedure were used.

2.7 HPLC analysis

For HPLC analysis, Holčapek et al., (1999) modified HPLC method and the Agilent 1100 High Performance Liquid Chromatograph, a Zorbax Eclipse XDB-C18 column: 4.4 m x 150 mm x 5 μm, Agilent technologies, Wilmington, USA and an UV/ViS detector were used. The flow rate of binary solvent mixture (methanol, solvent A, and 2-propanol/n-hexane, 5:4 by volume, solvent B) was 1 ml/min, with a linear gradient (from 100% A to 40% A+ 60% B in 15 min). The column temperature was held constant at 40°C. The components were detected at 205 nm. The free fatty acids (FFA), methyl esters (ME), monoacylglycerols (MAG), diacylglycerols (DAG) and triacylglycerols (TAG) were identified by comparing the retention times of the lipid components with those of standards. The samples of the reaction mixture were dissolved into a mixture of 2-propanol:n-hexane, 5:4 v/v and filtered through 0.45 μm Millipore filters.

2.8 GC analysis

For GC analysis, fatty acids methyl esters were prepared. The lipid were alkaline hydrolyzed and methylated by methanol and BF_3 as catalysts. The final fatty acids methyl esters concentration was about 8 mg/ml in heptane. For obtaining a methyl esters GC spectra, the HP 5890 SERIES II GAS-CHROMATOGRAPH, HP with FID detector and 3396 A HP integrator was used. Column was ULTRA 2 (25m x 0.32mm x 0.52 μm) (Agilent Technologies, Wilington, USA), injector temperature of 320°C, and injector volume of 0.4 μl. The carrier gas was He at a constant flow rate of 1 ml/min. The flame ionization detector was at 350°C and split ratio was 1:20. Oven temperature was initially 120°C and was maintained at 120°C, for 1 min, then increased by 15°C/min until 200°C, increased by 3 °C/min until 240°C, increased by 8°C/min until 300°C and maintained at 300°C for 15 min. The fatty acids were identified by comparison of retention times of the lipid components with those of standards.

2.9 Energetic value

Based on total carbohydrates (CHC), protein (PC) and lipid content (LC), the energetic value (EV) of wheat flour and white-soybean flour mixtures was calculated as:

$$EV = (CHC + PC) \cdot 17 + LC \cdot 37 \tag{3}$$

2.9.1 Statistical analysis

STATISTICA, version 5.0 software was used to perform the statistical analysis: the means and standard deviations, the correlation coefficients and cluster analysis. The means and standard deviations were obtained by Descriptive Statistics, marking the Median & Quartiles and Confirm Limits for Means. In order to classify wheat flour and wheat-soybean mixtures, the cluster analysis and the Euclidean method with the complete linkage was used.

3. Results and discussion

3.1 The main components content

The results of wheat and soybean flour moisture, starch, protein, ash, lipid, gluten and carbohydrates content are showed in Table 1. Data are presented as means of three determinations with standard deviation. Based on these data, considering the soybean flour portion in mixtures and compared to wheat flour, it is evident the moisture (from 12.8 to 11.6%), the starch (from 76.6 to 56.8%), gluten (23.9 to 16.7%) and carbohydrates content (from 78.8 to 62.8%) decreased, while the protein (from 8.6 to 20.0%), ash (from 0.48 to 2.08%) and lipid (1.2 to 7.2%) content increased with increasing soybean flour portion in mixtures.

Taking into account the protein, carbohydrates and lipid content, based on formula (3) for energetic value determination, it was obtained that the soybean flour increased energetic value in flour mixtures (from 1530 obtained for wheat flour, to 1544, 1554, 1579, 1625, 1674 kJ/100g, when soy flour portion was 3, 5 10, 20 and 30%, respectively) and the maximal increasing of 9.4% was in mixture with soybean flour portion of 30%.

Content (g/100g)	Moisture	Starch	Protein	Ash	Lipid	Gluten	Carbo-hydrates
Wheat flour	12.8±0.6	76.6±1.2	8.6±0.34	0.48±0.04	1.2±0.05	23.9±0.4	78.8±2.2
Soybean flour	8.7±0.6	10.7±0.6	46.7±0.6	5.80±0.6	21.2±0.6	-	25.4±0.6
3%	12.7	74.6	9.7	0.64	1.8	23.2	77.2
5%	12.6	73.3	10.5	0.75	2.2	22.7	76.1
10%	12.4	69.9	12.4	1.01	3.2	21.5	73.5
20%	11.9	63.4	16.2	1.54	5.2	19.1	68.1
30%	11.6	56.8	20.0	2.08	7.2	16.7	62.8

Table 1. The main components content in wheat and soybean flour and their mixtures

3.2 Rheology properties

Farinograph, extensograph and amylograph data of flours and flour mixtures with different portions of soybean flour are given in Table 2. The results showed the farinograph data depended on the soybean flour portion in mixtures. The water absorption increased from 53.4 to 64.2% with increasing soybean flour portion in mixtures. It is well known, that the main dough component in wheat flour responsible for water absorption is gluten. The soy flour is without gluten but had even more than 5 times higher protein content than wheat flour (46.7 to 8.6 g/100g). The higher absorption ability could be due to soy protein components, such as globulins, which interacted with gluten protein in the composite dough (Maforimbo et al., 2008). The same effect of soy flour on water absorption value was reported by Ribotta et al., 2005, when heat-treated full-fat flour, enzyme-active defatted flour and soy protein isolates were used for wheat flour substitution in portion from 3 to 12% and enzyme-active full-fat flour in portion from 5 to 12%. The differences in dough development time and dough stability among flour mixtures with different soybean flour portions, ranged from 1.3 to 7.3 min and 0.7 to 2.9 min, respectively and both values for mixtures were higher than for wheat flour where value for dough development time was 1 min and for dough stability was 0.1 min. The effect of delaying dough development time and dough stability by addition the mostly of investigated soy flours was obtained in Ribotta et al. (2005) experiments, too.

Degree of dough softening was increased with increasing soybean flour portion in mixtures, from 45 to 66 BU and all values were lower than value for wheat flour, i.e. than 90 BU. The same effect of soybean flour on this farinographic characteristic was obtained by Ribotta et al. (2005), for all investigated samples, except for soy protein isolate Samsory 90 HI and portion of 5, 10 and 12%, when dough softening value was higher. According to appropriate triangle area on farinograph curves, the quality number, known as Honkocy number could be in the range from 0 to100, and quality groups are A_1 (area of 0-1.4 cm^2), A_2 (area of 1.5-5.5 cm^2), B_1 (area of 5.6-12.1 cm^2), B_2 (area of 12.2-17.9 cm^2), C_1 (area of 18.0-27.4 cm^2) and C_2 (area of 27.5-50.0 cm^2) (Djaković, 1980). The soybean flour addition in all investigated portions had positive influence on quality number and quality group which was B1 instead B2, which was for wheat flour.

Data obtained on extensograph, showed that dough with soybean flour portion of 20 and 30% had lower values for energy and dough resistance in comparison to dough made of wheat flour only. The extensibility of dough with soybean flour ranged from 125 to 89 EU and those values were lower than extensibility of dough with wheat flour only. The ration number, R/Ex, varied depending on soybean flour portion and ranged from 2.76 to 3.42. Based on curve of volume versus ratio number (Djaković, 1980), obtained for round bread, the bread volume was predicted to be in range from 576 to 557 cm^3 and it was lower than bread volume obtained from wheat flour only which was 581 cm^3.

Farinograph data						
	Wheat flour	3%	5%	10%	20%	30%
WA (ml/100g)	53.4±1.4	53.3±1.4	53.8±1.3	53.9±1.6	61.9±1.6	64.2±1.7
DT (min)	1±0.1	1.3±0.1	1.3±0.1	6.5±0.3	6.8±0.3	7.3±0.4
DSt (min)	0.1±0.1	0.7±0.2	0.8±0.2	1.8±0.4	2.3±0.5	2.9±0.5
DSf (BU)	90±1.6	45±1.4	50±1.4	65±1.7	65±1.7	66±1.5
QN	52.8±1.4	57.7±1.4	66.7±1.4	61.7±1.4	64.2±1.4	68.8±1.4
Group	B2	B1	B1	B1	B1	B1
Extensograph data						
E (cm^2)	67.8±1.3	70.5±1.5	74.2±1.5	79.0±1.6	57.1±1.4	43.4±1.2
R (EU)	315±5.5	345±6.0	350±6.0	355±5.5	345±5.5	305±5.5
Ex (EU)	126±3.0	125±3.0	123±3.0	117±2.5	101±2.5	89±2.0
R/Ex	2.50±0.2	2.76±0.3	2.84±0.4	3.03±0.4	3.41±0.4	3.42±0.4
V (cm^3)	581±10	576±10	574±10	566±10	559±10	557±10
Amylograph data						
T$_{max}$ (°C)	81.2±0.5	81.5±0.5	82.0±0.5	85.0±0.5	86.5±0.5	88.8±0.5
η$_{max}$ (AU)	630±20	315±15	250±10	180±10	120±10	90±5

Table 2. Rheological properties of wheat flour (WF) and wheat–soy bean flour mixtures

By amylograph data, dough with soybean flour had higher gelatinization temperature (in range 81.5 to 88.8°C) than dough with wheat flour only (81.2°C), and this value was higher when soybean flour portion was higher. Based on Stevenson et al. (2006) results, gelatinization temperature for wheat-soybean flour mixtures value could be lower. They found the gelatinization temperature of soybean starches were lower compared to wheat starch, due to the short amylopectin branch-chains and positive relationship between gelatinization temperature and amylopectin branch-chains. The reason why our gelatinization temperature values were higher maybe a different behaviour of soy starch in combination with wheat starch, which was present in wheat-soybean flour mixtures. The maximal pasta viscosity decreased from 315 to 90 AU when soybean flour portion was increased. The lowest pasta viscosity was when the soybean flour portion was 30% (w/w) and it was seven times lower than maximal pasta viscosity value for wheat flour only. The low peak viscosity of soybean starch could be due to short amylopectin branch-chain which has been correlated in wheat starches (Sasaki & Matsuki, 1998; Shibananuma et al., 1996).

3.3 Lipid profile

The lipid profile of wheat flour and soybean flour obtained by HPLC analysis and based on these results the lipid profile of wheat-soybean flour mixtures, is presented in Table 3. The content of components was determined by measuring the peak area at 1.76 min for free fatty acids, peak area at 2.15 min for methyl esters, peaks area in the range of 3.44-4.58 min for monoacylglycerols, peaks area in the range of 5.28-8.68 min for diacylglycerols and peaks area in the range of 10.91-15.81 min for triacylglycerols (Holčapek et al., 1999).

Flour	FFA (g/100g)	ME (g/100g)	MAG (g/100g)	DAG (g/100g)	TAG (g/100g)
Wheat	11.9±0.4	23.2±1.3	2.5±0.2	12.2±0.9	50.2±1.5
Soybean	25.9±1.2	1.6±0.3	0.8±0.1	2.3±0.4	69.4±1.6
3%	12.3	22.5	2.4	11.9	50.8
5 %	12.6	22.1	2.4	11.7	51.2
10 %	13.3	21.0	2.3	11.2	52.1
20 %	14.7	18.9	2.2	10.2	54.0
30 %	16.1	16.7	2.0	9.2	55.9

Table 3. The lipid profile of wheat and soybean flour and their mixtures obtained by HPLC

Fatty Acid content in g/100g	Wheat flour	Soybean flour	3%	5%	10%	20%	30%
Myristic ($C_{14:0}$)	0	0.11±0.01	<0.035	<0.035	<0.035	<0.035	<0.035
Palmitic ($C_{16:0}$)	19.45±0.45	10.35±0.6	19.17	18.99	18.54	17.63	16.71
Linoleic ($C_{18:2}$)	57.91±0.72	55.23±1.2	57.83	57.77	57.63	57.37	57.11
Oleic ($C_{18:1}$)	20.23±0.21	13.07±0.8	19.20	19.87	19.52	18.79	18.08
Linolenic ($C_{18:3)}$	0	13.46±0.9	0.40	0.67	1.34	2.69	4.04
Stearic ($C_{16:0}$)	1.36±0.14	5.23±0.8	1.48	1.56	1.76	2.18	2.58
Nonadecanoic ($C_{19:0}$)	0	0.52±0.09	<0.053	<0.053	<0.053	0.10	0.16
Arachidic ($C_{20:0}$)	0	0.60±0.1	<0.061	<0.061	<0.061	0.12	0.18
Behenic ($C_{22:0}$)	0.26±0.06	0.48±0.06	0.27	0.27	0.28	0.30	0.33
ND RT 25.96	0	0.17±0.04	<0.052	<0.052	<0.052	<0.052	<0.052
Lignocerinic ($C_{24:0}$)	0	0.55±0.01	<0.056	<0.056	<0.056	0.11	0.17
Phthalic acid	0.68±0.08	0	<0.069	<0.069	<0.069	0.14	0.20
TU	78.14	81.46	77.43	78.31	78.49	78.85	79.23
TMU	20.23	13.07	19.20	19.87	19.52	18.79	18.08
TPU	57.91	68.39	58.23	58.44	58.97	60.06	61.15
TS	21.07	17.84	20.92	20.82	20.58	20.44	20.13
TU/TS	3.71	4.58	3.70	3.76	3.81	3.86	3.94

Table 4. Fatty acids composition of wheat and soybean flour and their mixtures obtained by GC

The lipid from soybean flour had higher free fatty acids content (25.9 g/100g lipid) and triacylglycerols content (69.4 g/100g lipid) than lipid from wheat flour (11.9 and 50.2 g/100g lipid, respectively), while the contents of methyl esters, monoacylglycerols and diacylglycerols were lower than appropriate contents in wheat flour, even 14.5, 3.1 and 5.3 times, respectively. The content of free fatty acids and triacylglycerols in mixtures had the same changing tendency as soybean flour portion: it increased when the soybean flour portion in flour mixtures increased, while the contents of monoacylglycerols and diacylglycerols had opposite changing tendency: they decreased when the soybean flour portion in flour mixture increased. The properties of dough such as water absorption, development time, dough stability and gelatinization temperature had the same dependency on soybean flour portion as the content of free fatty acids and triacylglycerols and these components seemed to have a proper influence on the mentioned dough properties. Further, the content of monoacylglycerols and diacylglycerols had a proper influence on dough energy, resistance, extensibility and gelatization maximum. Some of these dependences were confirmed by determination of correlations coefficients.

The fatty acids composition of flours obtained by GC analysis and based on these results, the fatty acids composition in flour mixtures is presented in Table 4. The lipid contained: linoleic ($C_{18:2}$), α-linolenic ($C_{18:3}$), oleic ($C_{18:1}$), palmitic ($C_{16:0}$), stearic ($C_{18:0}$), arachidonic ($C_{20:4}$), behenic ($C_{22:0}$), nonadecanoic ($C_{19:0}$), γ-linolenic, lignocerinic, myristic ($C_{14:0}$) and several non-determined components (ND). GC analysis showed the lipid from wheat flour contained 78.14 g/100g of the total unsaturated fatty acids, consisting of linoleic (57.91 g/100g) and oleic acid (20.23 g/100g). The total polyunsaturated fatty acids content in lipid was 57.91 g/100g and it was from linoleic acid. The monounsaturated fatty acids content was 20.23 g/100g, from oleic acid, while the total saturated fatty acids content was 21.07 g/100g where the main fatty acids were palmitic and stearic acid with the content of 19.45 and 1.36 g/100g, respectively. The ratio of total unsaturated to total saturated fatty acids content was 3.7.

The lipid from soybean flour contained 81.46 g/100g of the total unsaturated fatty acids, composed of linoleic, oleic and linolenic acid with content of 55.23, 13.07 and 13.46 g/100g of lipid, respectively. The palmitic and stearic acid were the main saturated fatty acids with content of 10.35 and 5.23 g/100g, respectively, while the content of other detected saturated fatty acid was 0.69 g/100g. The myristic, linolenic, nonadecanoic, arachidic and lignocerinic fatty acids were detected only in soybean flour, while phthalic acid was detected only in wheat flour. Based on this composition and content of fatty acids, the ratio of total unsaturated to total saturated content in soybean flour was 4.58 and it was higher than ratio in wheat flour. Such fatty acids composition had influence on fatty acids composition in wheat-soybean flour mixtures: the content of total saturated fatty acids decreased when soybean flour portion in flour mixture increased, and all values were lower than the content of total saturated fatty acids in wheat flour; the content of total unsaturated fatty acids increased when soybean flour portion in flour mixture increased and all values (except in mixture with portion of soybean flour of 3%) were higher than the content of total saturated fatty acids in wheat flour. Finally, the ratio of total unsaturated to total saturated fatty acids content was higher in flour mixtures (except in mixture with portion of soybean flour of 3%) than in wheat flour.

3.4 Statistical analysis data

The correlation coefficients between the rheological properties (water absorption, dough stability, Ex, gelatinization temperature and gelatization maximum) and the content of some lipid components (free fatty acids, monoacylglycerols, diacylglycerols, triacylglycerols, palmitic (Pal), linoleic (Lin) and oleic (Ole) acid) in wheat flour and wheat-soybean flour mixtures are presented in Table 5.

	WA	DSt	Ex	T_{max}	μ_{max}	FFA	MAG	DAG	TAG	Pal	Lin
DSt	0.88										
Ex	-0.94	-0.84									
T_{max}	0.91	0.98	-0.90								
μ_{max}	-0.64	-0.86	0.67	-0.78							
FFA	0.96	0.96	-0.93	0.98	-0.77						
MAG	-0.75	-0.78	0.91	-0.83	0.71	-0.83					
DAG	-0.81	-0.81	0.94	-0.86	0.73	-0.86	-0.97				
TAG	0.94	0.94	-0.85	0.94	-0.72	0.97	-0.69	-0.72			
Pal	-0.89	-0.89	0.97	-0.93	0.78	-0.94	0.95	0.98	-0.84		
Lin	-0.45	-0.47	0.71	-0.53	0.51	-0.51	0.85	0.88	-0.30	0.77	
Ole	-0.85	-0.85	0.90	-0.85	0.78	-0.88	0.88	0.84	-0.81	0.92	0.65

Table 5. Correlation coefficients between rheological properties and lipid components content (correlations are significant at $p < 0.05$, N=12)

The sample size was twelve (N=12 (6x2): wheat flour and five wheat-soybean flour mixtures with minimal and maximal obtained value). As there were many correlations, only the one which had absolute value 0.85 and above 0.85 were taken into consideration. There were 54.5% of correlations, among which 30.3% were proper, and 24.2% were opposite correlations. The correlations can be divided into three groups: correlations between rheological properties, between rheological properties and lipid components content and between lipid components content. In the first group, there are the correlations where high water absorption value is associated with high dough stability and gelatinization temperature and low extensibility, high dough stability is associated with high gelatinization temperature and low gelatization maximum, and correlation between low gelatization maximum and high extensibility. In the second group of correlations, the high content of free fatty acids caused high water absorption, dough stability, and gelatinization temperature but low extensibility as well as the high monoacylglycerols and diacylglycerols content was proper correlated with extensibility. Also, when triacylglycerols content was higher, the water absorption, dough stability and gelatinization temperature were higher and extensibility was lower. The palmitic and oleic acid had opposite effect on the rheological properties, such as water absorption, dough stability and gelatinization temperature and proper effect on extensibility. The linolenic acid content was not associated with any rheological properties. Among lipid components content, there were proper correlations between free fatty acids and triacylglycerols content, between monoacylglycerols and diacylglycerols contents on one side and palmitic, linoleic and oleic acid content on the other side. It can mean that monoacylglycerols and diacylglycerols are mainly consists of palmitic, linoleic and oleic acids. The higher oleic acid content was associated with the higher palmitic acid content and it was only correlation among fatty acids.

By cluster analysis, based on multiple variables, wheat and wheat-soybean flour mixtures were classified into groups. Number of variables was six: wheat and five wheat-soybean flour mixtures (3, 5, 10, 20 and 30% w/w of soybean flour portions); number of cases i.e. parameters were eight: water absorption, dough stability, free fatty acids, diacylglycerols, triacylglycerols, palmitic, linolenic and oleic acid content. Linkage distances were obtained and presented by dendrogram in Fig. 1.

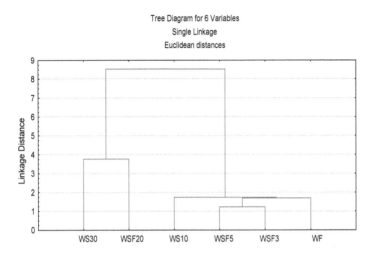

Fig. 1. Dendrogram obtained for wheat flour and wheat-soybean flour mixtures

The cluster analysis based on flour rheological and lipid characteristics, shows the linkage distance between wheat flour and flour mixtures increases when soybean flour portion in mixtures increases. The mixtures with soybean flour portion of 3 and 5% (w/w) are joined with wheat flour at the same distance level of 1.6 and make the first group. The mixtures with soybean flour portion of 10 % (w/w) are joined with wheat flour at the distance level of 1.7 and could be added to the first group. The mixture with soybean flour portion of 20 and 30% (w/w) is joined with wheat flour at distance level of 8.6 and make the second group. This means that soybean flour portion of 30% could be used to more enrich dough with soybean protein and the main rheological properties remain satisfactory as at portion of 20%. This provides a possibility of soybean flour being included in portions even higher than 30%, so future work could include this investigations as well as investigations to examine what happens with lipid components during dough mixing and baking: does their content and composition stay the same as in flour mixture or there occur changes.

4. Conclusion

The soybean flour addition increased the protein content up to 20.0%, the ash content up to 2.08% and lipid content up to 7.2%, while decreased starch, gluten and carbohydrates content for 19.8, 7.2 and 16 g/100 g flour mixture, respectively. Dough rheological properties and lipid profile depend on soybean flour portion. The soybean flour addition had positive influence on the quality number and group and extended duration of dough stability. The dough water absorption and the degree of softening increased with increasing soybean flour portion. The dough with soybean flour portion of 20 and 30% had lower values for energy in comparison to dough made of wheat flour only and it could be economically important. Values of gelatinization temperature for dough with soybean flour were higher than dough with wheat flour, maybe due to a specific behaviour of soy starch in combination with wheat starch. The maximal pasta viscosity decreased when soybean flour portion increased, even seven times when the soybean flour portion was 30%, probably due to short amylopectin branch-chain in soybean starch. All the wheat-soybean flour mixtures had higher free fatty acids and triacylglycerols content than wheat flour. Wheat-soybean flour mixtures had higher content of stearic and behenic acid compared to wheat flour, had almost the same content of linoleic acid as wheat flour and contained linolenic acid which was absent in wheat flour. The ratio of total unsaturated to total saturated fatty acids content was higher in flour mixtures than in wheat flour. The rheological properties of dough, such as water absorption and stability, had the same dependency on soybean flour portion in mixtures as the content of free fatty acids, while the content of triacylglycerols had the same dependency as water absorption and dough development time. In the same way, the content of monoacylglycerols and diacylglycerols had influence on dough extensibility and all these dependences were confirmed by statistical analysis with positive correlation coefficient value, higher than 0.85. The cluster analysis showed that the mixtures with soybean flour portions of 20 and 30% (w/w) were joined with wheat flour at the same distance level, so the soybean flour portion of 30% could be used to enrich even more dough with soybean protein and the main rheological properties remain satisfactory as with portion of 20%.

5. Acknowledgement

This work was supported under the project No.OI 172047 by the Ministry of Science of the Republic of Serbia.

6. References

Anjum, F.M.; Khan, M.I.; Butt, M.S.; Hussain, S. & Abrar, M. (2006) Functional properties of soy hulls supplemented wheat flour. *Nutrition and Food Science*, Vol.36, No.2, pp. 82-89, ISSN 0034-6659

Auerman, L.J., (1979). *Technology of bread production* (Ed Petar Spasic), Faculty of Technology, Novi Sad, Serbia

Bond, B.; Fernandez, D.R.; VanderJagt, D.J.; Williams, M.; Huang, Y.S.; Chuang, L.; Millson M.; Andrews, R. & Glew, R.H. (2005). Fatty acid, amino acid and trace mineral analysis of three complementary foods from Jos Nigeria. *Journal of Food Composition and Analysis*, Vol.18, No.7, (November, 2005), pp. 675-681, ISSN 0889-1575

Brandon, D. & Friedman, M. (2002). Immunoassays of soy proteins. *Journal of Agricultural and Food Chemistry*, Vol.50, No.22, (September, 2002), pp. 6635-6642, ISSN 1520-5118

Bressani, R. (1975). Nutritional contribution of soy protein to food systems. *Journal of the American Oil Chemists' Society*, Vol.52, No.4, (April, 1975) , pp. 254-262, ISSN 0003-021x

Djaković, Lj. (1980). *Wheat flour*, (Third edition), Faculty of Technology, Novi Sad, Serbia

Dobraszczyk, B.J. & Morgenstern, M.P. (2003). Rheology and bread making process. *Journal of Cereal Science*, Vol.38, No.3, (November, 2003), pp. 229-245, ISSN 07335210

Gujral, H.S.; Guardiola, I.; Carbonell, J.V. & Rosell, C.M. (2003). Effect of cyclodextrinase on dough rheology and bread quality from rice flour. *Journal of Agricultural and Food Chemistry*, Vol.51, No.13, (May, 2003), pp. 3814-3818, ISSN 0021-85661

Holčapek, M.; Jandera, P.; Fisher, J. & Prokeš, B. (1999). Analytical monitoring of biodiesel by high-performance liquid chromatography with various detection methods. *Journal of Chromatography A*, Vol.858, No.1, (October, 1999), pp. 13-31, ISSN 1873-3778

Jong, G.; Slim, T. & Creve, H. (1968). Bread without gluten. *Baker's Digest*, Vol.42, pp. 24-27, ISSN 0191-6114.

Liu, K. (1997). *Soybean: chemistry, technology and utilization*, Springer, ISBN 0-8342-1299-4

Mashayekh, M.; Mahmoodi, M.R. & Entezari, M.H. (2008). Effect of fortification of defatted soy flour on sensory and rheological properties of wheat bread. *International Journal of Food Science and Technology*, Vol.43, No.9, (September, 2008), pp. 1693-1698, ISSN 1365-2621

Maforimbo, E.; Skurray, G.; Uthaykumaran, S. & Wringly, C. (2008). Incorporation of soy proteins into the wheat-gluten matrix during dough mixing. *Journal of Food Science*, Vol.47, No. 2, (March, 2008), pp. 380-385 ISSN 0022-1147

Mohamed, A.A.; Shogren, R.L. & Sessa, D.J. (2006). Low carbohydrates bread: Formulation, processing and sensory quality. *Food Chemistry*, Vol.99, No.4, pp. 686-692, ISSN 0308-8146

Moore, M.; Tilman, T.S.; Dockery, P. & Arendt, E.K. (2004). Textural Comparisons of Gluten-Free and Wheat-Based Dough, Batters and Breads. *Cereal Chemistry*, Vol.81, No.5, (September/October, 2004), pp. 567-575, ISSN 0009-0352.

Olguin, M.C.; Hisano, N.; D·Ottavio, E.A.; Zingale, M.I.; Revelant, G.C. & Calderari, S.A. (2003). Nutritional and antinutritional aspects of an Argentinean soy flour assessed on weanling rats. *Journal of Food Composition and Analysis*, Vol.16, No.4, (August, 2003), pp. 441-449, ISSN: 0889-1575

Ortega-Garcia, J.O; Medina-Juarez L.A.; Gamez-Meza, N. & Noriega-Rodrigez, J.A. (2005). Optimisation of bleaching conditions for soybean oil using response surface methodology. *Food Science Technology International*, Vol.11, No.6, (December, 2006), pp. 443-450, ISSN 1532-1738

Qian, H.; Zhou, H. & Gu, J. (2006). Effect of different kinds of soy protein powder on bread-making properties. *Nongye Gongcheng Xuebao/Transactions of the Chinese Society of Agricultural Engineering*, Vol.22, No.10, (October, 2006), pp. 233-236, ISSN 10026819

Pérez, G.T.; Ribotta, P.D.; Steffolani, M.E. & Leon, A.E. (2008). Effect of soybean proteins on gluten depolymerization during mixing and resting. *Journal of the Science of Food and Agriculture*, Vol.88, No.3, (February, 2008), pp. 455-463, ISSN 0022-5142

Ribotta, P.D.; Arnulphi, S.A. & León, A.E. (2005). Effect of soybean addition on rheological properties quality of wheat flour. *Journal of the Science of Food and Agriculture*, Vol.85, No.11, (August, 2005), pp. 1889-1896, ISSN 1097-0010

Ribotta, P.D.; Pérez, G.T.; Aňo'n, M.C. & Leon, A.E. (2010). Optimization of additive combination for improved soy-wheat bread quality. *Food and Bioprocess Technology*, Vol.3, No.3, (June, 2010), pp. 395-405, ISSN 1935-5149

Roccia, P.; Ribotta, P.D.; Pérez, G.T. & León, A.E. (2009). Influence of soy protein on rheological properties and water retention capacity of wheat gluten. *LWT Food Science and Technology*, Vol.42, No.1, pp. 358-362, ISSN 1096-1127

Sabanis, D. & Tzia, C. (2009). Effect of rice, corn and soy flour addition on characteristics of bread produced from different wheat cultivars. *Food and Bioprocess Technology*, Vol.2, No.1, (March, 2009), pp. 68-79, ISSN 1935-5149

Sasaki, T. & Matsuki, J. (1998). Effect of wheat structure on swelling power. *Cereal Chemistry*, Vol.75, No.4, (May/June, 1998), pp. 385-391, ISSN 0009-0352

Schober, T.J.; O·Brien, C.M.; McCarthy, D.; Dardnedde, A. & Arendt, E.K. (2003). Influence of gluten-free flour mixes and fat powders on the quality of gluten-free biscuits. *European Food Research and Technology*, Vol.216, No.5, (May, 2003), pp. 369-376, ISSN 1438-2385

Shibanuma, Y.; Takeda, Y. & Hizukuri, S. (1996). Molecular and pasting properties of some wheat starches. *Carbohydrate Polymers*, Vol.29, No.3, (March, 1996), pp. 253-261, ISSN 0144-8617.

Stevenson, D.G.; Russell, K.D. & Jay-lin, J. & George, E.I. (2006). Structures Functional Properties from seeds of three soybeans (*Glycine max* (L.) Merr.) Varietes. *Starch*, Vol.58, No.10, (October, 2006), pp. 509-519, ISSN 1521-379x

Trajković, J.; Baras, J.; Mirić S. & Šiler, S. (1983). *Analysis of viands*, Technology and Metallurgy, Beograd, Serbia

Yoshida, H.; Kanerei, S.; Yuka, T. & Mizushina, Y. (2006). Regional characterization of tocopherols and distribution of fatty acids within soybean seeds (*Glycine max* L.). *Journal of Food Lipid*, Vol.13, No.1, (March, 2006), pp. 12-26, ISSN 1745-4522

8

Protein Sources in Ruminant Nutrition

Monica I. Cutrignelli, Serena Calabrò,
Raffaella Tudisco, Federico Infascelli and Vincenzo Piccolo
Department of Animal Science and Food Control,
University of Naples Federico II, Naples,
Italy

1. Introduction

Since 2001 the European Commission banned the use of meat and bone meal and its by-products in diets for livestock animals (EC directive 999/2001) in order to assure consumer safety on animal products. Consequently, soybean meal became the most utilised protein source in the intensive livestock systems.

Moreover, the proteins of this source are low degradable in the rumen and well proportioned to the non structural carbohydrates (NSC).

Soybean meal solvent extract (s.e.) is a by-product of oil industry, where soybean seeds are treated with organic solvents (e.g. hexane) and subsequently with high temperature. For this reason soybean meal has been banned in the organic livestock (EC directive 2092/1991; EC directive 834/2007).

Even if in Europe the high part of soybean is imported, soybean solvent extract represents the less expensive protein source for its high crude protein content (44-50 % as fed). However soybean meal costs and availability are strongly related with the price development of agricultural commodities on the world market (Jezierny et al., 2010). Factors which may influence world market prices include variations in population and economic growth, changes in consumer's product preferences, but world market prices are also dependent on weather conditions (Gill, 1997; Trostle, 2008).

Finally, another factor has to be evaluated that is the genetically modification (GM) of soybean. Indeed, public concerns are increasing in GM food consumption due to the fact, even if for several years no direct evidence that it may represent a possible danger for health has been reported, recently, a number of papers have been published with controversial results.

Thus, the search for alternative protein sources has led to an increasing interest in the use of grain legumes, as they supply the important source of plant protein.

The botanical family of grain legumes is known as *Fabaceae*, also referred to as *Leguminosae*. Grain legumes are cultivated primarily for their seeds which are harvested at maturity, and which are rich in protein and energy. The mature dry seeds of grain legumes are used either as animal feed ingredient or for human consumption (Singh et al., 2007). Beans, lentils and chickpeas are utilised exclusively for human nutrition, while the other grains are used in animal feeding too.

In Italy grain legumes cultivation is progressively increased due the presence of new cultivars more hardy and productive. These new cultivars were selected principally in

France and are characterised by lower water requirements, higher production and higher resistance to the parasitic infestations and to the adverse environmental conditions.

Generally, legumes are characterised by their ability to use atmospheric nitrogen as a nutrient due to the symbiosis with nitrogen-fixing bacteria from the *Rhizobium* species (Sprent and Thomas, 1984; Zahran, 1999). Therefore, unlike other cultivated plants, legume crops need less nitrogen fertiliser for optimal growth, and the use of legumes in crop rotation systems reduces the need of nitrogen fertiliser in subsequent crops (López-Bellido et al., 2005). Nitrogen benefits in legume-cereal rotation systems have been attributed not only to the transfer of biologically fixed nitrogen (Díaz-Ambrona and Mínguez, 2001; Evans et al., 2001), but also to lower immobilisation of nitrate in the soil during the decomposition of legumes compared to cereal residues (Green and Blackmer, 1995), also termed as the nitrogen-sparing effect. Thus, nitrogen benefits may result from a combination of legume nitrogen sparing effects and the bacterial nitrogen fixation (Chalk et al., 1993; Herridge et al., 1995). In addition, crop rotation and intercropping with legumes may provide successful strategies for weed suppression (Liebman and Dyck, 1993; Bulson et al., 1997). Weed growth and development may be disrupted due to varying cultivation conditions prevailing for the different crops used (e.g. fertiliser requirements, planting or maturation dates), thereby preventing domination of only a few weed species (Froud-Williams, 1988; Liebman and Janke, 1990). Due to these crop effects, cultivation of grain legumes is an important part of crop rotation, particularly in organic farming (Badgley et al., 2007).

In animal nutrition, grain legumes are mainly used as protein supplements, but also as a valuable energy source, due to their partly high contents of starch (faba bean, peas) and lipids (lupins) (Gatel, 1994; Bach Knudsen, 1997; Salgado et al., 2002a).

However, the use of grain legumes in animal nutrition has been hampered due to partially high concentrations of secondary plant metabolites, also referred as antinutritional factors (ANFs), including condensed tannins, protease inhibitors, alkaloids, lectins, pyrimidine glycosides and saponins. Possible negative effects of these secondary plant metabolites include, for example, feed refusals (tannins, alkaloids), reduced nutrient digestibility (tannins, protease inhibitors, lectins) or even toxic effects (alkaloids) (Rubio and Brenes, 1995; Lallès and Jansman, 1998; Huisman and Tolman, 2001).

The objectives of the following chapters are the comparison of chemical composition as well as the nutritive value of soybean, soybean meal solvent extract and several legume grain (e.g.: peas, lupine, faba bean). In addition, in order to evaluate the opportunity of soybean replacement with grain, the results of *in vitro* studies are described and the influence of protein sources on meat quality are discussed.

2. Nutritional characteristics of soybean meal solvent extract and legume grains

The chemical composition of some grain legumes in comparison to soybean and soybean meals s.e are pictured in Table 1.

Soybeans (*Soja hispida*) are characterized by high protein (380 g/kg dry matter) and lipid (200 g/kg dry matter) concentrations, which provide high energy density (1.1-1.2 UFL/kg dry matter). In ruminant feeding, soybean integral seeds could represent the only source of protein supplementation for cattle fattening. However, in dairy cows it is preferable not to exceed the dry matter administration of soybean seeds as it may modify fatty acid profile of milk fat and worse butter consistency and conservation.

Soybean meal s.e. proteins, due to the heat treatment, are medium to low degradable in the rumen (Chaubility et al., 1991; Infascelli et al., 1995). In addition, soybean meal shows an elevated ratio of protein/non structural carbohydrates.

Legume grains are characterised by high energy density allowed to the high protein, starch and/or fat concentrations, as more than sufficient is their calcium concentration.

The proteins of legume grain are highly degradable in the rumen and digestible in the intestine. Notwithstanding, large part of legume grain shows anti-nutritional factors (i.e. lecithin, trypsin inhibitors, tannins, saponin, phytase), that are inactivated by the enzymes produced by the bacteria present in the rumen. Within the grain legumes, lupins have higher amounts of crude protein (324–381 g/kg dry matter), compared to faba beans (301 g/kg dry matter) and peas (246 g/kg dry matter) (Degussa, 2006). Jezierny et al. (2007) reported similar contents of crude protein in different batches of lupins, faba beans and peas averaging 387, 308 and 249 g/kg dry matter, respectively. In comparison to soybean meal, faba beans and peas contain between 45 to 55% and lupins (L. albus) even up to 70% of its crude protein content (Degussa, 2006).

The ether extract content in peas and faba beans is generally rather low compared to lupins; crude fat contents of faba beans and peas range from 15 to 20 g/kg dry matter, thus being in a similar range as values for soybean meal (15–28 g/kg dry matter) (DLG, 1999; Jezierny et al., 2007). In lupins, the crude fat content varies between cultivars, with values of about 57 g/kg dry matter (L. luteus, L. angustifolius) to 88 g/kg dry matter (L. albus) (DLG, 1999).

The carbohydrate fraction includes the low molecular-weight sugars, starch and various non-starch-polysaccharides (NSP) (Bach Knudsen, 1997). The NSP and lignin are the principal components of cell walls and are commonly referred to as dietary fibre (Theander et al., 1989; Canibe and Bach Knudsen, 2002). Generally, faba beans and peas are rich in starch (422–451 and 478–534 g/kg dry matter, respectively) (DLG, 1999; Jezierny, 2009), whereas lupins have comparatively low levels of starch (42–101 g/kg dry matter) (DLG, 1999; Jezierny, 2010). However, it needs to be emphasized that the determination of starch in grain legumes may be confounded by the analytical method used (Hall et al., 2000).

Faba beans and peas contain rather low amounts of fibre fractions in comparison to lupins (Bach Knudsen, 1997; Jezierny, 2009), and, with regard to lignin content, faba beans and L. angustifolius have similar amounts of lignin (1 to 7 and 6 to 9 g/kg dry matter, respectively), whereas the lignin content in peas is of minor importance (0.4–3 g/kg dry matter) (Salgado et al., 2002a; Jezierny, 2010).

The NSP fraction of faba beans consists mainly of cellulose (89–115 g/kg dry matter), with lower levels of hemicellulose (21–57 g/kg dry matter) (Salgado et al., 2002a,b; Jezierny, 2009).

Hemicellulose contents in peas range from 23 to 95 g/kg dry matter and cellulose contents range from 52 to 77 g/kg dry matter (Salgado et al., 2002a,b; Jezierny, 2009).

Lupins contain high levels of NSP, with contents of cellulose generally being higher than hemicellulose (131 to 199 vs. 40 to 66 g/kg dry matter) (Bach Knudsen, 1997; Salgado et al., 2002a,b; Jezierny, 2009), and they also have considerable amounts of oligosaccharides (Bach Knudsen, 1997; Salgado et al., 2002a).

Comparing the different batches in each species, lupin show lower variability than faba bean and peas, probably because the genetic selection in this species was addressed principally on the reduction of secondary plant metabolites (Colombini, 2004) than to the improving of chemical characteristics.

The protein of faba beans and peas contains similar or even higher proportions of lysine (70 and 80 g/kg crude protein, respectively), when compared to protein from soybean meal s.e. (69 g/kg crude protein) or lupins (51 to 54 g/kg crude protein) (Degussa, 2006).

Comparing the different batches in each species, lupin show lower variability than faba bean and peas, probably because the genetic selection in this species was addressed principally on the reduction of secondary plant metabolites (Colombini, 2004) than to the improving of chemical characteristics.

Comparing the different batches in each species, lupin show lower variability than faba bean and peas, probably because the genetic selection in this species was addressed principally on the reduction of secondary plant metabolites (Colombini, 2004) than to the improving of chemical characteristics.

Protein sources	DM	CP	EE	Ash	NDF
Soybean	93.30	33.2	25.0	5.57	14.5
Soybean meal s.e.	89.80	41.3	1.36	6.00	22.2
Lupin Lublanc	93.26	36.92	6.78	4.33	4.33
Lupin Lutteur	94.31	35.30	5.73	4.19	4.19
Lupin Multitalia	93.97	36.67	9.54	3.92	3.92
Faba bean Irena	89.22	25.62	1.05	3.99	20.94
Faba bean Lady	88.48	25.17	1.02	4.21	25.82
Faba bean Scuro di Torre Lama	90.55	26.91	0.90	4.32	21.75
Faba bean Chiaro di Torre Lama	90.80	24.69	1.01	4.19	21.23
Faba bean ProtHABAT69	90.60	28.69	1.10	4.51	18.21
Faba bean Sicania	90.29	26.52	0.95	3.89	21.43
Peas Alembo	88.49	31.27	0.73	3.94	21.60
Peas Alliance	89.20	28.47	0.56	3.82	20.40
Peas Attika	89.68	25.04	0.81	4.01	18.09
Peas Corallo	88.80	28.50	0.80	3.69	21.82
Peas Iceberg	90.08	27.30	0.78	4.12	22.75
Peas Ideal	89.94	28.28	0.88	4.17	18.37
Peas Spirale	93.00	28.74	0.55	4.30	19.05

Table 1. Mean values of chemical composition (%dry matter) of different protein sources (Calabrò et al., 2001; Calabrò et al., 2009; Calabrò et al., 2010).

Comparing the different batches in each species, lupin show lower variability than faba bean and peas, probably because the genetic selection in this species was addressed principally on the reduction of secondary plant metabolites (Colombini, 2004) than to the improving of chemical characteristics.

The protein of faba beans and peas contains similar or even higher proportions of lysine (70 and 80 g/kg crude protein, respectively), when compared to protein from soybean meal (69 g/kg crude protein) or lupins (51 to 54 g/kg crude protein) (Degussa, 2006).

The proportion of threonine in grain legume protein (38 to 42 g/kg crude protein) is similar to that in soybean meal (45 g/kg soybean meal) (Degussa, 2006), however, there is a severe deficiency in the sulphur containing AA methionine + cystine, while tryptophan is marginally deficient to fulfill nutrient requirements for pigs (20 to 50 kg body weight) (NRC, 1998; Degussa, 2006). In fact, apart from L. albus, the seeds of faba beans, peas and lupins contain less than 50% of these AA in comparison to soybean meal (Table 2), thus constraining the use of grain legumes as sole protein source in pig diets.

	Vicia faba	Pisum sativum	Lupinus albus	Lupinus angustifolius	Lupinus luteus	SBM
CP	301	246	381	324	361	541
Indispensable AA						
Arginine	26.4	21.0	39.3	33.5	38.0	39.7
Histidine	7.8	6.1	9.3	8.8	9.7	14.4
Isoleucine	11.8	10	15.3	12.7	14.2	24.3
Leucine	21.4	17.4	27.5	21.5	24.1	40.9
Lysine	18.4	17.3	18.2	15	16.3	33.1
Methionine	2.2	2.2	2.5	2.0	2.0	7.3
Phenyl-alanine	12.6	11.7	14.9	12.5	13.6	27.2
Threonine	10.5	9.1	13.3	10.9	11.9	21.3
Tryptophan	2.6	2.2	3.0	2.6	3.0	7.4
Valine	13.3	11.4	14.5	12.5	13.6	25.5
Dispensable AA						
Alanine	11.9	10.5	12.5	10.9	11.8	23.3
Aspartic acid	31.6	28.2	38.5	31.5	35.1	62.0
Cystine	3.5	3.5	6.7	4.3	4.8	8.0
Glutamic acid	46.9	40.0	79.3	65.6	72.5	97.6
Glycine	12.2	10.6	15.0	13.4	14.3	23.0
Proline	11.8	10.2	15.3	13.5	14.3	27.5
Serine	14.1	11.5	19.0	15.3	17.0	27.3

Table 2. Amino acid contents of grain legumes compared to soybean meal (g/kg dry matter) (Jezierny et al., 2010) SBM= soybean meal; CP= crude protein; AA= amino acids.

As concerns the fatty acid profile, the rather high proportion of essential unsaturated fatty acids of some grain legumes, e.g. some Vicia species (Akpinar et al., 2001) or L. albus (Erbas, et al., 2005) may be attractive both from the human and animal nutrition perspective (Bézard et al., 1994), while adverse effects of unsaturated fatty acids on meat quality should be taken into account (Wood et al., 2003). For example, in faba beans a ratio of saturated to unsaturated fatty acids of 40–60 has been reported (Akpinar et al., 2001), whereas in L. albus, a ratio of saturated, monounsaturated and polyunsaturated fatty acids of 13.5 to 55.4 to 31.1 has been established (Erbas, et al., 2005).

As concerns the mineral composition of soybean and grain legumes, only few data are reported in literature.

The calcium concentration ranges between 1.0 g/kg (faba beans) and 1.9 g/kg (L. angustifolius). Phosphorus concentration varies between 4.2 g/kg (L. angustifolius) and 7.6 g/kg (L. luteus). No extreme differences in trace mineral concentrations occurred except for the manganese concentration of L. albus, which containes approximately 10 times more manganese than the other legume grains (Brand et al., 2004).

Concerning the trace elements, Cabrera et al (2003) reported that in legumes their levels ranged from 1.5–5.0 µg Cu/g, 0.05–0.60 µg Cr/g, 18.8–82.4 µg Fe/g, 32.6–70.2 µg Zn/g, 2.7–45.8 µg Al/g, 0.02–0.35 µg Ni/g, 0.32–0.70 µg Pb/g and not detectable–0.018 µg Cd/g. In nuts, the levels ranged from 4.0–25.6 µg Cu/g, 0.25–1.05 µg Cr/g, 7.3–75.6 µg Fe/g, 25.6–69.0 µg Zn/g, 1.2–20.1 µg Al/g, 0.10–0.64 µg Ni/g, 0.14–0.39 µg Pb/g, and not detectable–0.018 µg Cd/g. The authors found a direct statistical correlation between Cu–Cr, Zn–Al and Cr–Ni (P<0.05), and Al–Pb (P<0.001).

Sankara Rao and Deosthale (2006) comparing for their total ash, calcium, phosphorus, iron, magnesium, zinc, manganese, copper, and chromium contents five grain legumes (indian legumes, chick pea, pigeon pea, green gram, and black gram), found significant varietal differences only for chromium content in black gram. The cotyledons of these legumes were significantly lower in calcium content as compared to the whole grains. The authors concluded that for human nutrition, differences in mineral composition of whole grain and cotyledons were marginal except for calcium. These legumes as whole grain and cotyledons, appeared to be significant contributors to the daily requirements of magnesium, manganese, and copper in the diet.

As concerns the mineral composition of soybean and grain legumes, only few data are reported in literature.

The calcium concentration ranges between 1.0 g/kg (faba beans) and 1.9 g/kg (*L. angustifolius*). Phosphorus concentration varies between 4.2 g/kg (*L. angustifolius*) and 7.6 g/kg (*L. luteus*). No extreme differences in trace mineral concentrations occurred except for the manganese concentration of *L. albus*, which containes approximately 10 times more manganese than the other legume grains (Brand et al., 2004).

Concerning the trace elements, Cabrera et al (2003) reported that in legumes their levels ranged from 1.5–5.0 µg Cu/g, 0.05–0.60 µg Cr/g, 18.8–82.4 µg Fe/g, 32.6–70.2 µg Zn/g, 2.7–45.8 µg Al/g, 0.02–0.35 µg Ni/g, 0.32–0.70 µg Pb/g and not detectable–0.018 µg Cd/g. In nuts, the levels ranged from 4.0–25.6 µg Cu/g, 0.25–1.05 µg Cr/g, 7.3–75.6 µg Fe/g, 25.6–69.0 µg Zn/g, 1.2–20.1 µg Al/g, 0.10–0.64 µg Ni/g, 0.14–0.39 µg Pb/g, and not detectable–0.018 µg Cd/g. The authors found a direct statistical correlation between Cu–Cr, Zn–Al and Cr–Ni (P<0.05), and Al–Pb (P<0.001).

Sankara Rao and Deosthale (2006) comparing for their total ash, calcium, phosphorus, iron, magnesium, zinc, manganese, copper, and chromium contents five grain legumes (indian legumes, chick pea, pigeon pea, green gram, and black gram), found significant varietal differences only for chromium content in black gram. The cotyledons of these legumes were significantly lower in calcium content as compared to the whole grains. The authors concluded that for human nutrition, differences in mineral composition of whole grain and cotyledons were marginal except for calcium. These legumes as whole grain and cotyledons, appeared to be significant contributors to the daily requirements of magnesium, manganese, and copper in the diet.

3. *In vitro* evaluation of soybean and legume grains

The nutritive value of a feed is assessed by its chemical composition, digestibility and level of voluntary intake. Feed evaluation methods are use to express nutritive value of feed. It is basically description of feeds interns that allow for a prediction of the performance of animals offered the feeds (Medsen et al., 1997).

Several methods are used in feed evaluation such as chemical analysis, rumen degradability measurement using the nylon bag technique, digestibility measurement and feed intake prediction. However, the tables supply mean values, which cannot be used for individual lots, and all the *in vivo* techniques are very expensive and time consuming while accuracy may be low.

The *in vitro* gas production technique (IVGPT, Theodorou et al., 1994) has proved to be a potentially useful technique for ruminant feed evaluation (Herrero et al., 1996; Getachew et al., 2004), as it is capable of measuring rate and extent of nutrient degradation (Groot et al., 1996; Cone et al., 1996).

To evaluate a feedstuff by IVGPT, it is incubated at 39°C and under anaerobiosis condition with buffered rumen fluid and gas produced is measured as an indirect indicator of fermentation kinetics. During the incubation the feedstuff is first degraded and the degraded fraction may either be fermented to produce gas (CO_2 and methane) and fermentation acids, or incorporated into microbial biomass.

The IVGPT is considered the most complete *in vitro* technique, because it allows to estimate the fermentation kinetics and contemporary gives information on the fermentation products (degradability of dry matter, volatile fatty acids).

The IVGPT has been used by our research group for many years in order to investigate feed fermentation kinetics.

In particular, in this chapter we report some results obtained incubating different protein sources (i.e. soybean and legumes grains) in order to compare their fermentation kinetics (Calabrò et al., 2001a), to assess the effect of some technological treatments (i.e. crushing and flaking) on the carbohydrates fermentation kinetics (Calabrò et al., 2001b) and to test different legumes grain cultivars (Calabrò et al., 2009).

Calabrò et al. (2001a), in order to study the *in vitro* fermentation characteristics and kinetics proposed the following protocol: the samples (about 1.00 g), ground to pass a 1 mm screen, were incubated in triplicate at 39°C in 120 ml serum bottles under anaerobic conditions. Rumen liquor for the *inoculum* was collected from four buffaloes fed a standard diet, and immediately transported to the laboratory where it was homogenised and filtered.

The gas measurements was made at 2-24 time intervals using a manual a pressure transducer (figure 1).

The cumulative gas produced at each time was fitted to the Groot et al. (1996) model which estimates the asymptotic value (A, ml/g), the time after incubation at which A/2 is formed (B, h), the time to reach the maximum rate (tmax, h) and the maximum rate (Rmax, ml/h).

At the end of incubation, the degraded organic matter (dOM, %) was calculated as a difference between incubated and residual OM (filtering the bottle content through pre-weighed glass crucibles and burning at 550°C for 3 hours), and pH and volatile fatty acid concentration (VFA, mM/g) were determined, using a pH-meter and a gas chromatography, respectively.

Several concentrate ingredients such as cereals and grain legumes, used in ruminant diets in order to increase production levels, were evaluated. In particular barley, maize, hard wheat, soft wheat, oats, faba bean and pea were used as test substrates.

Carbohydrates fractionation was carried out according to Cornell Net Carbohydrate and Protein System (CNCPS, Sniffen *et al.*, 1992).

The Cornell Net Carbohydrate and Protein System was developed to predict requirements, feed utilization, animal performance and nutrient excretion for dairy and beef cattle and sheep, using accumulated knowledge about feed composition, digestion, and metabolism in supplying nutrients to meet requirements.

Fig. 1. Pressure transducer.

The CNCPS partitions crude protein into fractions A, B, and C, depending on their rate and extent of degradability in the rumen (NRC, 2001). Fraction A represents the non-protein N (NPN) (ammonia, peptides, amino acids) and is considered to be completely soluble; fraction B, subdivided into B1, B2, and B3, consists of true protein with progressively declining ruminal degradability. Fraction C is unavailable true protein. Broadly, these crude protein fractions are categorized into rumen degradable protein (RDP) and rumen undegradable protein (RUP). The rumen degradable protein meets protein requirements for ruminal microbial growth and protein synthesis.

Once reaching the rumen, feed and protein degradation is a function of microbial activity. Rumen microbial activity, growth and protein synthesis is primarily limited by the rate and extent of carbohydrate fermentation in the rumen. Consequently, dietary fiber fractions in the forage determine the animal response to feed.

Microbial protein and rumen undegradable protein reaching the small intestine are absorbed to meet the ruminant's protein requirement. When rumen degradable protein exceeds the capacity of the rumen microbes to assimilate it, ammonia builds up in the rumen. This is followed by absorption of ammonia into the blood, conversion into urea by the liver, and excretion in the urine. The conversion of ammonia to urea costs the dairy cow energy that could otherwise be used for milk production. This loss of dietary crude protein and energy reduces the utilization efficiency of rumen degradable protein and therefore, reduced ruminant production (NRC, 2001). It also causes a negative energy balance that leads to a reduced fertility.

In the study of Calabrò et al. (2001a), the CNCPS carbohydrate fractions (table 3) was consistent with the values reported by Sniffen et al. (1992).

Oats (table 4) had the lowest potential gas production (A: 251 ml/g, P<0.0) and a very fast fermentation process (evidenced by low B and high RM). This result was probably due to its high soluble sugars contents (A fraction). Interestingly, oat values were very high compared to in situ observations. As reported by Van Soest et al., (1992), maize proved to be the slowest because of its high starch content (B1 fraction), which degraded slowly. Hard and soft wheat showed very similar fermentation characteristics according to their chemical composition. Barley has fermentation kinetics between those of wheat and maize. Results related to these last three grains were similar to the results obtained in vitro using the IGPT (Mould et al., 2005).

	Barley	Maize	Oats	Hard wheat	Soft wheat	Faba bean	Pea
CP (% DM)	11.0	8.79	12.2	14.4	13.5	29.8	23.9
A	3.80	0.49	12.6	2.40	5.00	6.00	18.0
B1	66.4	75.4	45.6	67.6	65.7	43.6	40.8
B2	11.4	6.14	16.0	9.86	9.64	9.99	7.92
C	3.60	2.57	6.77	2.64	2.16	6.41	4.18

Table 3. Chemical composition and CNCPS fraction of the tested grains.

Legumes were characterised by high gas production alongside a rather slow fermentative process. However, interpretation of faba bean and pea data was complicated by their high protein content, whose degradation can interact with gas production (Schofield., 2000).
Organic matter degradability for all tested grains was always higher than 95%.
At 72 h the pH always remains good to guarantee microbial activity.
Oats reached the maximum rate before all the other grains (figure 2) and remained the highest until 12 incubation hours, subsequently decreasing below the soft wheat level. Until 6 hours of incubation barley was faster than the two wheat, later becoming slower until the end; up to this time the three sharp curves were practically overlapping, suggesting that the fermentation trend was similar for the three grains. Maize, faba bean and pea always showed the slowest fermentation rate with overlapping trends: fermentation slowly reaches its maximum rate and slowly decreases. The results of this investigation evidenced the validity of the IVGPT in describing the kinetics fermentation of the examined grains.
The range of rates obtained with the GPT (oats>wheat>barley>maize>legumes) approximately reflects that reported by other authors cited by Sniffen et al. (1992): wheat>oats>barley>maize.

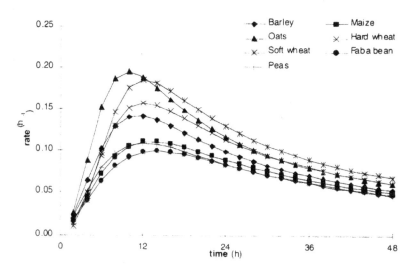

Fig. 2. Gas production fractional rate plots of different grains.

In the second investigation Calabrò et al. (2001b) evaluated the effects of flaking and crushing on some cereal and legume grains, using the *in vitro* cumulative gas production technique (Theodorou *et al.*, 1994).

The grains used in animal nutrition are commonly subjected to different technological treatments (i.e.: grounding, crushing, flaking, extrusion, cooking, micronization). Flaking is a hydrothermal-mechanical process which causes starch gelatinization and increases the enzymatic hydrolysis rate of the polysaccharide, which in turn favours the rumen microbial attack. Crushing differs from flaking in the shorter exposure to vapour. Also in this case, the obtained product presents a larger surface area compared to the primary grain, and a modified starch, which promotes faster fermentation kinetics.

Several investigations to evaluate the effect of flaking on organic matter degradability in the rumen, indicated the considerable influence of the primary grain (Arieli *et al.*, 1995; Bittante *et al.*, 1989); by contrast, little was known about the effects of crushing.

		VC_{SOI}	A	B	RM	dOM	Y	VFA
		ml/g	*ml/g*	*h*	*h-1*	*%*	*ml/g*	*mM/g*
Maize	wholegrain	340	362	14.9	0.078	92.9	363	99.1
	flaked grain	338	347	13.2	0.096	96.6	343	83.3
	whole + flaked	339[A]	355[A]	14.1[B]	0.087[B]	94.8	353[A]	91.2[A]
Barley	wholegrain	304	323	10.2	0.125	90.6	341	92.4
	flaked grain	346	369	11.7	0.104	97.9	359	94.8
	whole + flaked	325[B]	346[A]	11.0[C]	0.115[A]	92.6	350[A]	93.6[A]
Soybean	wholegrain	204	236	15.5	0.051	96.7	209	71.6
	flaked grain	225	280	19.8	0.039	86.8	257	76.0
	whole + flaked	215[C]	258[B]	17.6[A]	0.045[C]	91.8	233[B]	73.8[B]
Main effect								
grain		***	***	***	***	n.s.	***	**
treatment		***	**	**	n.s.	n.s.	**	n.s.
Interaction								
grain. x treat.		***	***	***	***	***	***	n.s.
Var.er.		87.9	135	1.21	0.00009	10.5	206	103

OMCV, *cumulative gas production related to incubated OM at 120 h; A, potential gas production; B, time at which A/2 is produced; Rmax, maximum fermentation rate; Y, yield related to the degraded OM; VFA, volatile fatty acids related to degraded OM.*
a,b,c: P<0.05; A,B: P<0.001ns, not significant;***: p < 0.001.

Table 4. *In vitro* fermentation characteristics of whole and flaked grains.

In this experiment three whole and flaked grains (barley, maize, soybean) and four whole and crushed grains (barley, oats, maize, faba bean) were used as test substrates.

Fermentation characteristics (Table 4 and 5) were affected by the grain type and treatment; in addition, the two factors interacted with each other.

		OMC V	A	B	Rmax	dOM	Y	VFA
		ml/g		H	h^{-1}	%	ml/g	mM/g
Oats	whole grain	321	334	10.6	0.107	87.8	362	90.6
	crushed grain	272	311	10.3	0.106	86.9	332	90.5
	whole +crushed	296[B]	322[B]	10.4[C]	0.0106[Aa]	87.3[aCD]	347[B]	90.5[AB]
Faba bean	whole grain	335	360	14.4	0.071	96.7	344	94.3
	crushed grain	325	367	14.8	0.062	89.9	359	79.3
	whole +crushed	330[A]	364[A]	14.6[A]	0.066[B]	93.3[B]	352[B]	86.8[AbB]
Maize	whole grain	324	348	12.2	0.099	97.6	331	90.6
	crushed grain	327	350	13.4	0.092	96.7	338	100
	whole +crushed	326[Ab]	349[A]	12.8[B]	0.096[AbC]	97.2[A]	335[B]	95.5[aA]
Barley	whole grain	310	327	8.68	0.138	81.5	398	77.3
	crushed grain	370	386	9.59	0.128	90.1	395	91.7
	whole +crushed	340[Aa]	357[A]	9.13[D]	0.133[D]	85.8[bD]	397[A]	84.5[B]
Main effect								
	Grain	***	***	***	***	***	***	***
	Treatment	n.s.	n.s.	*	n.s.	n.s.	n.s.	n.s.
Interaction								
	gran x treat	***	***	n.s.	n.s.	***	***	n.s.
	Var. er	204	398	0.73	$1 \bullet 10^{-1}$	1.85	437	48.19

OMCV, cumulative gas production related to incubated OM at 120 h; A, potential gas production; B, time at which A/2 is produced; Rmax, maximum fermentation rate; Y, yield related to the degraded OM; VFA, volatile fatty acids related to degraded OM.
a,b,c: P<0.05; A,B: P<0.001ns, not significant;***: p < 0.001.

Table 5. *In vitro* fermentation characteristics of whole and crushed.

Flaking increased the potential gas production (A) for soybean and barley (P< 0.01), and slowed their fermentation kinetics (increased t/1/2 and decreased RM values). By contrast, maize flakes showed lower gas production (P<0.05) and the fermentation process was faster (P<0.01), compared to whole grain.

Organic matter degradability decreased after the treatment for soybean (P<0.01), while it increased for barley (P<0.01) and, not significantly, for maize. The slower fermentative process for barley flakes (increased t/1/2and decreased RM values) agrees with the results of other authors on treated cereals for both *in vitro* (Bittante et al., 1989) and *in situ* experiments (Arieli et al., 1995).

Comparing each grain with the respective flakes, OM degradability did not strictly follow the trend of potential gas production, confirming that not all the degraded OM is fermented. This result holds mainly for soybean flakes, which had a higher A (P<0.01) and a lower dOM (P<0.01) compared to the whole grain.

Such decreased OM degradability was probably due to the thermal treatment causing a decrease in the rapidly degradable fraction (Sarubbi, 1999).

Overall, the two cereals showed higher gas and VFA production, as well as a faster fermentative process.

Crushing showed slighter effects compared to flaking, and also the results of the grain x treatment interaction were less important.

The different technological condition of the two treatments (less drastic for crushing) may well have contributed to this trend. In particular, crushed barley presents, with respect to the whole grain, the same behaviour as the flakes (higher A and dOM, P<0.001; higher $t/1/2$ and lower RM, not significant).

Interestingly, there was a decrease in faba bean degradability due to the treatment (P<0.01), which agrees with the results of a contemporaneous *in situ* trial (Sarubbi, 1999). Besides, faba bean also showed a lower VFA concentration, which was probably also caused by the considerable decrease in their protein degradability (P<0.001) as found *in situ* (Sarubbi, 1999). As usually observed *in vivo* in young bulls fed a cereal-rich diet, pH values were always quite low.

In the third study Calabrò et al. (2009) evaluated the fermentation characteristics of different cultivars of grain legumes using IVGPT.

Three grain legumes were tested: lupine (*Lupinus* spp.) (Lublanc, Luteur, Multitalia), faba bean (*Vicia faba* L.) (Chiaro di Torre Lama, Irena, Lady, ProtHABAT69, Scuro di Torre Lama, Sicania) and peas (*Pisum sativum*) (Alembo, Alliance, Attika, Corallo, Iceberg, Ideal, Spirale).

The fermentation characteristics are reported in table 6.

The values of pH ranged between 6.35 and 6.72, indicating a normal pattern of fermentation, and were consistent to the crude protein content.

As regards faba bean, the cultivar "Scuro di Torre Lama" showed significantly (P<0.01) lower values of dOM and OMCV than the other 5 cultivars.

In the case of lupine the cultivar "Lublanc" had lower (P<0.01) OMCV than the other 2 cultivars and for peas the cultivar "Spirale" produced less gas and showed a faster kinetics than the other 6 cultivars.

As expected, the OM degradability resulted very high in any case. However, comparing the pools of the grain legumes, dOM was in each case lower than that of soybean meal. OMCV was significantly (P<0.01) higher for pea than faba bean (330 vs. 316 ml/g, P <0.05) and lupine (330 vs. 258 ml/g, P <0.01). Gas production of peas was always higher than that of lupine, faba bean and also soybean meal according to the results of Buccioni et al. (2007) who studied the *in vitro* fermentation of soybean meal, faba bean and pea, and found in the latter the best balance between energy and nitrogen inputs.

The slower fermentation kinetics of faba bean may be due to the content in polyphenols while that of lupine may be caused by the very low starch content (INRA, 1988).

From the data obtained, the authors concluded that the tested grain legumes show only few differences compared to soybean meal (higher dOM and lower OMCV), consequently they may be considered in replacing, totally or partially, soybean.

The reported results are of particular importance as well as highlight the differences in dietary and nutritional characteristics of soybean and legume grains, provide data on the effects of the treatments of these feedstuffs and the differences among the cultivars present on the European market.

With this information, the nutritionist may from time to time choose the most suitable protein source in order to satisfy animal requirements ensuring the simultaneous availability of nitrogen and energy for the bacteria present in the rumen with beneficial effects on livestock production and environmental impact.

Cultivar	pH	dOM	OMCV	Yield	A	B	tmax	Rmax
		%	ml/g	ml/g	ml/g	h	h	ml/h
Faba bean								
Irene	6.46	92.9	370	397	328	22.9	12.42	9.14
Lady	6.35	93.3	354	363	333	24.5	15.67	9.26
Scuro di Torre Lama	6.49	87.8	308	351	269	22.0	15.39	8.84
Chiaro di Torre Lama	6.41	91.8	348	379	310	23.1	13.20	8.69
ProtHABAT69	6.47	93.8	359	383	303	20.2	12.49	10.02
Sicania	6.40	92.9	324	349	299	21.0	12.83	9.71
MSD	0.135	3.19	60.1	49.3	60.1	3.90	3.65	1.89
Lupine								
Lublanc	6.63	93.4	256	279	283	26.1	10.25	6.73
Luteur	6.69	92.4	275	298	309	25.6	5.58	7.83
Multitalia	6.72	91.2	273	297	303	27.0	8.48	7.08
MSD	0.219	5.40	26.6	91.5	13.9	45.8	15.6	4.74
Peas								
Alembo	6.57	99.0	406	410	361	20.6	12.52	11.73
Alliance	6.49	99.3	397	396	358	20.1	11.99	11.72
Attika	6.57	98.4	397	404	360	20.5	11.82	11.46
Corallo	6.53	98.9	393	394	365	22.3	11.42	10.38
Iceberg	6.55	98.8	381	385	347	21.0	12.45	10.86
Ideal	6.58	97.0	371	383	336	20.7	13.24	11.06
Spirale	6.58	98.8	344	343	310	17.1	10.66	12.14
MSD	0.188	2.81	52.5	53.5	3.68	74.1	5.22	2.47
Faba bean[1]	6.52Ab	90.9b	368B	405A	321a	21.1ab	12.4	10.0Ab
Lupine[1]	6.64B	91.8ab	284C	309B	293b	24.4a	9.03	7.42B
Peas[1]	6.60a	95.1a	394A	413A	336a	18.2b	11.4	12.6Aa
Soybean meal[2]	6.73	96.5	295	306	323	18.7	6.01	10.67
MSE	0.001	2.31	56.2	30.5	101	4.40	2.99	0.56

MSD: Minimum Significant Differences for P<0.01. MSE: Mean Square Error.,In the column A,B,C: P<0.01; a,b,c: P<0.05. [1]Data obtained from the grain legumes incubated in vitro as a pool. [2]Data not statistically assessed.

Table 6. Fermentation characteristics of the different grain legume cultivars and soybean meal.

4. Effect of different protein sources on animal performance

Several researches have been carried out in order to compare the nutritional characteristics of soybean solvent extract and legume grains, as faba bean, peas and lupine, for ruminant feeding.

Di Francia et al. (2007) evaluated the effect of partial replacement of soybean cake with extruded peas in the diet of lactating buffalo cows on milk yield and quality over the first 100 days of lactation. Their results showed that peas could represent an attractive GMO free protein feed when approaching the problem of the choice of a protein source alternative to soybean in diet formulation for buffalo cows raised in organic farms.

On the other hand Morbidini et al (2005) investigating on the effect of two different fattening diets with different protein sources - soybean meal and flaked faba bean - on slaughtering performance and carcass quality in light Apennine and Italian Merino lambs, observed that the use of faba beans lightly depressed growth performance (Morbidini et al., 2004) and slaughtering weight, even if did not affect the carcass quality.

In order to make a contribution on this issue, our research group conducted a trial on the effect of protein source on growth performance and meat quality of Marchigiana young bulls (Cutrignelli et al 2008 a, b).

The trial was carried out on a farm situated at 700 m a.s.l. in Campania Region (Southern Italy), where 12 weaned young bulls (129 d of age) were equally divided into two groups. Each animal was placed in individual box up to the slaughtering weight (620 kg).

The groups were fed diets with the same protein and energy concentrations and the same forage/concentrate ratios (F/C), but differing in protein source: faba bean (*Vicia faba minor* L.) vs soybean meal (*Soja hispida*).

All the animals were regularly weighed until the body weight (BW) of 620 kg fixed in advance as slaughter weight, was reached. All animals were slaughtered in an authorized slaughterhouse according to EU legislation (EU Regulation EC No 882/2004).

Live animals and carcasses were weighed and measured according to ASPA (1991). After 9 days of refrigeration at 4 ± 1 °C, dissection of the carcasses was carried out.

Samples of *Longissimus thoracis* (LT), *Semitendinosus* (ST), *Iliopsoas* plus *Psoas minor* (IP) muscles, perirenal (PF) and subcutaneous (SF) adipose tissues were collected and rapidly transported, upon refrigeration temperature, to the laboratories for the chemical analysis in order to evaluate the rheological (water holding capacity) and nutritional (chemical composition fatty acid profile, cholesterol and hydrossiproline contents) characteristics of meat.

The protein source did not affect any *infra vitam* parameters except the body weight (BW) at 180 d of age (173 vs. 186 kg, for group FB and SB, respectively; P<0.01).

The difference was probably due to the higher non-protein nitrogen (NPN) concentration of the faba bean than the soybean meal (about 12 vs 1.3% of crude protein, respectively). Indeed, in the months immediately following the weaning it should be preferable to administer diets with higher rumen undegradable protein content because at this age the rumen is not yet perfectly functional and microbial protein synthesis is less efficient. Furthermore, for the same reason, the animals in this period are probably unable to neutralize possible anti-nutritional factors of faba beans.

As respects the influence of the replacement of soybean meal solvent extract with legumes seeds, the literature results are contrasting. Moss et al. (1997) found no significant effects on weight gain and feed intake when soybean meal was replaced on an iso-protein basis by lupin seeds in diets for growing bulls (BW from 182 to 243 kg); similar results are reported by Kwak and Kim (2001) on Korean native bulls (BW from 247 to 427 kg) utilising two different concentrations (15 and 30%) of flaked lupin. Instead, according to our results, Murphy and McNiven (1994) found significantly higher weight gain in growing steers (BW from 235.2 to 343.7 kg) fed soybean meal vs raw or roasted lupin although the differences were not significant in the finishing phase (final BW 503.4 kg).

No significant differences between the groups were found for carcass measurements. In each case the carcass measurements of this trial ranged into the interval indicated by Keane (2003) for European/North American breeds.

No differences were found in dressing out, organs and tare incidence on net weight.

Protein source influenced neither body and carcasses conformations nor dressing out with the exception of the incidence of long bones showed a significant difference (6.2 vs 6.7 for faba and soybean, respectively; P<0.05).

The first quality meat cuts were acceptable in both groups (58.1 and 57.8%, for faba and soybean, respectively). Concerning the comparison between the two protein sources, only the incidence of long bones showed a significant difference (6.2 vs 6.7 for faba and soybean, respectively; P<0.05). It is important to underline that the sample cut measurements were contradictory and conflicted with the data obtained from total carcass dissection. The results of the sample cut dissection indicated a significant (P<0.05) difference between groups faba and soybean in meat incidence (69.4 vs 66.9, for faba and soybean, respectively); while no differences were found between groups at carcass dissection. Moreover, sample cut of soybean group showed in the meantime the smallest meat incidence and the highest LT area (88.4 vs. 84 for soybean and faba group, respectively).

The animals fed faba bean showed significantly higher water losses, measured with the compression method (WHC 7.6 vs. 5.7% for faba and soybean group, respectively; P<0.01).

Our grilling loss data (31.7 vs. 28.3 % for faba and soybean group, respectively; P>0.01) were higher than those reported by Sami et $al.$ (2004) in Simmental young bulls but they were in agreement with those reported by Pen et $al.$ (2005) cooking ST samples of Holstein steers in an oven, on the contrary our drip loss data resulted higher than that reported by these authors.

The chemical composition of LT was not statistically different between groups. Meat from both groups showed a very low fat content (<3%) and higher protein concentrations than the Holstein steers (Pen et $al.$, 2005), confirming the high quality of the Marchigiana meat.

Hydroxyproline (60.0 vs. 62.6 mg 100g^{-1} of meat, in group faba and soybean, respectively) contents were not influenced by protein sources. Regarding the differences registered for this parameter among the texted muscles (LT: $Longissimus$ $thoracis$; ST: $Semitendinosus$; IP: $Iliopsoas$ plus $Psoas$ $minor$) the IP samples, which correspond to the tenderloin, showed in all the groups significantly (P<0.01) lower hydroxyproline concentrations than the other two muscles; also between the LT and ST muscles the differences were statistically significant (P<0.05) being lower for the former.

Cholesterol content (56.3 vs.55.1 mg 100g^{-1} of meat, in group faba and soybean, respectively) was not influenced by protein sources according to the observations of Cutrignelli (2000) on Podolian young bulls and by Poli et $al.$ (1996) on Chianina young bulls.

Considering the differences among the muscles, cholesterol values were significantly (P<0.05) lower for IP. Cifuni et $al.$ (2004) found no differences in cholesterol contents among muscles, while Rusman et $al.$ (2003) found significant differences. This contradiction is probably due to the different muscles analysed in each experiment. As theorised by Wheeler et $al.$ (1987) the cholesterol content may be affected by the different physiological function of the muscles. In both groups, and especially for LT and ST muscles, the cholesterol contents were slightly higher than the value (less than 50 mg100g^{-1} of muscle) indicated by the Protected Geographical Indication (PGI) of the "Vitellone Bianco dell'Appennino Centrale" (Council Regulation EEC No 2081/92; Floroni, 2002). Nevertheless, our results are very close to those reported for Italian meat breeds (Poli *et al.* 1996; Cifuni *et al.*, 2004) and lower than those from other breeds (Migdal *et al.*, 2004).

Regarding the fatty acids profile in both groups and in each analysed tissues (intramuscular, perirenal and subcutaneous adipose tissues) palmitic (C16:0), stearic (C18:0) and oleic (C18:1) acids were the most widely represented fatty acids. In particular, in intramuscular fat, the sum of oleic, stearic and palmitic acids represents over 50% of total fatty acids, according to the observations of Cifuni *et al.* (2004) and Migdal *et al.* (2004).

Comparing the data of groups faba and soybean, the only significant (P<0.01) difference was for stearic acid being higher for bulls fed faba bean than for those fed soybean meal solvent extract. However this result did not significantly affect the SFA concentration, or AI and TI indexes (table 7).

Protein source did not affected the fatty acids composition of the analysed adipose tissues. This is probably due to the very low concentration of phospholipids in subcutaneous and perirenal fat tissue.

Comparing the fatty acid composition of the three adipose tissues the following differences were noted:

-   the erucic (C22:1), docosahexaenoic (DHA, C22:6, ω-3) eicosapentaenoic (C 20:5, ω-3) and docosapentaenoic (C 22:5, ω-3) acids, were present only in intramuscular fat, due to the lower level of intramuscular fat compared to the two adipose tissues. Lower fat content is associated with fewer and smaller adipocytes containing fewer triglycerides, accompanied by a relative increase in the proportion of phospholipids in total lipids and an increased PUFA content (Scollan *et al.*, 2006);

-   the proportion of total SFA was minimal in LT and maximal in SF (mean values 46.56 *vs* 54.50 and 59.35% of total fatty acids; in LT, PF and SF, respectively, Figure 1); the single saturated fatty acids amounts also showed a similar trend. These data are partly in contrast with those of Kim *et al.* (2004) who in Hanwoo cattle found a higher SFA proportion in perirenal than in subcutaneous fat. Even if the differences in the fatty acids profile between internal and external fat deposits have not been fully elucidated, Eguinoa *et al.* (2003) suggested that the lipogenic enzyme activities per cell could be influenced by several factors such as adipocite size, nutrient supply, etc.;

-   the proportion of total polyunsaturated fatty acids (PUFA) was higher in LT and lower in the perirenal adipose tissue (mean values 22.23 *vs* 11.36 *vs* 11.02% of total fatty acids, respectively in LT, SF and PF, Figure 1). This difference may be primarily ascribed to the lower concentration of ω-6 PUFA and to the absence DHA in SF and PF adipose tissues;

-   the atherogenic and thrombogenic indexes were considerably higher in SF and PF tissues than in LT. This is due to the high SFA concentration of SF and PF adipose tissues as well as to the high PUFA concentration of LT.

The LT fatty acid profile was similar to those reported by Raes *et al.* (2003) for *Longissimus lumborum* of Belgian Blue and Limousin beef (SFA: 338 and 506 mg 100g^{-1} edible portion; MUFA: 323 and 554 mg 100g^{-1} edible portion; PUFA: 195 and 195 mg 100g^{-1} edible portion; in Belgian Blue and Limousin bulls, respectively) and different from values found in Irish and Argentine beef. The latter showed significantly higher total intramuscular fatty acid content compared to the former, probably due to genetic selection.

This observation confirms that of Carnovale and Nicoli (2000) who concluded that Italian meat showed a favourable intramuscular fatty acid composition with high PUFA content. Muscle with a high percentage of unsaturated fatty acids (UFA) generally scored higher in taste panel evaluation (Westerling and Hedrick, 1979) and food with high UFA, especially PUFA, is good for human health (Rusman *et al.*, 2003). The ω-6/ω-3 ratio was higher than the value (less than 3) reported by Scollan *et al.* (2006) but lower than that registered by Warren *et al.* (2003) for steers fed corn silage and concentrates (8.9).

It has to be underlined that the fatty acid profile o the food arouse high interest in human medicine due to their influence on the functionality of the cardio-circulatory apparatus. A number of epidemiological researches put in evidence that diets with high content of saturated fatty acids (SFA) were associated with high levels of serum cholesterol (especially of low density lipoprotein, LDL) which appear important in atheroma. Successively, Ulbricht and Southgate (1991) reported that:

1.   diets high in C18:0, stearic acid, do not raise serum cholesterol;
2.   short-chain SFA (C 10 and below) likewise do not raise blood cholesterol, so the putative atherogenic SFA are C12:0 (lauric), C14:0 (myristic) and C16:0 (palmitic). Myristic acid is the most atherogenic, with about four times the cholestrol-raising potential of palmitic acid.

PUFA are considered protective factors: ω-6 fatty acids show mainly anti-atherogenic activity while ω-3 fatty acids have anti-thrombogenic activity. More recently, high prominence is attributed to the role developed by the MUFA, and particularly by the oleic acid that, reducing the oxidation of the cholesterol LDL, may slow the progression of atherosclerosis.

As was pointed out above, the P/S (poliunsaturated/saturated) ratio is not suitable measure of the atherogenicity or thrombogenicity of a diet or foods. Currently they are expressed as follows:

Index of atherogenicity: C12:0 + (4 x C14:0) + C16:0/ ω-3 + ω-6 + MUFA

Index of thrombogenicity: C14:0 + C16:0 +C18:0/(0,5 x C18:1) + (0,5 other MUFA) + 0,5(ω-6) + 3(ω-3) + (ω-3/ ω-6).

The AI of the meat in this trial was particularly interesting, rather lower than the data reported by Ulbricht and Southgate (1991) for raw minced beef and than those reported by Badiani *et al.* (2002) for cooked beef (0.72 and 0.77, respectively); our data were similar to those of Poli *et al.* (1996) on Chianina young bulls (AI: 0.58).

	FB	SB	Significance
SFA	365.1 ± 39.1	330.6 ± 26.9	Ns
MUFA	244.3 ± 39.5	271.7 ± 29.4	Ns
PUFA	186.6 ± 20.5	201.9 ± 15.7	Ns
ω-6	174.2 ± 19.9	188.3 ± 12.9	Ns
ω-3	12.39 ± 1.5	13.57 ± 0.8	Ns
AI	0.54 ± 0.04	0.49 ± 0.07	Ns
TI	1.46 ± 0.12	1.20 ± 0.09	Ns

FB: faba bean; SB: soybean meal solvent extract;AI: atherogenic index; TI thrombogenic index.
ns: not significant.

Table 7. Fatty acid profile of Longissimus thoracis muscle (mg 100g^{-1} of edible part).

However, the index of thrombogenicity in this trial was higher than the findings of the above-cited authors (1.27 and 1.30 for Ulbricht and Southgate, 1991 and Poli et al., 1996, respectively); only the TI (1.77) reported by Badiani et al. (2002) was similar to our data.

These results show that the faba bean could be used as an alternative protein source to soybean meal solvent extract as it did not affect the growth rate (body weight, daily weight gain and biological efficiency of growth) or the feed conversion indexes during the whole experimental period, and offers decided agronomical, economical and healthy advantages.

Nevertheless, in the first period after weaning the faba bean reduced the growth rate, probably due to the higher concentrations of NPN and anti-nutritional factors. It might be useful in this period to use this protein source associated with other richer in rumen undegradable.

Our results contribute to show that both protein sources (soybean meal and faba bean) could be utilised in the diet for young Marchigiana bulls ensuring the high quality of the meat obtained by this breed.

Although the meat of group fed faba bean had significantly higher concentrations of stearic acid compared to the level found in soybean group, neither the atherogenic and thrombogenic indexes, nor the cholesterol content were influenced.

As regards the effect of protein source on organolepitc characteristics, our results on hydroxyproline content and water holding capacity were conflicting. While the meat of the group fed soybean meal solvent extract showed a potential low tenderness (higher level of hydroxiproline) the water holding capacity measured by compression was lower for the group receiving faba bean.

From our results it is also possible to formulate a favourable assessment of the nutritional characteristics of the of meat Marchigiana young bulls. Indeed, the cholesterol values were very close to those indicated by the PGI of the "Vitellone Bianco dell'Appennino Centrale" and lower than those found in other breeds. Moreover, the fatty acids profile of LT confirms that the meat of the Italian breed specialised in meat production has higher unsaturated fatty acids concentration and lower saturated fatty acids levels, which in turn ensures medium-low atherogenic and thrombogenic indexes.

5. Conclusions

The protein source largely used in ruminant nutrition (soybean, soybean meal solvent extract and grain legumes) are characterised by several differences in chemical composition, in particular as concerns crude protein, ether extract and carbohydrates concentrations. Also the technological treatments as flaking and crushing could affect these parameters. In addition, mainly concerning the grain legumes, the contents of different nutrients is affected by the cultivar.

In order to better estimate the nutritive value of protein source, the in vitro gas production technique seems to be the most useful methods, as it allow to study also the kinetics of degradation in the rumen. The knowledge of the latter phenomenon is particularly useful, since the nutritionist may choose to treat a particular protein source in order to synchronize the availability of nitrogen and energy for rumen bacteria.

From our in vitro data, faba bean, lupin and peas, even if from different cultivars, may be considered in replacing, totally or partially, soybean as only few differences were found among these feeds.

In addition, our *in vivo* studies concerning the *infra vitam* and *post mortem* performances of young bulls fed either faba bean or soybean showed:

- no influence of protein source on growth rate (body weight, daily weight gain and biological efficiency of growth) and feed conversion indexes;
- no influence of protein source on the nutritional characteristics of meat. Indeed, even if the meat of group fed faba bean had significantly higher concentrations of stearic acid compared to the level found in soybean group, neither the atherogenic and thrombogenic indexes, nor the cholesterol content were influenced;
- conflicting results on hydroxyproline content and on water holding capacity. Indeed, while the meat of the group fed soybean showed a potential low tenderness (higher level of hydroxiproline), the water holding capacity measured by compression was lower for the group receiving faba bean.

From these results, it can be concluded that the use of grain legumes as a protein source in ruminant diets could be used as alternative to soybean meal. However, mainly in the intensive livestock system the soybean meal solvent extract represents the protein source par excellence for the high protein content (in particular undegradable fraction), the absence of anti-nutritional factors and the extremely favourable biological value of its proteins. Not insignificant, then is the economic assessment, soybean is, especially in the areas where the ruminants breeding has been particularly developed, the protein source with the quality/price ratio more favourable.

By contrast, it has to be underlined that biotech herbicide tolerant soybean continued to be the principal biotech crop in 2010, occupying 73.3 million hectares or 50% of global biotech area. Farm animals are currently fed soybean and soybean meal developed from genetic transformation. Europe is strongly dependent upon the American continent for its protein requirements amounting up to 90 to 95%.

Thus, the possible risk connected to genetically modified organisms use in animal breeding has led to the reconsideration of animal production processes with special reference to the use of alternative protein sources (e.g. faba beans, dried peas, lupine seeds, chickpeas) able to replace soybean. These legumes have agronomic importance because they improve soil fertility and reduce nitrogenous dressing, with positive effects on environmental pollution. Moreover, they need a limited initial investment for their modest requirements of chemical and energetic inputs and their short culture cycle.

6. References

A.S.P.A. (1991). Metodologie relative alla macellazione degli animali di interesse zootecnico ed alla valutazione e dissezione della loro carcasse. Ed. ISMEA, Roma.

Akpinar, N., Akpinar, M.A. & Türkoglu, S. (2001). Total lipid content and fatty acid composition of the seeds of some Vicia L. species. *Food Chem.*, 74, pp. 449–453.

Arieli, A., Bruckental, I., Kedar. & I., Sklan, D. (1995). *In sacco* disappearance Of starch nitrogen and fat in processed grains. *Anim. Feed sci. Techn.*, 51, pp. 287-295.

Bach Knudsen, K.E. (1997). Carbohydrate and lignin contents of plant materials used in animal feeding. *Anim. Feed Sci. Technol.*, 67, pp. 319–338.

Badgley, C., Moghtader, J., Quintero, E., Zakem, E., Chappell, M.J., Avilés-Vázquez, K., Samulon, A. & Perfecto, I., (2007). Organic agriculture and the global food supply. *Renew. Agric. Food Sys.*, 22, pp. 86–108.

Badiani, A., Stipa, S., Bitossi, F., Gatta, P.P., Vignola, G. & Chizzolini, R., (2002). Lipid composition, retention and oxidation in fresh and completely trimmed beef muscles as affected by common culinary practices. *Meat Sci.*, 60, pp. 169-186.
Bézard, J., Blond, J.P., Bernard, A. & Clouet, P., (1994). The metabolism and availability of essential fatty acids in animal and human tissues. *Reprod. Nutr. Dev.*, 34, pp. 539-568
Bittante, G., Chies, L., Fasone, V., Sinatra, M.C. & Gallo, L. (1989). Degradabilità e digeribilita *in vitro* della sostanza organica di cereali fioccati e/o macinati. *Zoot. Nutr. Anim.*, 15, pp. 613-626.
Brand, T.S., Brandt, D.A. & Cruywagen, C.W. (2004). Chemical composition, true metabolisable energy content and amino acid availability of grain legumes for poultry. *South African Journal of Animal Science*, 34, pp. 116-122.
Boccioni, A., Minieri, S., Petacchi, F. & Antongiovanni, M. (2007). Il favino (*Vicia faba minor*) e il pisello proteico (*Pisum sativum*) in sostituzione della farina di estrazione di soia nell'alimentazione zootecnica. Valutazione comparativa delle caratteristiche nutrizionali con metodi di laboratorio. *Arsia*, Toscana.
Bulson, H.A.J., Snaydon, R.W. & Stopes, C.E., (1997). Effects of plant density on intercropped wheat and field beans in an organic farming system. *J. Agric. Sci.*, 128, pp. 59-71.
Cabrera, C., Lloris, F., Giménez, R., Olalla, M. & López, M.C. (2003). Mineral content in legumes and nuts: contribution to the Spanish dietary intake. *The Science of total environment*, 308, pp. 1-14.
Calabrò et al. (2010) www.progettobufale.it/pubblicazionenew.html
Calabrò, S., Marchiello, M. & Piccolo, V. (2001a). *In vitro* fermentation of some cereals and legumes grains. *Proc. XIII Congr. Naz. ASPA*, Firenze (Italy), pp. 153-155.
Calabró, S., Piccolo, V., Bovera, F. & Sarubbi, F. (2001b). *In vitro* fermentation kinetics of crushed and flaked grains. *Proc. VI World Buffalo Congr.*, Maracaibo (Venezuela), pp. 478-484.
Calabrò, S., Tudisco, R., Balestrieri, A., Piccolo, G., Infascelli, F. & Cutrignelli, M.I. (2009). Fermentation characteristics of different grain cultivars with the *in vitro* gas production technique. *Ita, J. of Anim. Sci.*, 8, pp. 280-282.
Canibe, N., Bach Knudsen, K.E., (2002). Degradation and physicochemical changes of barley and pea fibre along the gastrointestinal tract of pigs. *J. Sci. Food Agric.*, 82, pp. 27-39.
Carnovale, E. & Nicoli, S., (2000). Changes in fatty acid composition in beef in Italy. *J. Food Compos. Anal.*, 13, pp. 505-510.
Chalk, P.M., Smith, C.J., Hamilton, S.D. & Hopmans, P., (1993). Characterization of the N benefit of a grain legume (Lupinus angustifolius L.) to a cereal (Hordeumvulgare L.) by an in situ 15N isotope dilution technique. *Biol. Fertil. Soils*, 15, pp. 39-44.
Chaubility, U.B., Gupta, R., Bird, S.H. & Mudgal, V.D., (1991). In sacco dry matter disappearance of different protein cakes in the rumen of male buffalo and their solubility in water. *Proceedings of the 3rd World Buffalo Congress*, Paestum (Italy), p. 165, Abstract.
Chumpawadee, S., Chantiratikul, A. & Chantiratikul, P. (2007). Chemical composition and nutritional evaluation of protein feeds for ruminants using an *in vitro* gas production technique. *Journal of Agricultural Technology* pp. 191-202
Cifuni, G.F., Napolitano F., Riviezzi, A.M., Braghieri, A. & Girolami, A., (2004). Fatty acid profile, cholesterol content and tenderness of meat from Podolian young bulls. *Meat Sci.*, 67, pp. 289-297.

Colombini, S.(2004). Resa proteica di alcune varietà di pisello, fava e lupino nell'Italia settentrionale. *Informatore agrario*, 60, pp. 73-75

Cone, J.W., van Gelder, A.H., Visscher, G.J.W. & Oudshoorn, L. (1996). Influence of rumen fluid and substrate concentration on fermentation kinetics measured with a fully automated time related gas production apparatus. *Animal Feed Science Technology*, 61, pp. 113-128.

Council Regulation (EC) No 834/2007 on organic production and labelling of organic products and repealing Regulation (EEC) No 2092/91 L 189/1- L 189/23.

Council Regulation (EEC) No 2092/91 of 24 June 1991 on organic production of agricultural products and indications referring thereto on agricultural products and foodstuffs L 198 , 22/07/1991 P. 0001 – 0015

Cutrignelli, M.I., (2000). Caratteristiche del taglio campione di mezzene di vitelloni Podolici sottoposti a differenti sistemi di allevamento. *Proc. Nat. Congr: Parliamo di... on Breeding in the third Millennium*, Fossano (CN), Italy, pp. 127-134

Cutrignelli, M.I., Piccolo, G., Bovera, F., Calabrò, S., D'Urso, S., Tudisco, R. & Infascelli, F. (2008a) Effects of two protein sources and energy level of diet on the performance of young Marchigiana bulls. 1. Infra vitam performance and carcass quality. *Ital.J.Anim.Sci.*, 7, pp. 259-270

Cutrignelli, M.I., Calabrò, S., Bovera, F., Tudisco, R., D'Urso, S., Marchiello, M., Piccolo, V. & Infascelli, F. (2008b) Effects of two protein sources and energy level of diet on the performance of young Marchigiana bulls. 2. Meat qualità. *Ital.J.Anim.Sci.*, 7, pp. 271-285.

Degussa, (2006). The amino acid composition of feedstuffs. 5th rev. ed., Degussa AG, Feed Additives, Hanau, Germany.

Di Francia, A., De Rosa, G., Masucci, F., Romano, R., Borriello, I. & Grassi, C. (2007). Effect of Pisum sativum as protein supplement on buffalo milk production. *Ital.J.Anim.Sci.*, 6, pp. 472-475

Díaz –Ambrona, C.H. & Mínguez, M.I. (2001). Cereal-legume rotations in a Mediterranean environment: biomass and yield production. *Field Crops Research*, 70, pp. 139- 51

DLG, (1999). Kleiner Helfer für die Berechnung von Futterrationen. Wiederkäuer und Schweine. DLG-Verlag, 10. Auflage, Frankfurt am Main, Germany.

Erbas, M., Certel, M. & Uslu, M.K., (2005). Some chemical properties of white lupin seeds (Lupinus albus L.). *Food Chem.*, 89, pp. 341–345.

European Commission directive, 999/2001. Regulation (EC) No. 999/2001 of the European Parliament and of the Council of 22 May 2001 laying down rules for the prevention, control and eradication of certain transmissible spongiform encephalopathies. *Official Journal of the European Communities* L 147/1–L 147/40.

European Commission, 1992. Council Regulation of 14 July 1992 concerning the protection of geographical indications and designations of origin for agricultural products and foodstuffs, 92/2081/EC. *Official Journal*, L 208, 24/07/1992, pp. 1-8.

European Commission, (2004). European Parliament and Council Regulation of the of 29 April 2004 concerning the official controls performed to ensure the verification of compliance with feed and food law, animal health and animal welfare rules, 882/2004/EC. *Official Journal*, L 165, 30/04/04, pp 1-141.

Evans, J., McNeil', A.M., Unkovich, M.J., Fettell, N.A. & Heenan, D.P., (2001). Net nitrogen balances for cool-season grain legume crops and contributions to wheat nitrogen uptake: a review. *Aust. J. Exp. Agric.*, 41, pp. 347–359.

Floroni, A. (2002). I.G.P. "Vitellone Bianco dell'Appennino Centrale". *L'Allevatore*, 58, pp. 12-13.

Froud-Williams, R.J., (1988). Changes in weed flora with different tillage and agronomic management systems. In: Altieri, M.A. & Liebman, M. (Eds.), *Weed Management in Agroecosystems: Ecological Approaches*. CRC, Boca Raton, Florida, pp. 213-236.

Gatel, F., (1994). Protein quality of legume seeds for non-ruminant animals: a literature review. *Anim. Feed Sci. Technol.*, 45, pp. 317-348.

Getachew, G., Robinson, P.H., DePeters, E.J. & Taylor, S.J. (2004). Relationships between chemical composition, dry matter degradation and *in vitro* gas production of several ruminant feeds. *Animal Feed Science Technology*, 111, pp. 57-71.

Gill, C., (1997). World feed panorama. High cost of feedstuffs: global impact, response. *Feed International*, 18, pp. 6-16.

Green, C.J. & Blackmer, A.M., (1995). Residue decomposition effects on nitrogen availability to corn following corn or soybean. *Soil Sci. Soc. Am. J.*, 59, pp. 1065-1070.

Groot, J.C.J., Cone, J.W., Williams, B.A. & Debersaques, F.M.A., (1996). Multiphasic analysis of gas production kinetics for *in vitro* fermentation of ruminant feedstuff. *Anim. Feed Sci. Tech.*, 64, pp. 77-89.

Herrero, M., I. Murray, R.H. Fawcett & Dent, J.B. (1996). Prediction of *in vitro* gas production and chemical composition of kikuyu grass by near-infrared reflectance spectroscopy. *Animal Feed Science Technology*, 60, pp. 51-67.

Herridge, D.F., Macellos, H., Felton, W.L., Turner, G.L. & Peoples, M.B., (1995). Chickpea increases soil-N fertility in cereal systems through nitrate sparing and N₂ fixation. *Soil Biol. Biochem.*, 27, pp. 545-551.

Huisman, J. & Tolman, G.H., (2001). Antinutritional factors in the plant proteins of diets for non-ruminants. In: Garnsworthy, P.C. & Wiseman, J. (Eds.), Recent Developments in Pig Nutrition, vol. 3. Nottingham University Press, Nottingham, pp. 261-322.

Infascelli, F., Di Lella, T. & Piccolo, V., (1995). Dry matter, organic matter and crude protein degradability of high protein feeds in buffaloes and sheep. *Zoot. Nutr. Anim.*, 21, pp. 89-94.

INRA (1988). Alimentation des bovins, ovins & caprins. Paris.

Jezierny, D., Mosenthin, R. & Bauer, E. (2010). The use of grain legumes as a protein source in pig nutrition: A review. *Animal Feed Science and Technology*, 157, pp. 111-128.

Jezierny, D., (2009). *In vivo* and *in vitro* studies with growing pigs on standardised ileal amino acid digestibilities in grain legumes. Ph.D. Thesis. University of Hohenheim, Stuttgart, Cuvillier Verlag Göttingen, Germany.

Jezierny, D., Mosenthin, R., Eklund, M. & Rademacher, M., (2007). Determination of standardized ileal digestibilities of crude protein and amino acids in legume seeds for growing pigs. *Proceedings of the 16th International Science Symposium on Nutrition of Domestic Animals*, Radenci, pp. 198-203.

Keane, M.G., (2003). Beef production from Holstein-Friesian bulls and steers of New Zeland and European/American descent, and Belgian Blue x Holstein-Fresians slaughtered at two weights. *Livest. Prod. Sci.*, 84, pp. 207-218.

Kim, C.M., Kim, J.H., Chung, T.Y. & Park, K.K., (2004). Effect of flaxseed diets an fattening response of Hanwoo cattle: 2 fatty acid composition of serum and adipose tissues. *Asian-Aust. J. Anim. Sci.*, 9, pp. 1246-1254.

Lallès, J.P. & Jansman, A.J.M., (1998). Recent progress in the understanding of the mode of action and effects of antinutritional factors from legume seeds in non-ruminant farm animals. In: Jansman, A.J.M., Hill, G.D., Liebman, M. & Dyck, E.,. Crop rotation and intercropping strategies for weed management. *Ecol. Appl.* 3, pp. 92-122.

Liebman, M. & Janke, R., (1990). Sustainable weed management practices. In: Francis, C.A., Flora, C.B. & King, L.D. (Eds.), Sustainable Agriculture in Temperate Zones. John Wiley&Sons, New York, pp. 111–143.

López-Bellido, F.J., López-Bellido, L. & López-Bellido, R.J., (2005). Competition, growth and yield of faba bean (Vicia faba L.). Eur. J. Agron., 23, pp. 359–378.

Medsen, J., T. Hvelplund & Weisbjerg, M.R. (1997). Appropriate method for the evaluation of tropical feeds for ruminants. Animal Feed Science Technology, 69, pp. 53-66.

Morbidini, L., Rossetti, E., Cozza, F. & Pauselli, M. (2005). Different protein source (soybean or faba bean) in postweaning diets for Apennine and Sopravissana (Italian Merino) light lamb: slaughtering performances. Ital.J.Anim.Sci., 4, pp. 360-362.

Morbidini, L., Pauselli, M., Rossetti, E. & Cozza, F., (2004). Diete post-svezzamento con differente fonte proteica (favino o soja) per la produzione di agnelli leggeri di razza Appenninica e Sopravissana: rilievi in vivo. Proc. Congr. "Parliamo di ... nuove normative in campo zootecnico", Cuneo, Italy, pp. 113-122.

Moss, A.R., Givens, D.I., Grundy, H.F. & Wheeler, K.P.A., (1997). The nutritive value for ruminants of lupin seeds from determinate plants and their replacement of soybean meal in diets for young growing cattle. Anim. Feed Sci. Tech., 68, pp. 11-23.

Mould, F.L., Kliem, K.E. Morgan, R. & Mauricio, R.M, (2005). In vitro microbial inoculum: A review of its function and properties Anim. Feed Sci. Technol., 123, pp. 31-50

Murphy, S.R. & McNiven, M.A., (1994). Raw or roasted lupin supplementation of grass silage diets for beef steers Anim. Feed Sci. Tech., 46, pp. 23-35.

NRC (2001). Nutrient Requirements of Dairy Cattle: Seventh Revised Edition, 2001. National Academy Press, Washington, D.C.

NRC, (1998). Nutrient requirements of swine. 10th ed., National Academy Press, Washington, D.C.

Pen, B., Oyabu, T., Hidaka, S. & Hidari, H., (2005). Effect of potato by-products based silage on growth performance, carcass characteristics and fatty acid composition of carcass fats in Holstein steers. Asian-Aust. J. Anim. Sci., 4, pp. 490-496.

Poli, B.M, Giorgetti, A., Bozzi, R., Funghi, R., Balò, F.& Lucifero, M., (1996). Quantity and quality of lipid fractions for human nutrition in Chianina muscles as influenced by age and nutritive level. In: G. Enne and G.F. Greppi (eds.) Food & Health: role of animal products, Proc. 31st Int. Symp. of Società Italiana per il Progresso della Zootecnia held in Milano, Italy, Elsevier publ., Paris, France, pp. 199-204.

Raes, .K., Balcaen, A, Dirinck, P., De Winne, A., Claeys, E., Demeyer, D. & De Smet, S., (2003). Meat quality, fatty acid composition and flavour analysis in Belgian retail beef. Meat Sci., 65, pp. 1237-1246.

Rubio, L.A. & Brenes, A., (1995). Utilization de leguminosas-grano en nutricion animal: problemas y perspectives. Proc. XI Curso de Especializacion FEDNA, Barcelona, Spain. www.cirval.asso.fr.

Rusman, S. & Setiyono, A.S., (2003). Characteristics of Biceps femoris and Longissimus thoracis muscles of five cattle breeds grown in a feedlot system. J. Anim. Sci., 74, pp. 59-65.

Salgado, P., Freire, J.P.B., Mourato, M., Cabral, F., Toullec, R. & Lallès, J.P., (2002a). Comparative effects of different legume protein sources in weaned piglets: nutrient digestibility, intestinal morphology and digestive enzymes. Livest. Prod. Sci., 74, pp. 191–202.

Salgado, P., Martins, J.M., Carvalho, F., Abreu, M., Freire, J.P.B., Toullec, R., Lallès, J.P. & Bento, O., (2002b). Component digestibility of lupin (Lupinus angustifolius) and pea (Pisum sativum) seeds and effects on the small intestine and body organs in anastomosed and intact growing pigs. *Anim. Feed Sci. Technol.*, 98, pp. 187–201

Sami, A.S., Augustini, C. & Schwarz, F.J., (2004). Effect of feeding intensity and time on feed on performance, carcass characteristics and meat quality of Simmental bulls. *Meat Sci.*, 67, pp. 195-201.

Sankara Rao, D.S. &, Deosthale & Y.G. (2006). Mineral Composition of Four Indian Food Legumes. *Journal of Food Science*, 46, pp. 1962–1963

Sarubbi, F. (1999). Studi sull'impiego di proteasi tioliche per la previsione della degradabilita ruminale delle proteine e sule modificazioni dell' attivita ruminale indotte nei bufali da razioni arricchite con *Aspergillus oryzae*. PhD Thesis, Univ. Napoli, 1-23. 1999.

Schofield, P. (2000). Gas production methods. In "Farm Animal Metabolism and Nutriion", J.P.F. D'Mello, ed. CAB International, Wallingford, Oxon, U.K, pp. 209-232.

Scollan, N., Hocquette, JF., Neuerbetg, K., Dannenberger, D, Richardson, I. & Moloney, A., (2006). Innovations in beef production system that enhance the nutritional and health value of beef lipids and their relationship with meat quality. *Meat Sci.*, 74, pp. 179-33.

Singh, R.J., Chung, G.H. & Nelson, R.L., (2007). Landmark research in legumes. *Genome*, 50, pp. 525–537.

Sniffen, C. J., J. D. O'Connor, P. J. Van Soest, D. G. Fox & J. B. Russell. (1992). A net carbohydrate and protein system for evaluating cattle diets: 11. Carbohydrate and protein availability. *J. Anim. Sci.*, 70, pp. 3562.

Sprent, J.I. & Thomas, R.J., (1984). Nitrogen nutrition of seedling grain legumes: some taxonomic, morphological and physiological constraints. *Plant Cell Environ.*, 7, pp. 637–645.

Theander, O., Westerlund, E., Åman, P. & Graham, H., (1989). Plant cell walls and monogastric diets. *Anim. Feed Sci. Technol.*, 23, pp. 205–225

Theodorou, M.K. (1994.) A new laboratory procedure for determining the fermentation Kinetics of ruminants feeds. *Ciencia e Investigcion Agraria*, 20, pp. 332-334

Trostle, R., (2008). Global agricultural supply and demand: Factors contributing to the recent increase in food commodity prices. A Report from the Economic Research Service. United States Department of Agriculture Economic Research Service, Washington, DC

Ulbricht, T.L.V. & Southgate, D.A.T., (1991). Coronary heart disease: seven dietary factors. *Lancet*, 338, pp. 985-992.

Warren, HE., Enser, M., Richardson, I., Wood, JD.& Scollan, N.D., (2003). Effect of breed and diet on total lipid and selected shelf-life parameters in beef muscle. *Proc Nat. Congr. BSAS*, York, UK. pp. 43.

Westerling, D.B. & Hedrick, H.B., (1979). Fatty acid composition of bovine lipids as influenced by diet, sex and anatomical location and relationship to sensory characteristics. *J. Anim. Sci.*, 48, pp. 1343-1348.

Wheeler, T.L., Davis, G.W., Stoeker, B.J.& Hatmon, C.J., (1987). Cholesterol concentration of Longissimus muscle, subcutaneous fat and serum of two beef cattle breed type. *J. Anim. Sci.*, 65, pp.1531-1537.

Wood, J.D., Richardson, R.I., Nute, G.R., Fisher, A.V., Campo, M.M., Kasapidou, E., Sheard, P.R. & Enser, M., (2003). Effects of fatty acids on meat quality: a review. *Meat Sci.*, 66, pp. 21–32.

Zahran, H.H., (1999). Rhizobium-legume symbiosis and nitrogen fixation under severe conditions and in an arid climate. *Microbiol. Mol. Biol. Rev.*, 63, pp. 968–989.

9

Soybean as a Feed Ingredient for Livestock and Poultry

H.K. Dei

Department of Animal Science, Faculty of Agriculture,
University for Development Studies, Tamale,
Ghana

1. Introduction

The need to meet animal protein demand of ever growing world population, currently at approximately 6.8 billion (US Census Bureau, 2010), is set to increase at an even greater rate as the economies of developing countries improve and their growing affluent populace alter their dietary habits. This means production of soybean, which is used extensively as animal feed, must increase beyond current production level of about 246 million metric tonnes (FAS/USDA, 2009).

Soybean (*Glycine max,* L) is not only a source of high quality edible oil for humans, but also a high quality vegetable protein in animal feed worldwide. Its universal acceptability in animal feed has been due to favourable attributes such as relatively high protein content and suitable amino acid profile except methionine, minimal variation in nutrient content, ready availability year-round, and relative freedom from intractable anti-nutritive factors if properly processed. Also, attention has been focused on soybean utilisation as an alternate protein source in animal diets due to the changing availability or allowed uses of animal proteins coupled with relatively low cost.

Despite soybean's pivotal role in animal production, it cannot be fed raw because there are a number of anti-nutritive factors (ANFs) present that exert a negative impact on the nutritional quality of the protein. The main ANFs are protease inhibitors (trypsin inhibitors) and lectins (Liener, 1994), which fortunately can be destroyed by heat treatment. The trypsin inhibitors cause pancreatic hypertrophy/hyperplasia with consequent inhibition of growth, while lectins inhibit growth by interfering with nutrient absorption (Liener, 1994). The elimination of these ANFs and those of less significance can be achieved through various processing methods. These methods have different impact on the nutritional quality of the products derived such as full-fat soybeans, soybean meal and soybean protein concentrates. Of these, soybean meal has been the major ingredient in both poultry and livestock diets.

This chapter discusses soybean production and consumption, primary soybean products and their nutritional value for feeding animals, anti-nutritive factors present and ways of eliminating them, and utilisation in animal feeds as well as future challenges of using soybeans as a major source of animal feed.

2. Soybean production and consumption

Soybean (*Glycine max*, L) is an annual crop that belongs to the Fabaceae or Leguminosae family. It originated from East Asia, but now grown over a wide geographical area worldwide with United States of America, Brazil and Argentina being the leading producers (Table 1). It is used primarily for production of vegetable oil and oilseed meal for animal feeding. The surge in the use of soybean meal in feeding animal as replacement protein source for animal protein feeds has been the main driving force in soybean production.

Table 1 shows the major soybean producing countries and their relative supplies. Generally, there has been an increase of supply with a slight depression in most producing countries between 2006 and 2008 cropping seasons. The US and China tend to consume virtually what they produce, while Argentina and Brazil are major exporters with exports largely to the EU (Table 2).

Major producing countries	2005/06	2006/07	2007/08	2008/09	2009/10 October
United States	83,507	87,001	72,859	80,749	88,454
Brazil	57,000	59,000	61,000	57,000	62,000
Argentina	40,500	48,800	46,200	32,000	52,500
China	16,350	15,967	14,000	15,500	14,500
India	7,000	7,690	9,470	9,100	9,000
Paraguay	3,640	5,856	6,900	3,900	6,700
Canada	3,161	3,460	2,700	3,300	3,500
Other	9,512	9,337	8,004	9,090	9,413
World Total	220,670	237,111	221,133	210,639	246,067

Source: FAS/USDA (2009)

Table 1. World soybean supply (million tonnes) and distribution.

Countries	2009/10		2010/11	
	Production	Consumption	Production	Consumption
China	37.42	35.82	41.71	40.36
US	37.31	27.22	35.41	27.58
Argentina	27.13	0.70	29.95	0.60
Brazil	24.41	12.80	25.42	13.38
EU	9.85	31.49	9.77	32.30
India	4.85	2.85	6.08	3.08
Others	20.58	49.60	21.30	51.20
World Total	161.63	159.77	169.64	167.89

Source: FAS/USDA (2009)

Table 2. World soybean meal production and consumption outlooks for 2010/11 in million tonnes.

3. Primary soybean products for animal feeding

Figure 1 shows a schematic processing of soybeans into various high quality protein products. The processes involved either reduce or eliminate the ANFs in the beans and improve the nutritional value substantially for all classes of animals. Several steps involved in processing these products can have either positive or negative effect on the quality of the protein depending on the conditions used in processing. The heat applied in processing is identified as the single most important factor that affects soybean meal protein quality. Proper processing conditions such as moisture content, heating time and temperature inactivate ANFs such as trypsin inhibitors and lectins, which results in improved performance when fed to monogastric animals (Araba, 1990). High processing temperatures of oilseeds has deleterious effects on proteins and amino acids due to formation of Maillard reaction products (Hurell, 1990) or denaturation (Parsons *et al.*, 1992).

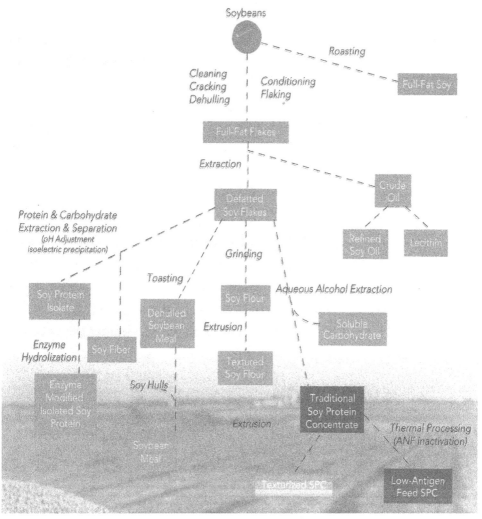

Fig. 1. Processing of soybeans into soybean products (USSEC, 2008).

3.1 Full-fat soybeans

These are whole soybeans in which the oil is not extracted. These products are produced by a variety of processes such as extruding (dry or wet), cooking/autoclaving, roasting/toasting, micronizing and jet-sploding to inactivate the ANFs. All of these processes have a different impact on the nutritive value of the products depending on heat damage or degree of inactivation of ANFs. Normally, soybeans are processed into defatted meals for feed formulation, particularly for poultry and pigs. However, the amount of full-fat soybeans used has been increasing in the livestock industry due to development of new varieties with limited number or levels of ANFs (Gu et al., 2010). Also, properly processed full-fat soybeans are a valuable feed ingredient for animal feeding because of their high energy content.

3.2 Soybean meal

Soybeans yield 18.6% of oil and 78.7% of soybean meal with the rest being waste (FEFAC, 2007). The oil can be extracted either mechanically or by solvent means. There are two main types of soybean meal. The dehulled soybean meal and soybean meal, depending on whether the testa (seed coat) is removed or not. Both products vary in their nutrient composition, but are quite high in protein content with a good amino acid balance except methionine, low in fibre, high in energy, and have little or no anti-nutritive factors when properly processed.

The amino acid profile of soybean meal is close to that of fishmeal, except methionine (INRA, 2004). This deficiency can easily be corrected in monogastric diets using synthetic source of methionine. Also, soybean meal is superior to other vegetable protein sources in terms of crude protein content and matches or exceeds them in both total and digestible amino acid content (Table 3a). Soybean meal protein digestibility in poultry is approximately 85% (Woodworth et al., 2001), ranging between 82% and 94% for individual amino acid digestibility. Among the vegetable protein sources, soybean meal is used to meet the animal's requirement for limiting amino acids in cereal-based (e.g. maize) diets (Table 3b), because it is usually the most cost-effective source of amino acids (Kerley and Allee, 2003).

The carbohydrates in soybean meal are incompletely digested by colonic microbiota in monogastrics (Kerley and Allee, 2003). Thus removal of raffinose and stachyose improved metabolisable energy content by 12% (Graham et al., 2002).

3.3 Soybean protein concentrate (SPC)

SPC is produced from the defatted flakes by the removal of the soluble carbohydrates. This can be achieved by two methods, either by ethanol extraction or enzymatic degradation (Figure 1). SPC is valuable as milk replacer feed for calves and as piglet pre-starter feed. This is because it contains only traces of the heat-stable oligosaccharides and the antigenic substances (Table 5). In milk replacer feed, it has been largely substituted for dried skim milk; whilst in pig starter feeds it can replace dried skim milk, whey powder and fishmeal.

3.4 Soybean oil

Soybean oil is produced primarily for human consumption. However, it has become a useful source of feed-grade fat for animals due to a need to formulate high-energy diets for modern breeds. Feed-grade soybean oil is popularly used in high energy diets, particularly for

	Soybean	Canola	Cottonseed	Palm kernel	Peanut	Sunflower
Crude protein	43.0	36.2	39.6	13.2	45.2	32.8
Amino acid						
Lysine	90.7	78.6	62.8	58.9	78.1	80.4
Methionine	90.6	88.6	71.9	83.7	85.6	91.2
Cystine	82.1	73.1	70.9	66.6	78.5	79.2
Threonine	84.1	77.6	67.2	69.2	83.8	83.7
Tryptophan	87.9	80.0	80.3	-	75.6	-
Arginine	91.1	90.6	85.3	88.6	89.6	93.1
Isoleucine	91.2	89.0	72.8	81.0	89.3	88.9
Leucine	90.7	94.1	74.8	85.0	89.7	88.7
Valine	88.9	87.8	76.3	80.1	88.9	85.8
Histidine	88.5	88.5	64.1	80.3	85.4	86.1
Phenylalanine	91.6	91.6	84.0	85.3	92.3	90.8

Source: Ajinomoto Heartland Lysine LLC Revision 7 (2006)

Table 3a. True digestibility (%) of essential amino acids in common oilseed meal proteins for poultry.

Amino Acid[1]	Soybean meal (475.0 g/kg CP)	Palm kernel meal (200.0 g/kg CP)	Maize (85.0 g/kg CP)
Arginine	73.3	135.0	44.7
Histidine	26.9	23.0	27.1
Isoleucine	44.6	32.0	34.1
Leucine	78.7	60.0	117.6
Lysine	62.3	36.0	30.6
Methionine	14.1	20.0	21.2
Cystine	15.2	15.0	21.2
Phenylalanine	49.3	39.0	44.7
Threonine	39.4	35.0	34.1
Tryptophan	15.6	10.0	7.1
Valine	46.7	57.0	47.0

[1]Data are adapted from Elkin (2002).

Table 3b. Comparative amino acid composition (g/kg protein basis) of soybean meal with palm kernel meal and maize.

poultry, because of its high digestibility and metabolisable energy content compared with other vegetable fats/oils (Table 4a). It is used widely in rations for broiler chickens and growing turkeys as a feed-grade fat to increase energy density of feeds and improve efficiency of feed utilisation (Sell *et al.*, 1978). The high energy value of soybean oil is attributed to its high percentage of (poly) unsaturated fatty acids (Table 4b), which are well absorbed and utilised as a source of energy by the animal (Huyghebaert *et al.*, 1988). Also, the high polyunsaturated fatty acids (PUFA) in soybean oil appears to have an energy independent effect on improving reproduction in dairy cattle (Lucy *et al.*, 1990; Kerley and Allee, 2003), and this has been attributed to the role of linoleic acid in reproduction (Staples *et al.*, 1998).

Source	Digestibility (%)		Metabolisable energy (MJ/kg)	
	3-4 weeks	>4 weeks	1-3 weeks	7.5 weeks
Soybean oil	96[+]	96[+]	38.5[*]	38.5[*]
Corn (maize) oil	84[+]	95[+]	-	41.3[#]
Lard	92[+]	93[+]	30.8[***]	-
Beef tallow	70[+]	76[+]	30.9[*]	32.9[*]
Menhaden oil	88[+]	97[+]	35.9[#]	37.6[#]
Palm oil	74[**]	-	27.7[*]	32.3[*]
Sunflower oil	85[****]	88[****]	-	40.4[#]

[+]Leeson and Summers (2001), [*]Wiseman and Salvador (1991), [**] Zumbado et al. (1999), [***]Huyghebaert et al. (1988), [****]Ortiz et al. (1998), [#]NRC (1994)

Table 4a. Comparison of digestibility and metabolisable energy values of triglycerides in broiler chickens fed soybean oil and selected dietary fats/oils.

Fatty acid	Soybean oil[2]	Palm oil[2]	Sunflower Oil[3]	Corn Oil[4]	Tallow[2]	Lard[5]	Poultry oil[4]
Lauric acid (C12:0)	1	3	-	-	2	-	-
Myristic acid (C14:0)	2	14	-	-	23	16	6
Palmitic acid (C16:0)	161	488	60	112	249	224	232
Palmitoleic acid (C16:1)	6	1	-	-	39	21	71
Stearic acid (C18:0)	61	55	64	21	206	177	64
Oleic acid (C18:1)	251	364	284	269	405	461	430
Linoleic acid (C18:2)	452	73	581	579	66	80	179
Linolenic acid (C18:3)	66	2	1	8	10	21	6
Arachidic acid (C20:0)	-	-	6	5	-	-	2

[1]Values may not total 1000 g due to trace amounts of other fatty acids not reported or rounding of figures
[2]Wiseman and Salvador (1991), [3]Ortiz et al. (1998), [4]Waldroup et al. (1995), [5]Huyghebaert et al. (1988)

Table 4b. Comparison of fatty acid composition of soybean oil with selected dietary fats/oils (g/kg total fatty acids)[1].

4. Chemical composition of commonly used soybean products in animal diets

There are variations in the reported chemical composition of soybean products that can be attributed to differences in processing methods (Table 5). Also, genetic variations have been observed in the soybean biotypes of *Glycine* (Yen et al., 1971; Gu et al., 2010), which may vary in their chemical compositions. The use of soybean products in non-ruminant diets can give reasonable performance only if diets are formulated correctly or their anti-nutritive factors removed. In this regard, nutrient levels, bioavailability, and anti-nutritive factors and their effects on animal performance must all be considered in determining the usefulness of any of the soybean products as a feed ingredient. Table 5 shows composition of some soybean products commonly used in animal feed. It is clear that soybean is a source of high protein content and quality as well as energy with little or no ANFs. It appears the quality of soybean proteins improves when subjected to multiple processing procedures. This is shown by increases in concentrations of limiting essential amino acids such as lysine and methionine for monogastric animals (Table 5). However, the cost of such improved products may limit their use in animal feeds.

	Full-fat soybean	Soybean Meal	Soy protein concentrate	Soy protein isolate
Dry matter	89.4	87.6 - 89.8	91.8	93.4
Crude protein	37.1	43.9 - 48.8	68.6	85.9
Crude fibre	5.1	3.4 - 6.3	1.7	1.3
Ether extract	18.4	1.3 - 5.7	2.0	0.6
Ash	4.9	5.7 - 6.3	5.2	3.4
NDF	13.0	10.0 - 21.4	13.5	-
ADF	7.2	5.0 - 10.2	5.4	-
ADL	4.3	0.4 - 1.2	0.4	-
Starch	4.7	3.3 - 7.0	-	-
Total sugars	-	9.1 - 9.3	-	-
Gross energy (MJ/kg)	20.95	17.22 – 17.41	17.89	22.45
Lysine	2.34	2.85 - 3.50	4.59	5.26
Methionine	0.52	0.62 - 0.80	0.87	1.01
Cystine	0.55	0.68 - 0.77	0.89	1.19
Tryptophan	0.49	0.56 - 0.74	0.81	1.08
Calcium	0.26	0.27 - 0.31	0.24	0.15
Phosphorus	0.57	0.64 - 0.66	0.76	0.65
Linoleic acid	9.7	0.6 - 2.9	-	-
Urease activity (pH-rise)	2.0	0.05 - 0.5	<0.05	<0.05
Trypsin inhibitor (mg/g)	45-50	1 – 8	2	<1
Glycinin (ppm)	180.000	66.000	<100	-
β-conglycinin (ppm)	>60.000	16.000	<10	-
Lectins (ppm)	3.5000	10 – 200	<1	0
Oligosaccharides (%)	14	15	3	0
Saponins (%)	0.5	0.6	0	0

Data are adapted from NRC (1994), INRA (2004), Peisker (2001)

Table 5. Per cent composition of some soybean products used in animal feed.

4.1 Anti-nutritive factors

Anti-nutritive factors are natural compounds in feedstuffs that impair utilisation of nutrients with consequent undesirable effects on animal performance. The ANFs in soybeans exert a negative impact on the nutritional quality for animals (Table 6). Fortunately, those ANFs with significant impact such as trypsin inhibitors and lectins are easily destroyed by heat. Of lesser significance are the anti-nutritional effects produced by relatively heat stable factors, such as goitrogens, tannins, phytoestrogens, oligosaccharides, phytate, and saponins (Liener, 994). Heat stable ANFs with the exception of oligosaccharides and the antigenic factors are low in soybeans and not quite likely to cause problems under practical feeding conditions. The removal of the oligosaccharides and antigens in the manufacture of soybean protein concentrates further improves the nutritional value.

Anti-nutritional factor	Mode of action	Method of detoxification
Protease inhibitors	Combines with trypsin or chymotrypsin to form an inactive complex and lower protein digestibility Causes hypertrophy of the pancreas Counteracts feedback inhibition of pancreatic enzyme secretion by trypsin	Heat treatment Germination Fermentation
Lectins (Phytohaemagglutinins)	Agglutinates red blood cells	Heat treatments
Anti-vitamin factors (rachitogenic factor and anti-vitamin B12 factor)	These factors render certain vitamins (e.g. vitamins A, B_{12}, D, and E) physiologically inactive	Cooking Supplementation of vitamins
Goitrogens	Enlargement of the thyroid	Heat treatment in some cases Administration of iodide
Metal-binding factors (phytate)	These factors decrease availability of certain minerals (e.g. P, Cu, Fe, Mn, Zn)	Heat treatment Addition of chelating agents Use of enzymes
Saponins	Bitter taste, hemolyze red blood cells	Fermentation
Estrogens	Cause an enlargement of the reproductive tract	
Cyanogens	Cause toxicity through the poisonous hydrogen cyanide	Cooking
Oligosaccharides	Impair digestion (e.g. intestinal cramps, diarrhoea, and flatulence)	Ethanol/water extraction
Antigens (glycinin and β-conglycinin)	Cause the formation of antibodies in the serum of calves and piglets. Prevent proliferation of beneficial bacteria in the GIT	Ethanol/water extraction

Sources: Liener (1977), Ensminger and Olentine Jr (1978), Peisker (2001)

Table 6. Anti-nutritive factors in soybeans.

Soybean meal contains high levels of phosphorus, but much of it is present in a complex form due to the presence of phytic acid. However, the use of phytase can increase phosphorus retention by 50% and reduction in excretion by 42% (Lei *et al.*, 1993).

5. Utilisation of soybean in animal production

The major farmed animal species diets containing soybean include poultry, pigs, cattle and aquatic. The global animal feed production by species a decade ago included pigs (31%), broiler (27%), dairy cattle (17%), beef cattle (9%), layer (8%), aquatic (5%) and 3% of others (Hoffman, 1999). Thus soybean meal is used relatively more in some types of animal feed than in others. The major aim is to provide high quality protein to poultry and pigs.

Of all plant protein sources, soybean cultivation alone occupies most land needed for production of animal products. For example, soybean meal is used extensively in animal compound feed in the United States (Table 7a) and European Union (Table 7b). The annual EU livestock consumption alone demands soybean acreage of 5.0 million hectares in Brazil and 4.2 million hectares in Argentina (Table 7c).

Species	Million metric tons	Percent of total
Poultry – broilers	12.36	44
Poultry – layers	1.88	7
Swine	6.69	24
Cattle – beef	3.45	13
Cattle – dairy	1.61	6
Pet animals	0.74	3
Aquaculture	0.18	1
Other	0.65	2
Total	27.56	100

Source: United Soybean Board (1999/2000)

Table 7a. Utilisation of soybean meal by livestock in the United States

Type of animal compound feed	Production volume (1,000 tonnes)	Estimated soy bean meal content (%)	Volume of soybean meal in compound feed (1,000 tonnes)
Cattle – meat	12,148	13.9	1,683
Cattle – dairy	27,852	10.4	2,893
Pigs	51,440	28.8	14,815
Poultry – broilers	30,929	36.8	11,389
Poultry – layers	15,532	22.4	3,477
Other animals (e.g. sheep, goats, ducks, etc)	9,522	16.6	1,577
Total	147,423	24.3	35,834

Source: PROFUNDO (2008)

Table 7b. Soybean meal used in types of animal compound feed in the European Union-27.

Livestock product	Soybean Equivalent[1] (1,000 tonnes)	Acreage (ha)	Country of origin	Soybean Equivalent (1,000 tonnes)[1]	Acreage (ha)
Beef and veal	1,557	595,519	United States	2,102	781,256
Milk	621	237,642	Canada	463	182,290
Pork	10,341	3,956,061	Argentina	11,450	4,240,559
Poultry meat	7,934	3,035,314	Brazil	12,789	4,995,608
Eggs	3,247	1,242,109	Paraguay	585	263,553
Cheese	1,156	442,402	Uruguay	53	26,319
Other products	2,764	1,057,330	Other countries	180	76,791
Total	27,620	10,566,377	Total	27,621	10,566,377

Source: PROFUNDO (2008) 1,000 tonnes of soybean meal = 771 tonnes of soybeans.

Table 7c. Soybean acreage needed for livestock consumption in the European Union-27 and by country of origin.

6. Future challenges of soybean utilisation in animal diets

Future challenges confronting soybean utilisation in animal diets have been discussed by Kerley and Allee (2003). The major challenges include the following:

- Increased demand for vegetable oil for biodiesel production may in turn reduce overall production of soybean in favour of other oilseed crops that produce more oil per acre. For instance, soybean produces about 36 litres of oil per acre compared to 72 litres of safflower, 84 litres of sunflower and 108 litres of canola (United Soybean Board, 2011). Even though the nutritional values of meals from these oilseeds are lower than that of soybean, the increased value of the oil may shift production to these crops at the expense of soybean.

- Competition between the bio-fuel industry and animal agriculture has increased the prices of feed ingredients with consequent increase in feeding cost. Also, by-products from ethanol and biodiesel production (e.g. distillers dried grains with soluble) are now competing with maize and soybean meal for their place in animal diets.

- Demands on animal production exerted by environmental regulations as a result of nitrogen waste, malodour and excretion of phosphorus into the environment by the use of soybean in diets.

- Pressures to improve nutritional value of soybean through breeding to modify aspects such as anti-nutritive factors, fatty acid profile, and oligosaccharide or protein synthesis in order to allow greater levels of soybean meal in animal diets.

7. Conclusion

Soybean is the major vegetable protein source in the animal feed industry. Its universal acceptability in animal feed is as a result of important attributes such as relatively high protein content and suitable amino acid profile except methionine, minimal variation in nutrient content, ready availability year-round, and relative freedom from intractable anti-nutritive factors if properly processed, limited allowable uses of animal proteins in feed and its relatively low cost. Therefore, its production and consumption will continue to grow as a preferred source of alternate high quality protein in animal diets.

Commonly used soybean products as protein source in animal feed are soybean meal, full-fat soybean and soybean protein concentrates, which are obtained through various heat processing methods that reduce anti-nutritive factors present such as trypsin inhibitors and lectins. Of these products, soybean meal is most preferred due to its relatively low cost. It is used extensively in feeds for poultry, pigs and cattle.

Soybean is also a major source of vegetable fat in animal feed. Feed-grade soybean oil is popularly used in high energy diets, particularly for poultry, because of its high digestibility and metabolisable energy content compared with other vegetable fats/oils.

Soybean production and utilization for animal feed is bound to face future challenges as a result of increased demand of vegetable oil for biofuel production; of which soybean is less competitive. There is also increased research to use co-products from biofuel production as substitutes for soybean meal in animal diets. Thus, there is a need to overcome these and other challenges in order not to jeopardise cheap meat production for ever increasing world population.

8. References

Ajinomoto Heartland Lysine LLC Revision 7. True digestibility of essential amino acids in poultry. http://www.lysine.com/new/Technical%20Reports/Poultry/PoultryDigTableV7.pdf.

Araba, M. and N.M. Dale. 1990. Evaluation of protein solubility as an indicator of under processing soybean meal. *Poultry Science,* 69:1749-1752

Elkin, R.G. 2002. Nutritional components of feedstuffs: a qualitative chemical appraisal of protein. In: J.M. McNab and K.N. Boorman, eds. *Poultry Feedstuffs: Supply, Composition and Nutritive Value, pp. 57-86.* CAB International, UK

Ensminger, M.E. and C. G. Olentine Jr. 1978. *Feeds and Nutrition Complete, 1st ed.* Ensminger Publishing Co., California, USA.

FAS/USDA (Foreign Agricultural Service/United States Department of Agriculture). 2009. World soybean supply and distribution. FAS/USDA, Washington, D.C.

FEFAC (European Feed Manufacturers Federation). 2007. Industrial compound feed production, FEFAC Secretariat General, Brussels, May 2007.

Graham, K.K., Kerley, M.S., Firman, J.D., and G.L. Allee. 2002. The effect of enzyme treatment of soybean meal on oligosaccharide disappearance and chick growth performance. *Poultry Science, 81: 1014-1019.*

Gu, C., Pan, H., Sun, Z. and G. Qin. 2010. Effect of soybean variety on antinutritional factors content, and growth performance and nutrients metabolism in rat. *International Journal and Molecular Science, 11: 1048-1056.*

Hoffman, P. 1999. Prospects for feed demand recovery. International Feed Markets '99, October 1999, Agra Europe.

Hurrell, R.F. 1990. Influence of the Maillard Reaction on nutritional value of foods. In: *The Maillard Reaction in foods processing, human nutrition and physiology, pp 245-358.* Birkhauser Verlag, Basel, Switzerland.

Huyghebaert, G., G.D. Munter and G.D. Groote. 1988. The metabolisable energy (AMEn) of fats for broilers in relation to their chemical composition. *Animal Feed Science and Technology, 20: 45-58.*

INRA (Institut Scientifique de Recherche Agronomique). 2004. *Tables of composition and nutritional value of feed materials, 2 ed.* (Sauvant, D., Perez, J.M. and Tran, G., eds.), p.186. Wageningen Academic Publishers, Netherlands.

Kerley, M.S. and G.L. Allee. 2003. Modifications in soybean seed composition to enhance Animal feed use and value: Moving from dietary ingredient to a functional dietary component. *AgBioForum, 6 (1&2)14-17.*

Lei, X.G., Ku, P.K., Miller, E.R. and M.T. Yokoyama. 1993. Supplementing corn-soybean meal diets with microbial phytase linearly improves phytate phosphorus utilisation by weanling pigs. *Journal of Animal Science, 71:3359-3367.*

Leeson, S. and J.D. Summers. 2001. *Nutrition of the chicken, 4th ed.* University Books, Ontario, Canada.

Liener, I.E. 1977. Removal of naturally occurring toxicants through enzymatic processing. In: R.E. Feeney and J.R. Whitaker (Eds.), *Food proteins: Improvements through chemical and enzymatic modifications (p.7-57).* Advances in Chemistry Series 160. Washington, D.C.

Lucy, M.C., T.S. Gross and W.W. Thatcher. 1990. Effect of intravenous infusion of a soybean oil emulsion on plasma concentration of 15-keto-13, 14-dihydro-prostaglandin F2a and ovarian function in cycling Holstein heifers. *In*: Livestock reproduction in Latin America. Internaqtional Atomic Energy Agency, Vienna, Austria.

National Research Council. 1994. *Nutrient Requirements of Poultry, 9th .rev. ed.* National Academy Press, Washington, DC.

Ortiz, L.T., Rebole, A., Rodriquez, M.L., Trevino, J., Alzueta, C and B. Isabela. 1988. Effect of chicken age on the nutritive value of diets with graded additions of full-fat sunflower seed. *British Poultry Science, 39: 530-535.*

Parsons, C.M., Hatsimoto, K., Wedeking, K.J., Han, Y. and D.H. Baker. 1992. Effect of over-processing on availability of amino acids and energy in soya bean meal. *Poultry Science, 71,* 133-140.

Peisker, M. 2001. Manufacturing of soy protein concentrate for animal nutrition. *Cahiers Options Mediterraneennes, 54: 103-107.*

PROFUNDO. 2008. Soy consumption for feed and fuel in the European Union (J.W. van Gelder, K. Kammeraat, and H. Kroes, eds.). The Netherlands.

US Census Bureau. 2010, World Population, International database, Washington, DC United Soybean Board 2011. Soybean meal-Demand, Soybean meal information centre, 2010, United Soybean Board, Washington, D.C.

USSEC (United States Soybean Export Council). 2008. Processing of soybeans into soybean products. Annual Report, 2008. Washington, D.C.

Sell, J.L., L.G. Tenesaca, and G.L. Bales. 1978. Influence of dietary fat on energy utilisation by laying hens. *Poultry Science, 58: 900-905.*

Staples, C.R., J.M. Burke, and W.W. Thatcher. 1998. Influence of supplemental fats on reproductive tissues and performance of lactating cows. *Journal of Dairy Science,* 81: 856-871

Waldroup, P.W., S.E Walkins, and EA. Saleh, 1995. Comparison of two blended animal-vegetable fats having low or high free fatty acid content. *Journal of Applied Poultry Research, 4:4143.*

Woodworth, J.C., Tokach, M.D., Goodband, R.D., Nelssen, J.L., O'Quinn, P.R., Knabe, D.A. and N.W. Said. 2001. Apparent ileal digestibility of amino acids and digestible and metabolisable energy content of dry extruded-expelled soybean meal and its effect on growth performance of pigs. *Journal of Animal Science, 79:1280-1287.*

Wiseman, J. and F. Salvador. 1991. The influence of free fatty acid content and degree of saturation on the apparent metabolisable energy value of fats fed to broilers. *Poultry Science,* 70, 573-582.

Yen, J.T., T. Hymowitz and A.H. Jensen. 1971. Utilization by rat of proteins from a trypsin-inhibitor variant soybean. *Journal of Animal Science, 33: 1012-1017.*

Zumbado ME, Scheele CW, C. Kwakernaak. 1999. Chemical composition, digestibility, and metabolizable energy content of different fat and oil by-products. *Journal Applied Poultry Research, 8:263–271.*

10

Soybeans (*Glycine max*) and Soybean Products in Poultry and Swine Nutrition

Leilane Rocha Barros Dourado[1,2], Leonardo Augusto Fonseca Pascoal[1,3],
Nilva Kazue Sakomura[1,4], Fernando Guilherme Perazzo Costa[1,3]
and Daniel Biagiotti[1,2]
[1]Department of Animal Science,
[2]Federal University of Piauí,
[3]Federal University of Paraíba and
[4]São Paulo State University
Brazil

1. Introduction

Soy is a legume and has been successfully cultivated around the world. Today, the world's top producers of soy are the United States, Brazil, Argentina, China and India. According Brazilian Association of Vegetable Oil Industries (Abiove), the Brazil is responsible for some 28 percent of the world's soybean production, with the estimate of a production of 57 million tons. The Brazil is the world's second largest producer and exporter of soybeans, soybean meal and soybean oil. The soybean complex, which gathers the productive chain of soybean, soybean meal and soybean oil, is the main item in the country's Trade Balance. Other activity that involves the use of soy products (oil) is the production of biodiesel.

In fact, so much in the Brazil as in most of the countries of the World, the soy represents one of the largest oilseeds of the world and to main source of vegetable protein for the poultry and swine feeding.

2. The nutritional composition of the soybeans and soybean products used in the feeding of poultry and swine

Soybeans and soybean products are now used widely in animal feeding. The crop is grown as a source of protein and oil for the human market and for the animal feed market. Soybean meal is generally regarded as the best of plant protein source in terms of its nutritional value. Also, it has a complementary relationship with cereal grains in meeting the amino acids (AA) requirements of farm animals. Consequently, it is the standard to which other plant protein sources are compared (Blair, 2008).

Soybeans provide an excellent source of both energy and protein for poultry and swine. As with any ingredient, their usage rate depends upon economics, although in the case of soybeans such economics relate to the relative price of soybean meal and of supplemental fats. Soybeans contain about 38% crude protein, and around 20% oil (Leeson & Summers, 2008). However, soybeans contain compounds that inhibit the activity of the proteolytic enzyme trypsin. They also contain other antinutrients, including hemagglutinins or lectins, which contribute to reduce nutrient use. Nutritional composition of the soybeans and soybean products is affecting by percentage of anti-nutritional factors (ANFs), variety genetic, efficiency of the oil-extraction process and the amount of residual hulls present, the heat processing and other factors.

2.1 Anti-nutritional factors (ANFs) and nutritional quality

The nutritional quality of soybean products for poultry and swine feeding is determined not only by the quantity of nutrients (protein, amino acids, fat and others), but mainly by nutrients availability for the animals. According Durigan (1989), anti-nutritional factor whole substance is synthesized by normal plant metabolism, which may result from different mechanisms to reduce the efficiency of utilization of the diet. The ANFs present in soybean can to affect the nutrients availability for poultry and swine.

Most of the ANFs present in the raw soybeans, as protease inhibitors and lectins is heat-labile, but others as phytic acid and polysaccharides non starch (PNS) only decrease with enzyme addition in diet, because poultry and swine has no ability to produce enzymes to degrade the PNS.

From Liener (1994), soybeans contain some heat-labile protease inhibitors and hemagglutinins. Soy also contains factors that are relatively heat-stable, though of lesser significance, such as: Goitrogens: substances that cause goiters, an enlargement of the thyroid gland; Tannins: complex plant compounds that are often bitter or astringent; Flatus-producing oligosaccharides: carbohydrates of small molecular weight that cause flatulence (gas); Phytates: which bind minerals preventing absorption; Saponins and Antivitamins.

This ANFs of soybean can cause inhibition of growth, decreased feed efficiency, goitrogenic responses, pancreatic hypertrophy, hypoglycemia, and liver damage in nonruminant animals depending on species, age, size, sex, state of health and plane of nutrition (Palacios et al., 2004)

2.1.1 Proteases inhibitors

Proteases inhibitors are substances that ability has to inhibit the activity of certain digestive enzymes (Durigan, 1989). They are polypeptides of 181 and 71 amino acid residues, respectively, which form well-characterized stable enzyme inhibitor complexes with pancreatic trypsin on a one-to-one molar ratio. The content of soya bean trypsin inhibitors varies in different varieties of soya bean and germination process (Bau et al., 1997). These are known as the Kunitz inhibitor and the Bowman-Birk inhibitor which are active against trypsin, while the latter is also active against chymotrypsin (Liener, 1994), because Bowman (1944) identified a protein in soy can inhibit trypsin and chymotrypsin and subsequently purified by Birk et al (1961), called Bowman-Birk inhibitor. It is a heat-stable protein, due to its large number of sulfur bridges. Later Kunitz (1945) identified and crystallized other protein, Kunitz trypsin inhibitor, which strongly inhibited the activity of digestive enzyme trypsin.

These protease inhibitors interfere with the digestion of proteins, resulting in decreased animal growth. Protease inhibitors stimulate protein synthesis and enzyme secretion from the pancreas. Inhibition of proteolysis, the presence of undigested protein in the intestinal tract, and a decreased release of amino acids in raw soy diets induce a compensatory reaction in the pancreas and a general stimulatory effect on other endogenous secretions

(Rackis & Gumbmann, 1981). The compensatory effect of the pancreas is effective since that the urease activity of the soy is in up to 0,20 (Butolo, 2010). The effect 8of hypertrophy of pancreas followed by a stimulation of its secretory activity can also result in an endogenous loss of the pancreatic enzymes, trypsin and chymotrypsin which are rich in the sulphur-containing amino acids, and thus accentuating the deficiency of methionine, being the first limiting amino acid in soybean (Johri, 2005)

Coca-Sinova et al. (2008) had evaluated the coefficient of apparent ileal digestibility (%) of DM, N, energy, and amino acids (AA) of the diet with different soybean meal (SBM) origin in broilers of 21 d of Age. They observed that digestibility coefficients were higher for SBM contained lower levels of TIA - trypsin inhibitor activity (1,8 mg/g), when compared with SBM with higher levels of TIA (4,8mg/g).

2.1.2 Lectins (haemagglutinins)

Lectins are glycoproteins with the ability to bind carbohydrate-containing molecules on the epithelial cells of the intestinal mucosa, with the property of agglutinating the erythrocytes of higher animals (Liener, 2000).) The cells of the intestine in the presence of lectin, tend to collapse by reducing the absorption (Butolo, 2010). According Fasina et al. (2004), when lectins are ingested by animals, they can be degraded by intestinal digestive enzymes or survive intestinal digestion and bind to enterocytes on the brush border membrane (BBM). However if bind, lectins may cause antinutritional effects such as disruption of the intestinal microvilli, shortening or blunting of villi, impairment of nutrient digestion and absorption, increased endogenous nitrogen loss, bacterial proliferation, and increased intestinal weight and size (Pusztai, 1993 cited by Fasina et al., 2004).

2.1.3 Goitrogens

The soybean and its products have been considered goitrogenic in humans and animals (Doerge et al., 2002)., because the acidic methanolic extract of soybeans contains compounds that inhibit thyroid peroxidase-(TPO) catalyzed reactions essential to thyroid hormone synthesis (Divi et al., 1997). Pigs feeding goitrogens (0.075% 1-methyl-2-mercaptoimidazole or .5% potassium thiocyanate) produced symptoms of hypothyroidism in a relative short period of time, usually 3 to 4 weeks with pronounced growth depression, but when the goitrogens were withdrawn from the diet, there was a marked increase in growth rate. The effects of goitrogens are more common in humans, mainly infants (Shepard et al., 1960), than in pigs, because the soybeans used in feed for pigs is generally thermally processed. the goitrogens of soybean can removed by heat treatment (Liener, 1970; Zhenyu et al., 2000).

2.1.4 Tannins

Tannins are complex plant compounds that are often bitter or astringent, they are naturally-occurring plant polyphenols which combine with proteins and other polymers such as cellulose, hemicellulose and pectin, to form stable complexes (Mangan, 1988). Egounlety et al. (2003) observed that the hulls were much richer in tannins than the whole soybean (2. 31 x 1.52 mg catechin equivalent/g). Soaking soybean for 12–14 h reduced the tannin content by 54.6%. No tannin was detected in dehulled and cooked and in fermented Soybean.

In contrast to the position with ruminant animals where tannins in the diet may have considerable benefits, and in plants where tannins give partial protection against predators, in simple-stomached animals, including man, tannins in the diet are generally undesirable, because they present effects as decrease of protein digestibility and reduction of the animal growth(Mangan, 1988).

2.1.5 Saponins

Saponins are steroid or triterpenoid glycosides, common in a large number of plants and plant products that are important in human and animal nutrition (Francis et al., 2002). Saponins have long been known to cause lysis of the erthrocytes when given in vitro. The hemolytic activity of saponins, coupled with this cholesterol inhibition effect, has been extensively used as a means of detecting and quantifying saponins in plant material. Such saponins activity results from their affinity for membrane sterols (Yoshiki et al., 1998).

Most listings of soybean antinutritional factors in the past included saponins, although with little or no justification. Toxicity was attributed to them simply by analogy with saponins from other sources that, in deed, are toxic (Anderson et al., 1995). Soyabean saponins did not impair growth of chicks when added at five times the concentration in a normal soyabean-supplemented diet (Ishaaya et al., 1969 as cited in Francis et al., 2002). Acording Yoshiki et al. (1998) no hemolytic activity in soybean saponin was observed, while 80% methanolic soybean extract had strong activity. As a result of detailed studies, the hemolytic compounds in soybean are indentified as linoleic acid and lipoxin, wich are secondary metabolitic products from lipoxigenase.

2.1.6 Antivitamins

There exists a fairly large category of natural substances that interfere with the utilization of certain minerals and vitamins. As examples, isolated soya-bean protein has been shown to interfere with the availability of such minerals as zinc, manganese, copper and iron as well as vitamin D (Liner, 1970) or other diverse but ill-defined factors appear to increase the requirements for vitamins A, B12, D, and E (Liner, 1994).

Raw soybean contains an enzyme lypoxygenase which catalyses oxidation of carotene, the precursor of vitamin A and can be destroyed by heating soybeans for 15 min at atmospheric pressure. Autoclaving of soybean protein or supplementation with vitamin D3 for about 8-10 times can eliminate the rachitogenic activity (Johri, 2005). Fisher et al. (1969) reported anti-vitamin E activity of isolated soy protein for the chick.

2.1.7 Olygosaccharides and polysaccharides non starch

Soybean carbohydrates make up approximately 35% of soybean (SB) seed and 40% of soybean meal (SBM) dry matter (DM). Approximately half of these carbohydrates are nonstructural in nature, including low molecular weight sugars, oligosaccharides, and small amounts of starch, while the other half are structural polysaccharides, including a large amount of pectic polysaccharides (Karr-Lilienthal et al., 2005). The fibre component of the grain consists primarily of nonstarchpolysaccharides (NSP) which in cereals form part of the cell wall structure. In legumes, NSP also play a role as an energy storage material. The role of fibre in monogastric diets has attracted much attention in recent years, due to the facts that (a) the soluble NSP elicit anti-nutritive effects (Choct, 1997). The components of non-digestible carbohydrates of a feedstuff are NSP, consisting of water insoluble cellulose and water soluble gums, hemicelluloses, pectic substances and mucilages (Mekbungwan, 2007) . Polysaccharides are plymers of monosaccharides joined through glycosidic linkages and are defined and classified in terms of the following structural (Choct, 1997). Choct et al. (1995)

showed that the addition of 40 g/kg NSP to a commercial broiler diet decreased the weight gain, feed efficiency and apparent metabolizable energy (AME) by 28.6, 27.0 and 21.2%, respectively.

According Smits & Annison (1996) the physicochemical properties of non-starch polysaccharides (NSPs) are responsible for their antinutritive activities in the poultry and swine. In particular, soluble viscous NSPs depress the digestibilities of protein, starch and fat. It is suggested that the gut microflora can mediate the antinutritive effects of soluble and viscous NSP. On the other hand, insoluble and non-viscous NSPs may have a beneficial effect. Hetland et al. (2004) reported that digestibility of starch is higher and digesta passage rate faster when a moderate level of insoluble fibre is present in the diet. The effect of insoluble fibre on gut functions stems from its ability to accumulate in the gizzard, which seems to regulate digesta passage rate and nutrient digestion in the intestine. NSP content of soybean meal is approximately 61 and 103 g kg-1 (dry matter basis) for soluble NSP and insoluble NSP, respectively (Bach Knudsen, 1997).

Choct et al. (2010) reported in review that non-digestible oligosaccharides can be fermented throughout all sections of the gastrointestinal tract including the large intestine, and the effects are most variation depends of specie. Soy oligosaccharides increased microbial activities as indicated by the increased volatile fat acids (VFA) contents and can cause intestinal disorder.

The α–galactoside family of oligosaccharides cause a reduction in metabolizable energy with reduced fiber digestion and quicker digesta transit time. Birds do not have an α-1:6 galactosidase enzyme in the intestinal mucosa (Leeson & Summers, 2008). Enzyme addition in diet for poultry and swine is most utilized for improve the nutritional value of soybean products. Zanella et al. (1999) found better body weight gain for broilers fed soybean meal (45% CP), extruded soybeans (38% CP) and roasted soybeans with addition of blend enzyme (xylanase, protease and amylase) ,when compared to without enzyme..

2.1.8 Phytate

Three terminologies, namely phytate, phytin and phytic acid, are used in the literature to describe the substrate for phytase enzymes. The most commonly used term, phytate, refers to the mixed salt of phytic acid (myo-inositol hexaphosphate; IP6). The term, phytin, specifically refers to the deposited complex of IP6 with potassium, magnesium and calcium as it occurs in plants, whereas phytic acid is the free form of IP6 (Selle & Ravindran, 2007). Historically, phytates have been considered solely as antinutrients because they are known as strong chelators of divalent minerals such as Ca^{2+}, Mg^{2+}, Zn^{2+} and Fe^{2+}. Moreover, phytates are also capable of binding with starch and proteins while preventing their assimilation through the digestive system (Afinah et al., 2010). The soybean meal present 0.34% of phytate the soybean has 0.34% of phytate which represents approximately 60% of the amount of total phosphorus (Kornegay, 2000). To improve the utilization of phosphorus and other nutrients complexed to phytate has been used in the enzyme phytase in diets for poultry and pigs. In Brazil more than 50% of poultry diets were formulated using phytase.

2.2 Soybean processing

Soybean processing is useful and necessary to destroy or remove undesirable constituents, make nutrients more accessible or to improve palatability. However, processing toward these ends also leads to changes in the composition of the various soybean materials compared with whole soybeans. These changes may be intentional, as in the case of heating to diminish trypsin inhibitor activity (Anderson et al.,1995). Although exist other options for reducing the antinutritional effects of soybean products, the cheaper and more efficient is heat-processing (Pusztai et al., 1997).

The most common procedures have involved a combination or extraction, cooking and fermentation. With soybeans, moist-heat treatment is particularly effective in reducing trypsin inhibitor activity below biological threshold levels, as determined by short-term animal bioassay. With present day manufacturing processes, residual trypsin inhibitor activity in edible-grade soy protein products is about 5-20% of the activity originally present in raw soybeans (Rackis & Gumbmann, 1981).

The heat necessary to destroy trypsin inhibitors and other hemagglutinins found in raw soybeans is dependent upon exposure time, and so high temperatures for a shorter time period are as effective as lower temperatures for longer times (Leeson & Summers, 2008).

In the 1930's, soybeans were mechanically processed using hydraulic or screw presses, which squeezed out the oil of the heated or cooked soybeans. In the late 1940's and early 1950's, most of the industry converted to the solvent-extraction process, which removes more oil from the soybean. Today, more than 99 percent of the U.S. processing capacity is using the solvent extraction process produced in large crushing facilities that produce meals of consistent high quality (Johnson & Smith, sd).

According Soybean Meal INFOcenter, the soybean products appear with the initial processes that included cleaned and dehulled and other three processes is used to separate the soybean oil from the protein meal. The first is solvent extraction, which is the one used most commonly around the world, uses hexane to leach or wash (extract) the oil from flaked oilseeds. This method reduces the level of oil in the extracted flakes to one percent or less. After this continuous pressing is performed at elevated temperatures, using a screw press to express the oil from ground and properly conditioned soybeans. The pressed cake is reduced to between 4 percent and 6 percent oil content by this method. At the end, dydraulic or batch pressing, this is an intermittent pressing operation carried out at elevated temperatures in a mechanical or hydraulic press after the soybeans have been rolled into flakes and properly conditioned by heat treatment. It is the oldest known method of processing oilseeds. According Butolo (2010), the industrialization flow for obtaining soybean oil, soybean meal or full fat soybean is divided into four distinct phases, which are shown in Figure 1.

A great processing variety exists, in table 1 sowed the composition of soybean feed ingredient products.

The quality of soya meal is the result of many factors, including bean variety, origin and storage. The various processing steps employed from the time the bean is received can affect the quality of the resulting meal and oil obtained. Heat treatment of the meal is essential to optimize its protein quality. The variables of moisture, temperature and time are interrelated and are important to achieve proper cooking conditions. The magnitude of these variables must be determined for each plant, preferably using a biological assay for evaluation. Many in vitro tests designed to measure protein quality in soya have been proposed and evaluated (Wright, 1968). Simple crude protein or amino acid assays provide information on the protein, but do not provide useful information on the quality of the protein. Chemists have used trypsin inhibitor analyses, urease activity, protein solubility in potassium hydroxide, protein solubility in water and dye binding methods to assay for protein quality (Johnson & Smith, nd).

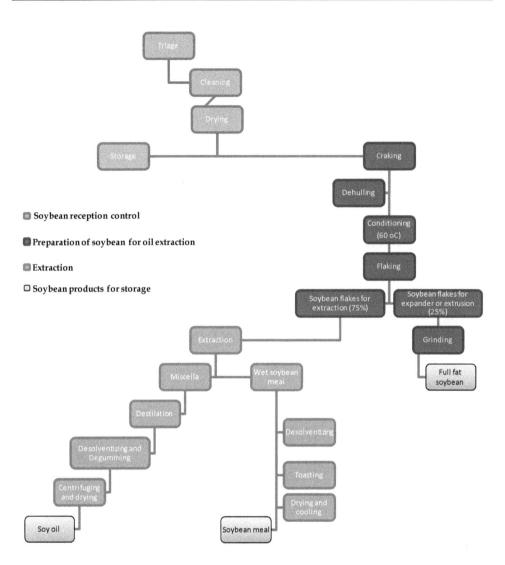

Fig. 1. The simple flowchart of soybean processing (Adapted of Butolo, 2010).

Inadequate heating fails to completely destroy the ANFs, which may have a detrimental impact on animal performance, while excessive heating reduces the availability of lysine via the Maillard reaction and possibly, to a lesser extent, of other amino acids (Caprita et al. 2010). While processed beans should be periodically tested for trypsin inhibitor or urease levels, a simple on-going test is to taste the beans. Under-heated beans have a characteristic 'nutty' taste, while over-heated beans have a much darker color and a burnt taste. The problem with overheating is potential destruction of lysine and other heat-sensitive amino acids (Leeson & Summers, 2008).

Nutrient profile	Full-fat extruded soybean	Soybean meal (48%)	Soybean protein concentrate	Soybean Hulls	Soy oil
Dry matter (%)	90.47	88.21	89.88	88.80	99.60
Gross energy (Kcal kg⁻¹)	4938	4164	4495	3854	9333
Starch (%)	6.70	3.00	-	-	-
Fat (%)	17.64	1.40	0.43	2.86	99.60
Crude fibre (%)	6.24	4.27	2.64	33.00	-
Crude protein(%)	37.00	47.90	62.92	13.50	-
Arginine(%)	2.71	3.50	5.32	0.81	-
Lysine (%)	2.23	2.92	4.07	0.89	-
Methionine + cystine (%)	1.08	1.37	1.90	0.39	-
Threonine (%)	1.47	1.86	2.60	0.51	-
Triptophan (%)	0.47	0.64	0.87	0.14	-
Calcium (%)	0.23	0.31	0.27	0.49	-
Av. Phosphorus(%)	0.17	0.21	0.27	0.05	-
Potassium(%)	1.67	2.11	2.18	-	-
Poultry					
Metabolizable energy (Kcal kg⁻¹)	3429	2302	2677	871	8790
Dig. Arginine(%)	2.54	3.31	5.12	0.64	-
Dig. Lysine (%)	2.02	2.70	3.75	0.59	-
Dig. Methionine + cystine (%)	0.93	1.21	1.67	0.21	-
Dig. Threonine (%)	1.29	1.66	2.29	0.25	-
Dig. Triptophan (%)	0.43	0.58	0.79	0.09	-
Swine					
Digestible energy (Kcal kg⁻¹)	4250	3540	4035	2370	8600
Dig. Arginine(%)	2.52	3.31		0.68	-
Dig. Lysine (%)	1.99	2.66		0.53	-
Dig. Methionine + cystine (%)	0.93	1.23		0.26	-
Dig. Threonine (%)	1.28	1.62		0.31	-
Dig. Triptophan (%)	0.40	0.57		0.09	-

Table 1. Nutrient profile of soybean products for poultry and swine nutrition (From Rostagno et al., 2005).

The most widely adopted method more economical and faster is the measurement of urease activity (urease test or urease index), however difficult the correlation of the protein solubility in KOH. Levels of the enzyme urease are used as an indicator of trypsin inhibitor activity. Urease is much easier and cheap to measure than is trypsin inhibitor and both molecules show similar characteristics of heat sensitivity. Two analytical methods for urease test were described, the first is based on ph difference in which 200 mg of sample (soybean or soybean product) is incubated in 10.0 ml of phosphate buffered urea solution at 300C for 30 minutes, after which the increase in pH units (ΔpH) from pH 7.00 is recorded (Palić et al. 2008; Butolo, 2010). The second is based a simple colorimetric assay in which urea-phenol-red solution is brought to an amber color by using either 0.1 N HCl or 0.1 N NaOH. About 25 g of soybean meal is then added to 50 ml of indicator in a petri dish. After 5 minutes, the sample is viewed for the presence of red particles. If there are no red particles showing, the mixture should stand another 30 minutes, and again if no red color is seen, it suggests overheating of the meal. If up to 25% of the surface is covered in red particles, it is an indication of acceptable urease activity, while 25 – 50% coverage suggest need for more detailed analysis. Over 50% incidence of red colored particles suggests an under-heated meal (Leeson & Summers, 2008). But urease index is not useful to determine excessive heat treatment since additional heating has no effect on the urease index (Caprita et al. 2010).

The KOH protein solubility test is based on the solubility of soybean proteins in a dilute solution of potassium hydroxide. The procedure involves the incubation of a sample with a 0.2% KOH solution for 20 min at room temperature. Following this incubation, the sample is centrifuged and the supernatant is analyzed for the protein concentration. The solubility of the protein, expressed as a percentage, was calculatedby dividing the protein content of the KOH extracted solution by the protein content of the original soybean sample (Caprita et al. 2010). KOH protein solubility is a better indicator of overprocessing than underprocessing of soybeans (Batal et al., 2000). In table 2 are shown the levels of urease activity and protein solubility in potassium hydroxide acceptable in most soybeans processing.

Protein Dispersibility Index (PDI) can be used to measure protein quality. According Butolo (2010) to determine the protein dispersibility index should be mixed 8g of soybean meal with 150ml water, then it is centrifuged to 8.500ppm for 10 minutes, filter and determining the soluble nitrogen by the Kjeldahl method. Batal et al. (2000) indicated that PDI demonstrates more consistent response to heating of soyflakes than did urease index or protein solubility in KOH, because the urease index is not linear and that it rapidly falls from approximately 2.0 units of pH change to near zero as SBM is heated contributes to the difficulty in determining a precise maximum acceptable.

Degree of soybean processing	Urease test (pH change)	KOH Protein solubility (%)
Under-processed	> 0.20	90
Normal	-	85
Adequately processed	0.05 - 0.20	77-80
Over-processed	-	< 77

Table 2. Globally accepted relation between the degree of soybean processing for urease activity and protein solubility in potassium hydroxide (from Palić et al. 2008; Butolo, 2010).

Nitrogen Solubility Index (NSI) is other methods can be used for determines protein quality. It uses a slow stirring technique. Nitrogen is extracted from the ground flour by placing approximately 1.5 g into a 200 ml beaker and adding 75 ml of 0.5% KOH. The sample is stirred 20 minutes at 120 rpm. PDI and NSI are a more consistent and sensitive indicator for monitoring both underheating and over-heating of SBM (Caprita et al. 2010).

It is very important whether the assessment process quality of soybeans for the animal is unlikely to have decreased performance.

2.3 Storage and genetic variety

The storage and genetic variation are factors that can alter the nutritional composition of soybean as well as the performance of monogastric animals. Narayan et al., (1988) found that chemical characteristics, moisture content, fat, water-soluble nitrogen (WSN), nitrogen solubility index (NSI), sugars, trypsin inhibitor activity, available lysine, pigment and lipoxygenase activity of seeds decreased during storage whereas non-protein nitrogen (NPN), extent of browning, free fatty acid (FFA) content and peroxide value are increased.

According Cromwell et al. (2002) the rate and efficiency of weight gain, scanned backfat and longissimus area, and calculated carcass lean percentage were not different (P > 0.05) for pigs fed diets containing conventional or genetically modified, herbicide (glyphosate)-tolerant soybean. For poultry, Taylor et al. (2007) concluded that the diets containing soybean meal produced from genetically modified (GM) glyphosate-tolerant were nutritionally equivalent to diets containing soybean meal produced from the control and conventional reference soybean varieties when fed to broilers.

Other genetically modified soybeans were study for Palacios et al. (2004) that compared the growth performance of chicks and pigs fed diets containing modified soybeans: Kunitz trypsin inhibitor-free (KF), lectin-free (LF), lectin and Kunitz trypsin inhibitor-free (LFKF), conventional soybeans (CSB), and commercially obtained, dehulled, solvent-extracted soybean meal (SBM). They verify that Chicks fed diets containing any of the raw soybean varieties gained less weight than did chicks and among the raw soybean treatments, there was a greater effect on growth performance by removing both lectins and Kunitz trypsin inhibitor (LFKF), than by removing each antinutritional factor separately. Feeding raw soybeans to chicks decreased average daily gain (ADG) by 49% for CSB, 37% for KF, 38% for LF, and 27% for LFKF compared with the ADG achieved by chicks fed SBM. For to pigs deprecreased ADG by 78% for CSB, 60% for LF, and 35% for LFKF compared with the ADG achieved by pigs fed the same variety but extruded. This results and others (Brune et al., 2010; Becker-Ritt et al., 2004; Vasconcelos et al., 1997) prove that variability in the amounts of these components (proteases inhibitors and lectins) can be affect by cultivars differences.

As observed, there seems great potential for reduction in content of anti-nutrients within GM soybeans, as studies have shown that the isogenic variant lacking the Kunitz trypsin inhibitor and other soybean variants low in Kunitz trypsin inhibitor are nutritionally superior to conventional raw soybeans but not as good as commercial soybean meal. Others genetic improvements in reducing the phytate-bound phosphorous, and reduction or elimination of oligosaccharide carbohydrates are the most important economical traits that are being researched

3. Soybean products for poultry and swine nutrition

3.1 Soybean meal and full fat soybean

Soybean meal is the most popular source of supplemental protein in livestock feeds (Table 3). That popularity derives from its nutrient content, its relative freedom from intractable antinutritional factors, and other issues (Pettigrew et al.2002).

Many studies are conducted comparing the inclusion of soybean meal with other soy products or other protein source, and where the animals fed soybean meal has a better perform in most cases. In general, full-fat soybeans may replace soybean meal in swine and poultry diets with similar performance anticipated. The decision on which soybean product to use needs to be based on the product's composition, availability and unit costs. Bertol et al. (2001) observed that substitution of 50% of soybean meal by full-fat extruded soybeans, texturized soybean protein and concentrated soybean protein in the weaning diet, promoted better performance, with additional 1 to 2 kg of body weight gains per piglet at the end of the nursery phase.

Protein Source	Million Metric Tons	Percent
Soybean meal	152.1	66.5
Rapeseed meal	30.7	13.5
Cottonseed meal	14.4	6.3
Sunflower meal	12.2	5.3
Palm kernel meal	6.1	2.7
Peanut meal	5.9	2.6
Fish meal	5.3	2.3
Copra meal	1.9	0.8
TOTAL	228.6	

* Soy Stats (2010)

Table 3. World Protein Meal Consumption*.

Micronizing is the name given to a cooking process that uses infrared rays to cook cereals and pulses at lower temperatures and for shorter times than other heating methods. Gas burners are used to generate the infrared rays that are absorbed by the products. The raw materials are passed under the burners on variable speed belts to achieve the desired level of "cook". The product is then passed through a roller mill to create flakes. These flakes can be used whole or ground into a meal (MMfeeds). The increase available energy and improve digestibility are both achieve due to the gelatinisation of starch molecules during the cooking process. Trindade Neto et al. (2002) observed that pigs fed micronized soybean takes more days to reach 50 and 90 kg of body weight when compared with those fed soybean meal.

3.2 Soybean hulls

Soybean hulls, due to their high fiber contents, are known to be poorly digested by non-ruminant animals. Recent studies, however, suggest that the hulls have potential as an alternative feed ingredient for swine and poultry. The soybean hulls can be included up to 10 and 12% for growing or finishing pig diets, respectively, replacing the wheat bran on a weight basis without any adverse effects on palatability of diets and animal performances (Chee et al., 2005). However, Moreira et al. (2009) not recommend the use of soybean hulls to piglets due reducing daily feed intake and daily weight gain for the animals fed feed containing soybean hull (15% inclusion in the diet) compared to the control feed without soybean hull.

Esonu et al. (2005) studing laying hens, found that inclusion of up to 20% soybean hulls, improves the Feed cost/dozen eggs, and when cellulolytic enzyme supplementation at 30% dietary level of soybean hull meal in layer diet could not significantly affect the performance of laying hens.

Currently, it is very common the use of soybean hulls in programmers of feed restriction and welfare of breeders and laying hens.

3.3 Soy protein isolates

Soybean protein concentrate (or soy protein concentrate) is the product obtained by removing most of the oil and water-soluble non-protein constituents from selected, sound, cleaned, dehulled soybeans. The traded product among 650 to 900 g/kg CP on a moisture-free basis. Soybean protein isolate (or soy protein isolate) is the dried product obtained by removing most of the non-protein constituents from selected, sound, cleaned, dehulled soybeans. Both soy protein concentrate and isolate have the potential to be used in poultry diets as a source of protein and AA (Blair, 2008). In Table 4 are showed broilers performance when feed with different soybean products in diets.

Protein source	Weight gain (g)
Casein	364 b
Soybean meal	405 a
Soy protein concentrate	356 b
Soy protein isolate	366 b

Batal & Parson (2003)

Table 4. Effect of protein sources on weight gain of chicks(week 0-3).

3.4 Soy oil

Soy oil has found many food uses due to its excellent nutritional qualities, widespread availability, economic value and wide-use functionality. Soy oil is a highly concentrated source of feed energy. Its caloric value is the major reason for its increased use.

Gaiotto et al. (2000) evaluated performance of broilers fed diets containing 4% supplemental fat from the sources: soybean oil (SOY4), beef tallow (TAL4), acidulated soapstock (SOAP4), mixtures 2%:2% (SOAP2/TAL2), (SOAP2/SOY2) and (SOY2/TAL2), and confirmed the superiority of soybean oil relative to the other fat sources fed to broiler and demonstrated that the quality of acidulated soapstock and beef tallow may be improved when used in 1:1 mixtures with soybean oil.

For laying hens Costa et al. (2008) evaluated soy and canola oil, and they observed better results for those characteristics were obtained as soybean oil increased. However, the egg mass conversion was negatively influenced by increase of canola oil. The addition of soybean oil promoted better performance as compared to canola oil.

For swine, Mascarenhas et al. (2002) evaluated the effects from two lipid sources (soybean oil and coconut oil) on performance from 60 to 100 kg boars and they observed that diets with coconut oil as lipidic source showed the best results of weight gain.

4. Conclusion

The benefits of the use of soybean and soybean products can be observed in the nutrition of poultry and swine, but it is very important to know the factors that affect the composition of the same ingredients for that may be included in adequate amounts without reducing animal performance.

5. References

ABIOVE. (Dez 2010). Brazilian Association of Vegetable Oil Industries. In : *About ABIOVE*. 17.12.2010. Available from : http://www.abiove.com.br/english/menu_us.html, Access in: march 2011.

Afinah, S. ; Yazid, A. M. ; Anis Shobirin, M. H. & Shuhaimi, M. (2010). Phytase: application in food industry. *International Food Research Journal*, vol. 17, pp. 13-21 , ISSN: 1985-4668

Anderson, R.L. & Wolf, W.J. (1995). Compositional changes in trypsin inhibitors, phytic acid, saponins and isoflavones related to soybean processing. *Journal of Nutrition*, vol. 125, pp.581–588, ISSN 1541-6100

Bach Knudsen, K.E. (1997). Carbohydrate and lignin contents of plant materials used in animal feeding. *Animal Feed Science and Technology*, vol. 67, pp. 319–338, ISSN: 0377-8401

Batal, A. B. & Parsons, C. M. (2003). Utilization of Different Soy Products as Affected by Age in Chicks. *Poultry Science*, vol. 82, pp.454–462, ISSN 0032-5791

Bau, H.; Villaume, C. ; Nicolas, J.P. & Jean, L.M. (1997). Effect of Germination on Chemical Composition, Biochemical Constituents and Antinutritional Factors of Soya Bean (Glycine max) Seeds. Journal of Science and Food Agriculture, vol. 73, pp. 1-9, ISSN : 1097-0010

Becker-Ritt, A. B. ; Mulinari, F. ; Vasconcelos, I. V. & Carlini, C. R. (2004). Antinutritional and/or toxic factors in soybean (Glycine max (L) Merril) seeds: comparison of different cultivars adapted to the southern region of Brazil. *Journal of the Science of Food and Agriculture*, vol. 84, pp.263–270, ISSN : 1097-0010

Bertol, T. M. ; Mores, N. ; Ludke, J. V. & Franke, M. R. (2001). Proteínas da Soja Processadas de Diferentes Modos em Dietas para Desmame de Leitões. *Brazilian Journal of Animal Science*, vol.30, no.1, pp.150-157, ISSN 1806-9290

Birk, Y. & Gertheler, A. (1961).Effects of mild chemical and enzimatic treatments of soybean meal and soybean trypsin inhibitors on their nutritive and biochemical properties. *Journal of Nutrition*, vol.75, pp. 379-387, ISSN 1541-6100

Blair, R. (2008). *Nutrition And Feeding Of Organic Poultry*, CAB International, ISBN 978-1-84593-406-4, Oxfordshire, England.

Bowman, D.W. (1944). Fractions derived from soybeans and navybeans whice retard tryptic digestion of casein. Proc. Soc. Exp. Biol. Med., New York, v.57, pp.139-140,

Brune, M. F. S. S. ; Pinto, M. O. ; Peluzio, M. C. G. ; Moreira,M. A. & Barros, E. G. (2010). Avaliação bioquímico-nutricional de uma linhagem de soja livre do inibidor de tripsina Kunitz e de lectinas. *Ciência e Tecnologia de Alimentos*, vol. 30, No. 3, pp. 657-663, ISSN : 0101-2061

Butolo, J. E. (2010). *Qualidade de ingredientes na alimentação animal*. CBNA, ISBN 85-902473-1-7, Campinas-São Paulo- Brazil

Căpriţă, R. ; Căpriţă, A. & Creţescu, I. (2010). Protein Solubility as Quality Index for Processed Soybean. *Animal Science and Biotechnologies*, vol. 43, pp.375-378.

Chee, K.M.; Chun, K. S.; Huh, B. D.; Choi, J. H. ; Chung, M. K.; Lee, H. S.; Shin, I. S. & Whang, K. Y. (2005). Comparative feeding values of soybean hulls and wheat bran for growing and finishing swine. *Asian-Australian Journal Animal Science*, vol. 18, No. 6, pp. 861-867, ISSN 1011-2367

Choct, M.; Hughes, R.J.; Wang, J.; Bedford, M.R. ; Morgan, A.J. & Annison, G. (1995). Feed enzymes eliminate antinutritive effect of non-starch polysaccharides and modify fermentationin broilers. Proceeding of Australian Poultry Science Symposium 7, pp. 121–125, ISSN 1034-6260, Sydney : University of Sydney.

Choct, M.; Dersjant-Li, Y. ; McLeish, J. & Peisker, M. (2010). Soy oligosaccharides and soluble non-starch polysaccharides: a review of digestion, nutritive and anti-nutritive effects in pigs and poultry. *Asian-Australian Journal Animal Science* , Vol. 23, No. 10, pp. 1386 – 1398, ISSN 1011-2367

Coca-Sinova, A.; Valencia, D. G. ; Jiménez-Moreno, E. ; Lázaro, R. & Mateos. G. G. (2008). Apparent Ileal Digestibility of Energy, Nitrogen, and Amino Acids of Soybean Meals of Different Origin in Broilers. *Poultry Science*, vol. 87, pp.2613–2623, ISSN : 0032-5791

Costa, F. G. P. ; Souza, C. J. ; Goulart, C. C. ; Lima Neto, R. C. ; Costa, J. S. & Pereira, W. E. (2008). Desempenho e qualidade dos ovos de poedeiras semipesadas alimentadas com dietas contendo óleos de soja e canola. *Brazilian Journal of Animal Science*, vol.37, no.8, pp.1412-1418, ISSN 1806-9290

Cromwell, G. L. ; Lindemann, M. D. ; Randolph, J. H. ; Parker, G. R. ; Coffey, R. D. ; Laurent, K. M. ; Armstrong, C. L. ; Mikel, W. B. ; Stanisiewski, E. P. & Hartnell, G. F. (2002). Soybean meal from Roundup Ready or conventional soybeans in diets for growing-finishing swine. Journal Animal Science, vol.80, pp.708–715, ISSN 1525-3163

Doerge, D. R. & Sheehan, D. M. (2002). Goitrogenic and Estrogenic Activity of Soy Isoflavones. *Environmental Health Perspectives*, Vol. 110, No. 3, pp. 349-353, ISSN : 0091-6765

Durigan, J.F. (1989). Fatores antinutricionais em alimentos. Anais do Simpósio Interfase Nutrição Agricultura, p.155-225, Piracicaba, São Paulo, Brazil.

Egounlety, M. & Aworh, O.C. (2003). Effect of soaking, dehulling, cooking and fermentation with Rhizopus oligosporus on the oligosaccharides, trypsin inhibitor, phytic acid and tannins of soybean (Glycine max Merr.), cowpea (Vigna unguiculata L. Walp) and groundbean (Macrotyloma geocarpa Harms). *Journal of Food Engineering*, vol. 56, pp. 249–254, ISSN : 0260-8774

Esonu, B.O. ; Izukanne, R.O. & Inyang, O.A. (2005). Evaluation of Cellulolytic Enzyme Supplementation on Production Indices and Nutrient Utilization of Laying Hens Fed Soybean Hull Based Diets. International. *Journal of Poultry Science*, vol. 4, no. 4, pp. 213-216, ISSN 1682-8356

Fisher,H. ; Griminger, P. & Budowski, P. (1969). Anti-vitamin E activity of isolated soy protein for the chick. Zeitschrift für Ernährungswissenschaft (European Journal of Nutrition), Vol.9, No. 4. pp. 271-278 , ISSN : 1436-6215

Francis, G. ; Kerem, Z. ; Makkar, H. P. S. & Becker, K. (2002). The biological action of saponins in animal systems: a review. *British Journal of Nutrition*, vol. 88, pp. 587–605, ISSN : 1475-2662

Gaiotto, J. B. ; Menten, J.F.M. ; Racanicci, A.M.C. &Iafigliola, M.C. (2000). Óleo de Soja, Óleo Ácido de Soja e Sebo Bovino Como Fontes de Gordura em Rações de Frangos de Corte. *Brazilian Journal of Poultry Science*. vol.2 no.3, pp. 219-227, ISSN 1516-635X

Hetland, H.; Choct, M. & Svihus, B. (2004). Role of insoluble non-starch polysaccharides in poultry nutrition. *World's Poultry Science Journal*, vol.60, pp.415-422, ISSN 0043-9339

Ilka M Vasconcelos, I. M. ; Siebra, E. A. ; Maia, A. A. B. ; Moreira, R. A. ; Neto, A. F. ; Campelo, G. J. A. & Oliveira, J.T. A. (1997). Composition, Toxic and Antinutritional Factors of Newly Developed Cultivars of Brazilian Soybean (Glycine max). Journal of Science and Food Agriculture, vol. 75, pp. 419-426, ISSN : 1097-0010

Johnson, L.; Smith, K. (Jan 2011). Soybean Processing. In: *Fact Sheet*, 19.01.2011, Available from: http://www.soymeal.org ISSN : 1065-3309

Johri, T.S. (2005). Endogenous and exogenous feed toxins. In: Poultry Nutrition Research in India and its Perspective. 19.01.2011, Available from: http://www.fao.org/docrep/article/agrippa/659_en-10.htm

Karr-Lilienthala, L.K.; Kadzereb, C.T. ; Grieshopc C.M. & Fahey Jr, G.C. (2005). Chemical and nutritional properties of soybean carbohydrates as related to nonruminants: A review. *Livestock Production Science*, vol. 97, pp. 1 –12, ISSN 0301-6226

Kunitz, M. (1945). Crystallization of trypsin inibitor from soybeans. *Science*, v.101, p.668-669, ISSN 1095-9203

Leeson, S. & Summers, J. D. (2008). *Comercial Poultry Nutrition*, Nottingham University Press, ISBN 978-1-904761-78-5, Guelph, Ontario

Liener, I.E. (1994). Implications of antinutritional components in soybean foods.*Critical Reviews in Food Science and Nutrition*, vol. 34, no. 1, pp. 31-67. ISSN : 1549-7852

Mascarenhas, A. G. ; Donzele, J. L. ; Oliveira, R. F. M. ; Ferreira, A. S. ; Lopes, R. S. & Tavares. S. L. (2002). Fontes e Níveis de Energia Digestível em Rações para Suínos Machos Inteiros dos 60 aos 100 kg. *Brazilian Journal of Animal Science*, vol.31, no.3, pp.1403-1408, ISSN 1806-9290

Mangan, J. L. (1988). Nutritional effects of tannins In animal feeds. *Nutrition Research Reviews*, vol. 1, pp. 209-231, ISSN : 1475-2700

Mekbungwan, A.   Application of tropical legumes for pig feed. (2007). *Animal Science Journal*, Vol. 78, pp. 342–350, ISSN 1525-3163

Moreira, I. ; Mourinho, F. L. ; Carvalho, P. L. O. ; Paiano, D. ; Piano, L. M. & Kuroda Junior, I. S. (2009). Avaliação nutricional da casca de soja com ou sem complexo enzimático na alimentação de leitões na fase inicia. *Brazilian Journal of Animal Science*, vol.38, no.12, pp.2408-2416, ISSN 1806-9290

Narayan, R.; Chauhan, G. S. & Verma, N. S. (1988). Changes in the quality of soybean during storage. Part 1—Effect of storage on some physico-chemical properties of soybean. *Food Chemistry*, Vol. 27, No. 1, pp. 13-23, ISSN 0308-8146

Palacios, M. F. ; Easter, R. A. ;   Soltwedel, K. T. ; Parsons, C. M. ; Douglas, M. W. ; Hymowitz, T. & Pettigrew, J. E. (2004). Effect of soybean variety and processing on growth performance of young chicks and pigs. Journal of Animal Science, vol. 82, pp.1108-1114, ISSN : 1525-3163

Palić, D. V. ; Lević, J. D. ; Sredanović, S. A. & Đuragić, O. M. (2008). Quality control of full-fat soybean using urease activity: critical assessment of the method. *Acta Periodica Technologica*. vol. 39, pp. 47-53, ISSN 1450-7188

Rackis, J. J. & Gumbmann, M. R. (1981). Protease inhibitors: physiological properties and nutritional significance. In: Antinutrients and Natural Toxicants in Foods. pp. 203-237, Food & Nutrition Press, Westport, CT. ISBN :

Selle, P.H. & Ravindran, V. (2007). Microbial phytase in poultry nutrition : Review Article. *Animal Feed Science and Technology*, Vol. 135, No. 1-2, pp. 1-41, ISSN: 0377-8401

Shepard, T. H. ; Pyne, G.E. ; Kirschvink, J.F. & McLean, M. (1960). Soybean Goiter — Report of Three Cases. *The New England Jounal of Medicine*, vol. 262, pp.1099-1103, ISSN : 1533-4406

Smits, C.H.M. & Annison, G. (1996). Non-starch plant polysaccharides in broiler nutrition – towards a physiologically valid approach to their determination. World's Poultry Science Journal, vol. 52, pp. 203-221, ISSN 0043-9339

Taylor, M. ; Hartnell, G. ; Lucas, D. ; Davis, S. & Nemeth, M. (2007). Comparison of Broiler Performance and Carcass Parameters When Fed Diets Containing Soybean Meal Produced from Glyphosate-Tolerant (MON 89788), Control, or Conventional Reference Soybeans. *Poultry Science*, vol.86, pp.2608–2614, ISSN 0032-5791

Wright, K. N. (1968). Soybean meal processing and quality control. *Journal of the American Oil Chemists' Society*, Vol. 58, No. 3, pp. 294-300, ISSN: 0003-021X

Yoshiki, Y.; Kudou, S. & Okubo, K. (1998). Relationship between chemical structures and biological activities of triterpenoid saponins from soybean (Review). *Bioscience Biotechnology and Biochemistry*, vol. 62, pp. 2291–2299, ISSN : 1347-6947

Zanella, I.; Sakomura, N. K.; Silverside, F. G. & (1999). Effect of enzyme supplementation of broiler diets based on corn and soybeans. *Poultry Science.*, v. 78, pp.561–568, ISSN 0032-5791

Zhenyu, G. ; Jianrong, L. ; Ping, Y. ; Xinle, L. & Peilin, C. (2000). Removal of Goitrogen in Soybean. Journal of the Chinese Cereals and Oils Association., vol. 01, ISSN : 1003-0174

11

Farming System and Management

Morio Matsuzaki
National Agricultural Reseach Center (NARC),
National Agriculture and Food Research Organization (NARO)
Japan

1. Introduction

In the cultivation of soybean, it is necessary to pay attention to nitrogen absorption and soil organism. Nitrogen of 11 - 31kg (an average of 16 kg) is necessary to produce soybean grains of 200 kg (Salvagiotti et al., 2008) because soybean has high grain protein content. Soybean and rhizobia (*Bradyrhizobium japonicum*) form symbiosis for N2 fixation (Gray & Smith, 2005). Soybean N2 fixation is approximately half of the soybean nitrogen uptake (Salvagiotti et al., 2008), and soybean absorbs the other half from fertilizer or soil. If there is much inorganic nitrogen in soil, the N2 fixation is suppressed (Ray et al., 2006), and the amount of fertilizer application for soybean is a little. Therefore, soybean yield probably depends on the quantity of soil organic nitrogen which is mineralized during crops growing period. This means nitrogen which a soil microbe holds, and it is called "biomass nitrogen" (Jenkinson & Parry, 1989).

Soybean is influenced by a biologic factor. Soybean forms symbiosis not only rhizobia but also arbuscular mycorrhizal (AM) fungi (Antunes et al., 2006; Troeh & Loynachan, 2003). The biologic factors such as nematodes, soil-borne diseases become the problem in soybean. In continuous cropping of soybean, soybean cyst nematode (SCN : *Heterodera glycines*) reduce soybean yield approximately 30% (Donald et al., 2006). Sudden death syndrome (SDS) due to the coinfection of SCN and *Fusarium solani* becomes the problem in U.S.A. (Rupe et al., 1997; Xing & Westphal, 2009).

Soybean secretes flavonoids such as daidzein or genistein, and they are key signal compounds for control of symbiosis with rhizobia and AM fungi (Antunes et al., 2006). Glycinoeclepin which kidney beans (*Phaseolus vulgaris*) secrete promotes the hatching of the SCN (Kushida et al., 2002). Crops influence soil organism by various compounds to secrete from root (Faure et al., 2009). Therefore, the growth and yield of soybean are probably influenced by the preceding crop. In soybean, yield decrease is remarkable by continuous cropping (Matsuda et al., 1980; Matsuguchi & Nitta, 1988).

The continuous cropping experiment with five crops including soybean was conducted in northern Japanese Hokkaido for 16 years (Memuro continuous cropping experiment). Organic matter application and soil fumigation were conducted in the experiment. Soybean continuous cropping will influence soil microbe (Kageyama et al., 1982; Matsuguchi & Nitta, 1988). Soil biomass nitrogen increases by organic matter application (Sakamoto & Oba, 1993). Soil fumigation promotes mineralization of soil nitrogen and suppresses nitrification

(Neve et al., 2004). Soil fumigation sterilizes Fungus (Asano et al., 1983) and nematodes. This experiment is a good example to study the influence of the soil microbe on soybean. About a subject picked up in this experiment, the knowledge of past were surveyed.

	C0	W15	W30	W50	B15	B30	B50	F30	F15	F0	R
Continuous cropping	O	O	O	O	O	O	O	O	O	O	
Rotation											O
Wheat straw manure		15	30	50				30	15		
Burk compost					15	30	50				
Soil fumigation								O	O	O	

* The application rate of organic matter is expressed in t/ha.
Soil fumigation (D-D) was applied from 1990 to 1995.

Table 1. Treatments in Memuro continuous cropping experiments.

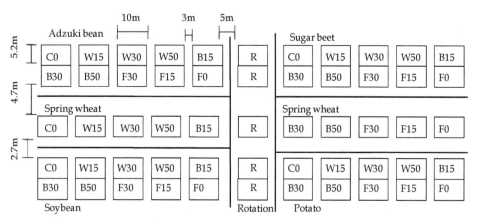

The cropping sequence of rotation plot is sugar beet – Potato - Adzuki bean - Spring wheat - Soybean ·

Fig. 1. The treatment plots in Memuro continuous cropping experiment.

2. Memuro continuous cropping experiment

2.1 Eexperimental design

Memuro continuous cropping experiment (42°53' N, 143°04' E) was conducted from 1980 to 1995. Soybean (*Glycine max*), adzuki bean (*Vigna angralis*), sugar beet (*Beta valgaris*), potato (*Solanum tuberosum*) and spring wheat (*Triticum aestivum*) were cultivated in the experiments. The eleven plots were established in each crop (Fig. 1). One plot was rotation plot, and other plots were continuous cropping plots (Table 1). Soybean was cultivated only chemical fertilizer in rotation plot (R) and the control plot in continuous cropping plots (C0). Wheat straw manure was applied in every year from 1980 at 15, 30, 50 t/ha, respectively (W15, W30, W50). Burk compost was applied in every year from 1981 at 15, 30, 50 t/ha, respectively (B15, B30, B50). D-D (1,3-dichloropropene) was applied from 1990 as a soil fumigation. In the soil fumigation plots, wheat straw manure was applied in every year from 1980 at 0, 15,30t/ha (F0, F15, F30).

2.2 The effect of continuous cropping to soybean cyst nematode (SCN)

The time course changes of egg density of SCN were showed in Fig. 2A. Closed symbols showed the value of rotation plot, and open symbols showed that of continuous cropping plots. The nematode susceptibility cultivar "Kitami-shiro" was used from 1980 to 1991, and nematode-resistant cultivar "Toyo-musume" was used from 1992. D-D was applied from 1990. In the continuous cropping plots, the egg density of SCN tended to decrease from 1985. Thereafter, the egg density in the rotation plot was higher than that in the continuous cropping plots.

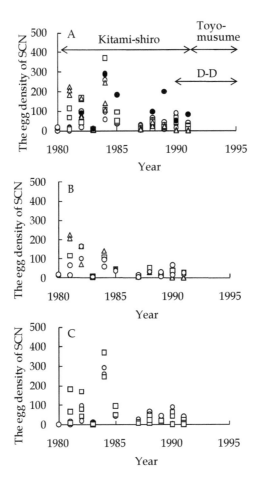

○ : C0, M15, M30, M50, □ : B15, B30, B50, △ : F0, F15, F30, ● : R.
Graph A: The changes of all plots.
Graph B: The changes of plots in which the egg density peaked at 1981.
Graph C: The changes of plots in which the egg density peaked at 1984.
The unit of the egg density of SCN is number / g dry soil.

Fig. 2. The time course changes of egg density of SCN.

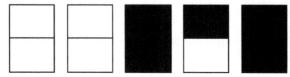

■:plots in graph B, □:plots in graph C.

Fig. 3. The placement of soybean continuous cropping plots in Graph B and C.

Fig. 2B and 2C show the time course changes of egg density of SCN in the continuous cropping plots. Fig. 2B shows the plots which the egg density of SCN was highest in 1981-1982. Fig. 2C shows the plots which the egg density of SCN was highest in 1984. The placement of each plots are showed in Fig. 3. The closed squares show the plots in Fig. 2B, and the open squares show the plots in Fig. 2C. The plots which belonged in Fig. 2B were located in the south side of experiment field, and the plots which belonged in Fig. 2C were located in the north side. The difference of time course changes of egg density probably depended on the position of plots. In both plots, the egg density of SCN decreased by 2 - 5 years continuous cropping.

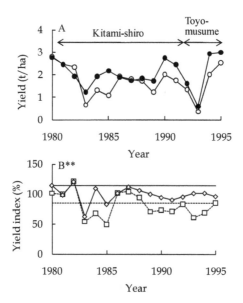

Graph A: The changes in yield of C0 and R.
 ○ : C0, ● : R.
Graph B: The time course changes in yield index.
 □ : C0 / R × 100.
◇ : The means of (W15, W30 W50, B15, B30 or B50) / R × 100,
Solid line: the value of ◇ in 1980.
Dotted line: solid line - least significant difference from Stutentized range.
** is significantly in 1%.

Fig. 4. The time course changes of yield and yield index of soybean.

2.3 The effects of treatments to soybean yield
2.3.1 Time course change
Soybean yield was measured from 1980 to 1995. The time course changes of yield of R and C0 plots were showed in Fig. 4A. In Hokkaido, soybean yield decrease by cool summer damage. It was a cool summer in 1983 and 1993, and soybean yield decreased. To examine the effects of continuous cropping and organic matter application, analysis of variance (ANOVA) was conducted for the soybean yield data. First, the yield of continuous cropping plots except soil fumigation plots (C0, W15, W30, W50, B15, B30, B50) were converted into the index by the yield of rotation plot (R). For the index of the continuous cropping + organic matter application plots (W15, W30, W50, B15, B30, B50), ANOVA was conducted as treatment replication (Fig. 4B). The solid line is a value of 1980, and the dotted line is solid line - least significant difference. In the year when index significantly decreased than that of 1980, soybean yield probably decreased by continuous cropping. For reference, the indexes of C0 were shown.

The mean of indexes of organic matter application plots did not decrease significantly except 1983. However, the indexes of C0 of 1983~1985 were lower than a dotted line. This time was almost coincided with the time when the egg density of SCN increased. The indexes of C0 increased again afterwards. The egg density of SCN decreased, too. Therefore, it is suggested that the yield decrease of continuous cropping was influenced by SCN. The organic matter application probably increased soybean yield.

2.3.2 Treatment effects
For the soybean yield data, ANOVA was conducted as year replication. Using a yield of all plots except soil fumigation plots (C0, W15, W30, W50, B15, B30, B50, R) of 1981~1995, the effects of continuous cropping and organic matter application were investigated. Using a yield of all plots in 1990~1991, the effects of continuous cropping, organic matter application and D-D on nematode susceptibility cultivar "Kitami-shiro" were investigated. Using a yield of all plots in 1992~1995, the effects of continuous cropping, organic matter application and D-D on nematode resistant cultivar "Toyo-musume" were investigated.

By ANOVA for the data of 1981-1995, yield decreased significantly by continuous cropping, and the decrease was approximately 20% (Fig. 5A). Organic matter application increased yield. By ANOVA for soil fumigation period (Fig. 5B, 5C), yield did not decrease significantly by continuous cropping. However, the trend in Fig. 5A was similar in those figures. Yield decreased by continuous cropping, and increased by organic matter application. By D-D, yield increased at the same level as the rotation plot. D-D might remove the effect of continuous cropping as a nematocide. However, the egg density of SCN declined before D-D application period. The effect of D-D was found to "Toyo-musume" that was nematode resistant variety. Therefore, it was suggested that D-D influenced the factor except SCN.

3. Factors to influence continuous cropping soybean

3.1 Soybean cyst nematode (SCN)
SCN forms the cyst containing a large number of eggs (Ichinohe, 1955a). Two or three generations of SCN can grow up in the soybean growing period of Hokkaido (Ichinohe, 1955a). SCN inhibits rhizobial adherence, too (Ichinohe 1955a). The damage of SCN is most remarkable if SCN invaded to soybean at 2-3 weeks after sowing (Okada, 1968). The damage of SCN is reduced by fertilization (Okada, 1966). SCN reduces the growth of soybean, but

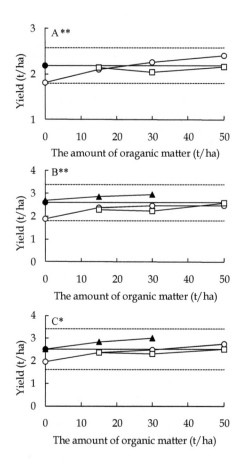

○ : C0, M15, M30, M50, □ : B15, B30, B50, ▲ : F0, F15, F30, ● : R.
Graph A : ANOVA for 1981 – 1995 (All period of continuous cropping),
Graph B : ANOVA for 1990 – 1991 (D-D for "Kitami-shiro"),
Graph C : ANOVA for 1992 – 1995 (D-D for "Toyo-musume").
Solid line : the value of R,
Dotted lines : solid line ± l.s.d. from Studentized range.
** is significantly in 1%, * is 5%.

Fig. 5. The treatment effects to soybean yield.

SCN cannot increase with poor growth soybean (Ichinohe, 1955b). Therefore, if the soybean growth is less, the density of SCN may be less. In Memuro continuous cropping experiment, the egg density of SCN was low in the cool summer damage year (1983).
Soil fumigation, organic matter application and cultivation of non-host crops or the resistant variety is effective for control of SCN. D-D etc are used for soil fumigation. Organic acid or ammonia released from organic matter suppresses a nematode (Oka, 2010). By organic matter including the chitinous substance, chitinase activity in the soil rises, and a nematode is suppressed (Akttar & Malik, 2000; Oka, 2010). Brassicaceae crops including glucosinolates

release isothiocyanates, and suppress a nematode (Oka, 2010). Probably Marigold suppresses a nematode by α- terthienyl (Oka, 2010).

Host crops of SCN is soybean, adzuki bean and kidney bean (Ichinohe, 1953). SCN is not parasitic on a non-leguminous crop. SCN is parasitic on other leguminous crops, but cannot become the adult (Ichinohe, 1953). After five years cultivation of corn (*Zea mays*) which is non-host crop, SCN increases by soybean cultivation (Porter et al., 2001). By the planting of the single resistant variety, the races adapted to the resistant variety increased (Shimizu & Mitsui, 1985). In contrast, the leguminous crop red clover (*Trifolium pratense*) may be used as trap crop (Kushida et al., 2002). Red clover hatch the egg of SCN, but the hatched larva cannot become the adult (Kushida et al., 2002). Therefore, after red clover cultivation, the density of SCN decreased (Kushida et al., 2002).

In Memuro continuous cropping experiment, SCN density decreased by 5 years continuous cropping of soybean. In other experiments, the density of SCN decreased by continuous cropping, too (Hashimoto et al., 1988). This phenomenon is called "SCN decline". SCN decreases in wheat-soybean double cropping (Bernard et al., 1996). It is suggested that these phenomena are caused because fungus or bacteria are parasitic on SCN. *Hirsutella* is parasitic on the second larva of SCN (Liu & Chen, 2000). *Fusarium* and *Verticillium* are parasitic on a female, cyst and egg of SCN (Bernerd et al., 1996; Sayre, 1986, Siddiqui & Mahmood, 1996). The nematode control using these microorganisms is possible. However, in Memuro continuous cropping experiment, soybean might be cropped continuously before the experiment station establishment. Long term continuous cropping may be needed to "SCN decline".

3.2 Nitrogen supply and soybean yield

In American Corn Belt, the potential yield of soybean is estimated at 6 - 8t/ha (Salvagiotti et al., 2008). Because soybean is crop which is high grain protein content, soybean need a large quantity of nitrogen. Nitrogen of 106 – 310 kg N / ha is necessary to get a yield of 2 t / ha (Salvagiotti et al., 2008). In the Tokachi district, the fixed nitrogen of soybean is 4 – 127 kg N /ha (Nishimune et al., 1983). There is negative correlation between amount of applied fertilizer and N2 fixation (Salvagiotti et al., 2008). Fertilizer nitrogen for soybean is less than 40kgN/ha. Therefore, soybean needs to absorb nitrogen from soil. Soybean yield will increase with soil nitrogen absorption.

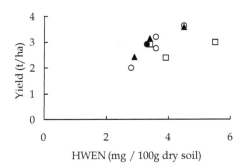

○ : C0, M15, M30, M50, □ : B15, B30, B50, ▲ : F0, F15, F30, ● : R.

Fig. 6. The relationship of the hot water extractable nitrogen (HWEN) of postharvest soil and soybean yield (1994).

Soil Nitrogen can be divided into inorganic, humus and biomass nitrogen (Jenkinson & Parry, 1989). Crops absorb inorganic nitrogen. The available nitrogen is organic nitrogen at sowing, but it is mineralized during a growing period. It is suggested that most of available nitrogen come from biomass nitrogen (Sakamoto & Oba, 1993). Organic matter application increases biomass nitrogen (Sakamoto & Oba, 1993). Available nitrogen and the heated water extraction nitrogen (HWEN) have corelation (Akatsuka & Sakayanagi, 1964). In Memuro continuous cropping experiment, the relationship of the HWEN of postharvest soil and soybean yield was investigated in 1994 (Fig. 6). With increase of the HEWN, soybean yield tended to increase. It is suggested that organic matter application increases biomass nitrogen, and contributes to yield increase.

D-D promotes mineralization from biomass nitrogen (Neve et al., 2004). D-D suppresses nitrification from ammonia nitrogen (Neve et al., 2004). Therefore, much ammonia nitrogen in soil will be kept by D-D. In the Memuro continuous cropping experiment, D-D was applied before one month of sowing. Therefore, it is thought that these effects influenced a soybean.

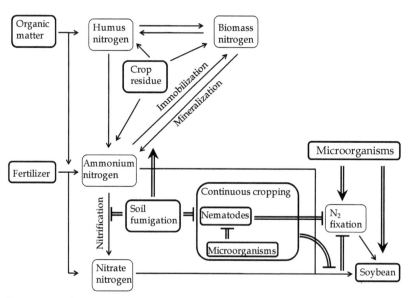

Fig. 7. The nitrogen flow and the factors influence to it.

4. Farming system and management to influence soybean yield

Farming system and management influencing soybean are described. Fig. 7 illustrates the results provided in Memuro continuous cropping experiment and the knowledge of the past.

4.1 Soil organic matter (SOM)

Soybean needs nitrogen absorption to get high yield. The Brazilian soybean absorbs approximately 100% of N need by N2 fixation, but in soybean of U.S.A. or Japan, fixed nitrogen is approximately 50% of N need (Graham & Vance, 2000). This may be connected with that soybean N2 fixation decrease at low temperature (F. Zhang et al., 1995). In other words, the nitrogen supply from soil probably become important so as to be a cold area.

4.1.1 Farming system

The soil organic matter (SOM) is broken down by the soybean planting (Cheng et al., 2003). This effect is called "priming effect". Soil carbon and nitrogen decrease in soybean continuous cropping in comparison with Gramineae - soybean rotation (Kelley et al., 2003; Wright & Hons, 2004). However, soybean-corn rotation can reduce the fertilizer nitrogen of 60kgN/ha/year in comparison with corn continuous cropping (Varvel & Wilhelm, 2003). The soil nitrogen mineralization quantity increases in soybean- corn rotation in comparison with soybean continuous cropping (Carpenter-Boggs et al., 2000). It is suggested that soybean breaks down SOM and will increase inorganic nitrogen, but soybean continuous cropping will cause a decrease of SOM.

In soybean rotation which incorporated alfalfa (*Medicago sativa*), the quantity of soil nitrogen mineralization increases greatly (Carpenter-Boggs et al., 2000). In soybean introducing to the permanent grass pasture, soybean yielded 3 t / ha at no chemical fertilizer (Diaz et al., 2009). One of the causes of these phenomena will be that pasture plant leaves much organic matter in soil.

4.1.2 Management

SOM increases by organic matter application. SOM is maintained by no-tillage (Wright & Hons, 2004). Because mineralization of soil nitrogen decreases by no-tillage, N2 fixation probably increases (van Kessel, 2000). These treatments are suggested to increase soybean yield.

4.2 Soybean cyst nematode (SCN)
4.2.1 Farming system

SCN inhibits the production of soybean. SCN does not increase by the cropping of non-host crop, but SCN increases by soybean cropping again (Asai & Ozaki, 1965). Gramineous crops such as corn and wheat are non-host crop of SCN. Soybean- corn rotation carried out in the northern part of U.S.A. (Varvel & Wilhelm, 2003; Xing & Westphal, 2009), but SDS by SCN and *Fusarium solani* occurs in this rotation (Rupe et al., 2003; Xing & Westphal, 2009). SDS can lead to defoliation of the leaflets, leaving the petioles attached to the plant after flowering (Rupe et al., 2003; Xing & Westphal, 2009). Pythium have a pathogenicity in soybean and corn and cause damping-off (B.Q. Zhang et al., 1998).

In the Southern U.S.A., soybean is cultivated by no-tillage in soybean - wheat double cropping (Bernard et al., 1996). No-tillage is used to corn, soybean, wheat and etc in U.S.A., and the cultivated area occupies 23% in U.S.A. (Triplett. Jr. & Dick, 2008). No-tillage reduces nematode density in soybean - wheat double cropping (Bernard et al., 1996). However, take-all (*Gaeumannomyces graminis*) of wheat cannot be reduced in soybean- wheat double cropping (Cook, 2003).

Some plants are able to control nematodes. Probably Marigold controls nematodes with chemical substances such as α -terthienyl (Oka, 2010). The Brassicaceae plants control nematodes with isothiocyanates which is broken down from glucosinolates (Oka, 2010). The leguminous crops such as red clovers are probably available as trap crop reducing the egg density of SCN (Kushida et al., 2003).

4.2.2 Management

Nematodes may be controlled by organic matter application. The organic matter including inorganic nitrogen or chitinous substance is effective for nematodes control (Akhtar & Malik, 2000; Oka, 2010). The application of organic matter including antagonism microorganism will be effective (Oka, 2010). However, these effects will vary according to materials or adjustment methods.

4.3 Arbuscular Mycorrhizal (AM) fungi

With soybean, plant growth promoting rhizobacteria (PGPR) such as rhizobia and *Bacillus* form symbiosis (Bai et al., 2003; Cattelan et al., 1999; Gray & Smith 2005), and arbuscular mycorrhizal (AM) fungi form symbiosis, too (Troeh & Loynachan, 2003). The AM fungi increase N2 fixation (Antunes et al., 2006). The symbiosis with AM fungi helps phosphorus acid absorption of the crop (Harrison, 1998). The crops doing N2 fixation need a lot of phosphoric acid (Graham & Vance, 2000), and the symbiosis with AM fungi will be effective for soybean. Soybean and corn form symbiosis with AM fungi (Troeh & Loynachan, 2003), and AM fungi probably increases in soybean - corn rotation. Sugar beet was non-host crop of AM fungi, and soybean growth after non-host crop such as sugar beet was suppressed (Karasawa, 2004).

5. Conclusion

Soybean forms symbiosis with rhizobia. However, in low temperature area, the nitrogen absorption of soybean may not be served only in N2 fixation, and soil nitrogen probably becomes important. Soybean stimulates the decomposition of SOM, and can absorb nitrogen from soil. The crops such as gramineous crop or pasture plant supply organic matter to soil. No-tillage maintains SOM. Organic matter application increases SOM. Soybean yield probably increases by the combination of these treatments.

SCN inhibits the nitrogen absorption of soybean. By cropping of non-host crops or resistant varieties, SCN does not increase, but the cyst of SCN does not decrease. Soybean is affected by not only SCN but also pathogenic fungi such as *Fusarium*. On the other hand, soybean growth is promoted by PGPR (rhizobia, *Bacillus* and etc) and AM fungi. In soybean, it is necessary to decide farming system and management while considering these organisms.

6. References

Akatsuka K. and Sakayanagi, M. (1964) Consideration on the Soil Nitrogen Test. Research Bulletin of Hokkaido National Experiment Station, Vol. 83, (March 1964) pp. 64-70, ISSN 0018-3415 (In Japanese)

Akhtar, M. and Malik, A. (2000) Roles of Organic Soil Amendments and Soil Organisms in the Biological Control of Plant-Parasitic Nematodes: A Review. Bioresource Technology, Vol. 74, No. 1, (August 2000) pp. 35-47, ISSN 0960-8524

Antunes, P.M., de Varennes, A., Rajcan, I. and Goss, M.J. (2006) Accumulation of Specific Flavonoids in Soybean (*Glycine max* (L.) Merr.) as a Function of the Early Tripartite Symbiosis with Arbuscular Mycorrhizal Fungi and *Bradyrhizobium japonicum* (Kirchner) Jordan. Soil Biology & Biochemistry, Vol. 38, No. 6 (June 2006) pp.1234–1242, ISSN 038-0717

Asai, K. and Ozaki, K. (1965) Studies on the Rotation Systems. Part 6. Behavior of Population Density of Soybean Cyst Nematode by Different Cropping Systems. Research Bulletin of Hokkaido National Experiment Station, Vol. 87 (December 1965) pp. 66-73, ISSN 0018-3415 (In Japanese)

Asano, M., Kato, T., Kinoshita, T. and Arisawa, M. (1983) Studies on the Behavior of Nutrients in Soil Resulting from Soil Disinfection (II) Effects of Soil Disinfection on Soil Microbes and the Growth of Fruit Vegetables in Greenhouse. Research Bulletin of the Aichi Agricultural Research Center, Vol. 15, (October 1983) pp. 216-222, ISSN 0388-7995 (In Japanese)

Bai, Y., Zhou, X. and Smith, D.L. (2003) Enhanced Soybean Plant Growth Resulting from Coinoculation of *Bacillus* Strains with *Bradyrhizobium japonicum*. Crop Science, Vol. 43, No. 5 (September 2003) pp. 1774–1781, ISSN 0011-183X

Bernard, E.C., Self, L.H. and Tyler, D.D. (1996) Fungal Parasitism of Soybean Cyst Nematode, *Heterodera glycines* (Nemata: Heteroderidae), in Differing Cropping-Tillage Regimes. Applied Soil Ecology, Vol. 5, No. 1, (January 1997) pp. 57–70, ISSN 0929-1393

Carpenter-Boggs, L., Pikul, J.L., Vigil, M.F. and Riedell, W.E. (2000) Soil Nitrogen Mineralization Influenced by Crop Rotation and Nitrogen Fertilization. Soil Science Society of America Journal, Vol. 64, No, 6 (November 2000) pp. 2038–2045, ISSN 0361-5995

Cattelan, A.J., Hartel, P.G. and Fuhrmann, J.J. (1999) Screening for Plant Growth–Promoting Rhizobacteria to Promote Early Soybean Growth. Soil Science Society of America Journal, Vol. 63, No. 6 (November 1999) pp. 1670-1680, ISSN 0361-5995

Cheng, W., Johnson, D.W. and Fu, S. (2003) Rhizosphere Effects on Decomposition Controls of Plant Species, Phenology, and Fertilization. Soil Science Society of America Journal Vol. 67, No. 5, (September 2003) pp. 1418-1427, ISSN 0361-5995

Cook, R. J. (2003) Take-All of Wheat. Physiological and Molecular Plant Pathology, Vol. 62, No. 3, (March 2003) pp. 73–86, ISSN 0885-5765

Diaz, D.A.R., Pedersen, P. and Sawyer, J.E. (2009) Soybean Response to Inoculation and Nitrogen Application Following Long-Term Grass Pasture. Crop Science, Vol. 49, No. 3, (May 2009) pp. 1058–1062, ISSN 0011-183X

Donald, P.A., Pierson, P.E., ST. Martin, S.K., Sellers, P.R., Noel, G.R., MacGuidwin, A.E., Faghihi, J., Ferris, V.R., Grau, C.R., Jardine, D.J., Melakeberhan, H., Niblack, T.L., Stienstra, W.C., Tylka, G.L., Wheeler, T.A. and Wysong, D.S. (2006) Assessing *Heterodera glycines* -Resistant and Susceptible Cultivar Yield Response. Journal of Nematolgy, Vol. 38, No. 1 (March 2006) pp. 76–82, ISSN 0022-300X

Faure, D., Vereecke, D. and Leveau, J.H.J. (2009) Molecular Communication in the Rhizosphere. Plant and Soil Vol. 321, No. 1-2, (August 2009) pp. 279–303, ISSN 0032-079X

Graham, P.H. and Vance, C.P. (2000) Nitrogen Fixation in Perspective: An Overview of Research and Extension Needs. Field Crops Research, Vol. 65, No.2-3, (March 2000) pp. 93–106, ISSN 1378-7290

Gray, E.J., and Smith, D.L. (2005) Intracellular and Extracellular PGPR: Commonalities and Distinctions in the Plant–Bacterium Signaling Processes. Soil Biology & Biochemistry, Vol. 37, No. 3, (March 2005) pp. 395–412, ISSN 0038-0717.

Harrison, M.J. (1998) Development of the Arbuscular Mycorrhizal Symbiosis. Current Opinion in Plant Biology, Vol. 1, No. 4, (August 1998) pp. 360–365, ISSN 1369-5266

Hashimoto, K., Inagaki, H., Momota, Y., Sakai, S., Nagasawa, T. and Kokubun, K. (1988) Variation in Populations of *Heterodera glycines*, and the Growth and Yield of Soybean Cultivars with Various Levels of Resistance Grown under Continuous Cropping. Bulletin of the Tohoku National Agricultural Experiment Station, Vol. 78, (August 1988) pp. 1–14, ISSN 0495-7318 (In Japanese)

Ichinohe, M. (1953) On the Parasitism of the Soy Bean Nematode, *Heterodera glycines*. Research Bulletin of Hokkaido National Experiment Station, Vol. 64, (February 1953) pp. 113-124, ISSN 0018-3415 (In Japanese)

Ichinohe, M. (1955a) Studies on the Morphology and Ecology of the Soy Bean Nematode, *Heterodera glycines*, in Japan. Report of Hokkaido Agricultural Experiment Station, Vol. 48, (March 1955) pp. 1–72, ISSN 0018-3407 (In Japanese)

Ichinohe, M. (1955b) A Study on the Population of the Soy Bean Nematodes (*Heterodera glycines*) 1. An Observation on the Relation between the Crop Damage and the Female Infestation. Research Bulletin of Hokkaido National Experiment Station, Vol. 68, (January 1955) pp. 67-70, ISSN 0018-3415 (In Japanese)

Jenkinson, D.S. and Parry, L.C. (1989) The Nitrogen Cycle in the Broadbalk Wheat Experiment: A Model for the Turnover of Nitrogen through the Soil Microbial Biomass. Soil Biology & Biochemistry, Vol. 21, No. 4 (1989) pp. 535-541, ISSN 0038-0717

Kageyama, K., Ui, T., Narita, Y. and Yamaguchi, H. (1982) Relation of *Pythium* spp. to Monoculture Injury of Soybean. Annals of the Phytopathological Society in Japan, Vol. 48, No. 3, (July 1982) pp. 333–335, ISSN 0031-9473 (In Japanese)

Karasawa T. (2004) Arbuscular Mycorrhizal Associations and Interactions in Temperate Cropping Systems. Research Bulletin of National Agricultural Research Center for Hokkaido Region, Vol. 179, (March 2004) pp. 1–71, ISSN 1347-8117. (In Japanese)

Kelley, K.W., Long Jr., J.H. and Tod, T.C. (2003) Long-Term Crop Rotations Affect Soybean Yield, Seed Weight, and Soil Chemical Properties. Field Crops Research, Vol. 83, No. 1 (June 2003) pp. 41–50, ISSN 1378-7290

Kushida, A., Uehara, T. and Momota, Y. (2002) Effect of Red Clover Hatching and Population Density of *Heterodera glycines* (Tylenchida: Heteroderidae). Japanese Journal of Nematology, Vol. 32, No. 2 (December 2002) pp. 69–76, ISSN 0919-6765

Liu, X.Z. and Chen, S.Y. (2000) Parasitism of *Heterodera glycines* by *Hirsutella* spp. in Minnesota Soybean Fields. Biological Control, Vol. 19, No. 2, (October 2000) pp. 161–166, ISSN 1049-9644

Matsuda, M., Horie, M., Honda, K. and Shimura, E. (1980) Studies on the Rotation of Upland Crops XI. Yield Trends of Crops in Continuous Cropping and Several Rotation Systems over 42-Year Experiments. Japanese Journal of Crop Science, Vol. 49, No. 4, (December 1980) pp. 548–558, ISSN 0011-1848 (In Japanese)

Matsuguchi, T. and Nitta, T. (1988) Organic Amendment to Upland Soil as a Conditioner of the Rhizosphere Ecosystem (Part 3) Effects of Organic Amendments on Root Development and the Rhizosphere Microflora of Monocultured Upland Crops. Japanese Journal of Soil Science and Plant Nutrition, Vol. 59, No.1. (February 1988) pp.1–11, ISSN 0029-0610 (In Japanese)

Neve, S. D., Csitári, G., Salomez, J. and Hofman, G. (2004) Quantification of the Effect of Fumigation on Short- and Long-Term Nitrogen Mineralization and Nitrification in Different Soils. Journal of Environmental Quality, Vol. 33, No. 5, (September/ October 2004) pp. 1647–1652, ISSN 0047-2425

Nishimune, A., Konno, T., Saito, G. and Fujita, I. (1983) The Relationship between Nitrogen Fixation and Grain Yields of Legumes on the Main Upland Soils in Tokachi, Hokkaodo. Research Bulletin of Hokkaido National Experiment Station, Vol. 137, (March 1983) pp. 81-106. ISSN 0367-5955 (In Japanese)

Oka, Y. (2010) Mechanisms of Nematode Suppression by Organic Soil Amendments – A Review. Applied Soil Ecology, Vol. 44, No. 2 (February 2010) pp. 101–115, ISSN 0929-1393.

Okada, T. (1966) Studies on the Parasitic Distribution of Nematodes. Part 3. Effects of Amounts of Fertilizer on the Growth of Soybeans and the Parasitic Distribution of Soybean Cyst Nematodes. Research Bulletin of Hokkaido National Experiment Station, Vol. 89, (March 1966) pp. 30-36, ISSN 0018-3415 (In Japanese)

Okada, T. (1968) Effects of the Inoculation Time of Soybean Cyst Nematodes, *Heterodera lycines* ICHINOHE, on the Growth of Soybean. Research Bulletin of Hokkaido National Experiment Station, Vol. 93, (October 1968) pp. 32-38, ISSN 0018-3415 (In Japanese)

Porter, P.M., Chen, S.Y., Reese, C.D. and Klossner, L.D. (2001) Population Response of Soybean Cyst Nematode to Long Term Corn–Soybean Cropping Sequences in Minnesota. Agronomy Journal, Vol. 93, No. 3, (May 2001) pp. 619–626, ISSN 0002-1962

Ray, J.D., Heatherly, L.G. and Fritschi, F.B. (2006) Influence of Large Amounts of Nitrogen on Nonirrigated and Irrigated Soybean. Crop Science, Vol. 46, No. 1, (January 2006) pp. 52-60, ISSN 0011-183X

Rupe, J.C., Robbins, R.T. and Gbur Jr., E.E. (1997) Effect of Crop Rotation on Soil Population Densities of *Fusarium solani* and *Heterodera gl'ycines* and on the Development of Sudden Death Syndrome of Soybean. Crop Protection, Vol. 16, No. 6 (September 1997) pp. 575–580, ISSN 0261-2194

Sakamoto, K. and Oba, Y. (1993) Relationship between Available N and Soil Biomass in Upland Field Soils. Japanese Journal of Soil Science and Plant Nutrition, Vol. 64, No. 1, (February 1993) pp. 42–48, ISSN 0029-0610 (In Japanese)

Salvagiotti, F., Cassman, K.G., Specht, J.E., Walters, D.T., Weiss, A. and Dobermann, A. (2008) Nitrogen Uptake, Fixation and Response to Fertilizer N in Soybeans: A Review. Field Crops Research, Vol. 108, No. 1, (July 2008) pp. 1-13, ISSN 1378-7290

Sayre, R.M. (1986) Pathogens for Biological Control of Nematodes. Crop Protection, Vol. 5, No. 4, (August 1986) pp. 268-276, ISSN 0261-2194

Shimizu, K. and Mitui, Y. (1985) Races of the Soybean Cyst Nematode in Tokachi Province, Hokkaido. Research Bulletin of Hokkaido National Experiment Station, Vol. 141, (January 1985) pp. 65–72, ISSN 0367-5995 (In Japanese)

Siddiqui, Z.A. and Mahmood, I. (1996) Biological Control of Plant Parasitic Nematodes by Fungi: A Review. Bioresource Technology, Vol. 58, No. 3, (December 1996) pp. 229–239, ISSN 0960-8524.

Triplett Jr., G.B. and Dick, W.A. (2008) No-Tillage Crop Production: A Revolution in Agriculture! Agronomy Journal, Vol. 100, Supplement, (May 2008) pp. 153–165, ISSN 0002-1962

Troeh, Z.I. and Loynachan, T.E. (2003) Endomycorrhizal Fungal Survival in Continuous Corn, Soybean, and Fallow. Agronomy Journal, Vol. 95, No. 1, (January 2003) pp. 224-230, ISSN 0002-1962

Varvel, G. E. and Wilhelm, W.W. (2003) Soybean Nitrogen Contribution to Corn and Sorghum in Western Corn Belt Rotations. Agronomy Journal, Vol. 95, No. 5, (September 2003) pp. 1220-1225, ISSN 0002-1962

van Kessel C. and Hartley, C. (2000) Agricultural Management of Grain Legumes: Has It Led to an Increase in Nitrogen Fixation? Field Crops Research, Vol. 65, No. 2-3, (March 2000) pp. 165–181, ISSN 1378-7290

Wright, A.L. and Hons, F.M. (2004) Soil Aggregation and Carbon and Nitrogen Storage under Soybean Cropping Sequences. Soil Science Society of America Journal, Vol. 68, No. 2 (March 2004) pp. 507 - 513, ISSN 0361-5995

Xing, L. and Westphal, A. (2009) Effects of Crop Rotation of Soybean with Corn on Severity of Sudden Death Syndrome and Population Densities of *Heterodera glycines* in Naturally Infested Soil. Field Crops Research, Vol. 112, No. 1, (April 2009) pp. 107-117, ISSN 1378-7290

Zhang, B.Q., Chen, W.D. and Yang, X.B. (1998) Occurrence of *Pythium* Species in Long-Term Maize and Soybean Monoculture and Maize/Soybean Rotation. Mycological Research, Vol. 102, No. 12, (December 1998) pp. 1450–1452, ISSN 0953-7562.

Zhang, F., Lynch, D. and Smith D. L. (1995) Impact of Low Root Temperatures in Soybean (*Glycine max* (L.) Merr.) on Nodulation and Nitrogen Fixation. Environmental and Experimental Botany, Vol. 35, No. 3 (July 1995) pp. 279–285, ISSN 0098-8472

12

Productivity and Nutritional Composition of *Lentinus strigosus* (Schwinitz) Fries Mushroom from the Amazon Region Cultivated in Sawdust Supplemented with Soy Bran

Ceci Sales-Campos, Bazilio Frasco Vianez
and Raimunda Liége Souza de Abreu
National Institute for Amazonian Research –INPA/Department of Forest Products – CPPF, Aleixo CEP-Manaus, AM Brazil

1. Introduction

The cultivation of edible mushrooms is a biotechnological process that uses various residues to produce food of high nutritional value. It is an activity of economical importance, in particular, the production of *Agaricus*, *Pleurotus* and *Lentinus* species (Guzmán et al., 1993). Nutraceutical properties of mushrooms are increasing its economic value. The cultivation of mushrooms can be a solution to problems of global importance, such as the lack of protein in developing countries and the possibility of environmental management. The use of organic materials for growing mushrooms is an indication of its extraordinary metabolic activity.

In the Amazon region there are large amounts of wood and agricultural residues whose potential has been underestimated. In the timber industry, raw material waste can be as high as 60%. In the agricultural industry there is no data on how much waste their activities produce. Vianez and Barbosa (2003), suggest several alternatives for the use of wood residues, including the use for the cultivation of edible mushrooms. This activity could contribute to a sustainable regional development. In this way, the objective of this study is to study the feasibility of using sawdust supplemented with soy bran for growing *L. strigosus*, a native mushroom of the Amazon region.

2. Literature review

The fungus proposed in the present study is a wild edible and a wood decomposer (white-rot fungus), whose domestication was sought for the production of mushroom. As there is no cultivation of this fungus with the proposed wild strain, for comparison in the literature review, a parallel association was made with the cultivation of species of edible fungi of related genera that have similar physiology and cultivation conditions, being considered mainly the genera *Lentinus* and *Pleurotus*.

2.1 History

Fossil finds have revealed that fungi exist since the Cretaceous period (approximately 130 million years ago), long before humanity (Chang, 1993). Fungi (mushrooms), also called macromycets, belong to the Fungi Kingdom, being known by man since the most remote period of human history. Edible mushrooms were first collected by man in China and dates from 5000-4000 BC. (Zhanxi and Zhanhua, 2001). It is estimated that the first cultivation of edible mushrooms in China started in the early 7th century, with the species *Auricularia auricula* (Chang and Miles, 1987). China is a country with a long tradition in cultivation and consumption of mushrooms and according to Zhanxi and Zhanhua (2001), it has more than ten species of fungi which are currently cultivated in several countries of the world. That country is a pioneer in the cultivation and consumption of edible and medicinal mushrooms, followed by Japan, Europe and The United States (Urben et al., 2001).

Edible mushrooms were described as the "food of the Gods" and as such, confirmed by Roman gourmets who appreciated them as a kind of spice. The Chinese considered them as the "elixir of life". The Greeks believed that the mushrooms were able to give strength to warriors in battles and the Egyptian pharaohs also nourished themselves on these spices (Chang and Miles, 1984). Mushrooms had a wide acceptance, and some species are considered as "Kings of the dining table" or "kitchen diamonds" (Zhanxi and Zhanhua, 2001).

The Greeks Euripides, Theophrastus and Plinio have described the consumption of edible mushrooms in their time (Guzmán et al., 1993). In some societies, the mushroom was a royal food, probably by its pleasant flavor and texture (Miles and Chang, 1997).

The Romans knew several edible and poisonous fungi. There is a story about the Emperor Julius Caesar who was very fond of *Amanita caesarea* mushroom, whose scientific name was a homage to him and for that reason, it became known as "Mushroom of the Caesars" (Guzmán et al., 1993).

According to Molena (1986), the species *Polyporus tuberoster* (*fungaie stone*) and *Polyporus corilinus* are among the first cultivated mushrooms, collected from the wood of hazels and eucalyptus. These fungi were consumed in 4-5 cm slices, and their production demanded about six months, yielding sometimes one or two mushrooms at a time. There was neither any knowledge about their nutritional requirements, nor about their growth cycle. The only thing that was known was that rubbing a mature mushroom on those woods, and leaving them in a wet environment during a particular period of the year could produce appreciable mushrooms (Molena, 1986).

During the Roman Empire the fungaie stone (stone that produces the mushroom) appeared in Italy, which was composed of a cluster of humus, leaves, twigs and limestone rocks, forming a compact mass, which was cut in blocks in the form of bricks and transported to the royal palaces. They were kept in a damp place and irrigated daily until harvest time to serve the senators and other members of the Roman aristocracy (Molena, 1986). In France, the mushroom cultivation began during the reign of Luis XIV, according to Molena (1986). However, the cultivation of *Agaricus bisporus*, the "Champignon de Paris", the most widely cultivated and commercialized species, has been produced since about 1650 (Delmas, 1978; Chang and Miles, 1984).

With the advances of knowledge and technology of mushroom cultivation, commercial production of dozens of species became viable in several countries in recent decades (Guzmán et al., 1993; Stamets, 1993; 2000; Vedder, 1996; Eira, 2000), reaching a production of approximately 4.3 million tons of edible mushrooms in 1991 (Miles and Chang, 1997). The world current production is around 6.2 million tons (Chang, 2003).

2.2 An overview of the commercial cultivation of edible fungi in the world and in Brazil

After World War II, the edible mushroom industry grew from 350,000 tons in 1965 to 4.3 million tons in 1991, from which 3.4 million tons belong to the six most worldwide important genera: *Agaricus, Pleurotus, Lentinula, Auricularia, Volvariella* and *Flammulina*. The major producers are China, Japan, USA and France (Miles and Chang, 1997). The most cultivated genera are *Agaricus, Pleurotus* and *Lentinula*. This increase was due to several factors, among them: a) the increase in the number of species on a commercial scale; b) the development of cultivation techniques using plastic bags, which allowed many wood decomposers edible fungi to be grown on lignocellulosic residues, preferably the cultivation on logs, reducing considerably the cultivation time; c) due to the marketing techniques highlighting the nutritional merits of mushrooms as an important part of the diet, so they wouldn't be marketed as simple accompaniments or delicacies, but as a food of high nutritional value (Miles and Chang, 1997).

The literature cites approximately 200,000 species of fungi existing in the world, from which, about 2,000 are potentially edible species. However, only 25 of them are commonly used as food, and fewer still are commercially cultivated (Chang and Miles, 1984; Chang, 1980; Bononi, 1999).

In the early 1980s, only *Agaricus bisporus* (Champignon de Paris) and other species of this genus and "shiitake" (*Lentinus edodes*, currently named *Lentinula edodes*) had a modern technology for commercial production, where 70% of the world production was represented by *Agaricus* and 14% by *Lentinula* (Chang and Miles, 1984). However, according to the same authors, the world's attention is turning to the development of new technologies for different species of worldwide known edible mushrooms, especially considering the difficulties of production in tropical and subtropical climates. Special technologies are being developed in several countries allowing the cultivation of: *Volvariella volvaceae* in China, Taiwan, Japan, Philippines and Indonesia; *Kuehneromyces mutabilis, Flamulina velutipes, Hypholoma capnoides* and *Coprinus comatus* in some countries of Europe and Asia; *Pleurotus ostreatus* in Italy, Hungary, West Germany, Mexico and Brazil (Chang and Miles, 1984; Guzman et al., 1993; Eira and Minhoni, 1997; Bononi et al., 1999; Zhanxi and Zhanhua, 2001; Urben et al., 2001). This way, the overview of the world production has changed suddenly, showing a considerable increase in cultivation and consumption of *Pleurotus* as reported by Eira (1997) adapted by Fermor (1993).

An adaptation based on Fermor (1993), made by Eira et al. (1997), the world production of cultivated mushrooms in the early 1990s was 1,424,000 tons for *Agaricus bisporus*, 900,000 tons for *Pleurotus* spp, 393,000 tons for *L. edodes* and 887,000 tons for other mushrooms, representing, respectively, 39.51%, 24.98%, 10.91% and 24.61%. The current trend is to increase production.

Concerning the production of mushrooms in Brazil, there is not a precise documentation that could allow us to determine when the cultivation of mushrooms started in the country (Fidalgo and Guimarães, 1985). Its popularization in the Center-South region of Brazil dates back to 50 years ago. Bononi (1999) reports that the cultivation of champignon (*Agaricus*) began in 1953, when the Chinese immigrants settled in Mogi das Cruzes and the Italian Oscar Molena in Atibaia, brought technology and imported strains of their countries. For Molena (1985), mushroom cultivation began in 1953 and developed after the poultry crisis in the period of 1955-1959, when breeders began to use chicken sheds for the cultivation of mushrooms, without proper technical conditions.

The commercial cultivation of edible fungi in Brazil is limited to *Agaricus bisporus* (champignon), *Lentinula edodes* (shiitake) and *Pleurotus* spp, known as oyster mushroom, giant mushroom or caetetuba (Bononi et al., 1999; Eira, 2000). Varieties or strains of mushrooms of the *Pleurotus* genus gave origin to the "hiratake" (mushrooms with very large basidiocarp, harvested in mature stage with opened basidiocarps, before they turn their edges upwards and with more than 5 cm in diameter) and the "shimeje" (with long stipes, harvested with their basidiocarps very young and dark, smaller than 5 cm, and can be harvested in bunches) (Eira and Minhoni, 1997).

There are few Brazilian research reports about the subject, and the Botanical Institute of São Paulo was one of the pioneers, creating a research center of edible mushrooms in Mogi das Cruzes in 1985 and a teaching, research and extension nucleus was created in the Faculty of Agronomic Sciences/UNESP in Botucatu, in 1986, named Module of Mushrooms (Eira, 2000). Other centers are springing up in many universities and research institutions.

The production of edible mushrooms in Brazil is difficult to be evaluated. Producers give preference (90%) to the cultivation of *A. bisporus*, (Bononi et al., 1999). Among producers, the majority, almost 90%, are from the East of Taiwan, China, Korea, Japan, working in small properties, in a family system, with all family members operating in all stages of cultivation, in a collective way. The region of the city of Mogi das Cruzes, São Paulo, is responsible for approximately 70% of the edible mushrooms commercialized in Brazil. The remainder are produced by other municipalities, most of them also in São Paulo, Ribeirão Pires, Suzano, Cabreúva, Atibaia, Mariporã, Sorocaba (Bononi et al., 1999; Souza, 2011). There are some important producers in Porto Alegre and some producing installations in southern Minas Gerais and Paraná States.

In 1990, the production of Mushrooms was only 3,000 tons according to Eira et al. (1997), being estimated at 10,000 tons per year until 1997. According to the APAN (Natural Agriculture Producers Association), the Brazilian production of shiitake in late 1995 among its associates, was approximately six tons per month. The official data are underestimated, because they include only the mushrooms marketed by CEAGESP (General Supply Center of São Paulo State) and those intended for export, which are recorded by the CACEX-Department of Foreign Trade (Eira et al., 1997; 2004; Bononi et al., 1999). It is known, however, that significant amounts are marketed directly by the producers with restaurants, pizzerias, snack bars and other establishments, as well as street markets.

Brazilian productivity of *Agaricus bisporus* "Champignon de Paris" in Mogi das Cruzes (in São Paulo State) until 2000 was of the order of 5 to 7 kg of fresh mushrooms/100 kg of moist substrate (4 to 6 kg of fresh mushroom/m²) (Eira, 2000). In Europe, however, in countries such as Belgium, Holland, Germany and France, the average productivity of mushroom at that time was 30 kg/100 kg of substrate.

Currently in Brazil, farms with more technology get on average a productivity (substrate conversion in mushrooms) for "Champignon de Paris" ranging from 18 to 24% in 20 to 30 days of the crop cycle. In more rustic crops this conversion varies from 12 to 15% in 70 to 90 days. While the numbers seem to have greatly increased, yet it is little when compared to Asian and European productions, where they manage 30 to 40% of conversion (Souza, 2011). Even today there are no official data concerning the production of mushrooms in Brazil, but some unofficial sources report that 12,050 tons a year of mushrooms "in natura" (table 1). Since 1995, there is an annual import of 12,000 tons per year, on average, most of it cooked *Agaricus bisporus*, to meet market demand. Therefore, it can be concluded that, Brazilian consumption is much higher than its production, reaching 24,050 tons per year. In this context Brazilian people consume around 130 g per capita (Souza, 2011). The world production is around 6.2 million tons (Chang, 2003).

Agaricus bisporus	8,000 ton
Pleurotus ostreatus	2,000 ton
Lentinula edodes	1,500 ton
Agaricus blazei	500 ton
Other species	50 ton

Table 1. Annual production of mushrooms in Brazil. Source (Souza, 2011).

2.3 The importance of fungi

The importance of fungi is unlimited in the terrestrial ecosystem and consequently in man's life. However, these organisms can be beneficial or not, according to the results of their actions. If we consider the decomposing action of fungi on food and the associated production of toxic substances (mycotoxins), the decomposition of other materials such as wood, pathogenicity caused to plants, animals and man, this is the negative aspect of it. On the other hand, if we consider the important role in the decomposition, which along with other microorganisms, participate in the mineralization of organic matter, as well as the symbiosis with plants in the process of mycorrhizae formation, bioremediation, biological control, food, and medicinal properties, one can see the positive side of these organisms.

In nature, the fungi do not participate only in the role of providing a food source for humans and other animals; they also play an important role in the cycling of carbon and other elements, by breaking the lignocellulosic residues and animal excrements which serve as a substrate for saprophytic fungi. This way, these decomposing agents play a very important environmental role along with other organisms, complementing the cycling of plants and animals. Simultaneously, they produce multiple enzymes that degrade complex substances that allow the absorption of soluble substances used for their own nutrition (Chang, 1993).

Trufem (1999) and Matheus and Okino (1999) highlight the importance of fungi in the context of biotechnology, where they are widely used in the food industry, pharmaceutical industry, bioremediation, in biosorption (removal of heavy and radioactive metals), in agriculture as arbuscular mycorrhizal fungi (AMF), where they are used in techniques that help the development of plants of economic interest, biological control, xenobiotics biodegradation, bioremediation of the soil, treatment of industrial effluents and bioconversion of lignocellulosic residues.

One of the most important processes from an economic point of view is the use of fungi in the conversion of lignocellulosic residues in edible mushrooms by fungus X substrate interaction, enabling the solid fermentation process, through enzymatic system of these microorganisms (Matheus and Okino, 1999).

The cultivation of edible mushrooms has become an increasingly important practice in modern society due to the biotechnological process of bioconversion of various residues in edible mushrooms or in dietary supplements of high nutritional value, enabling a more efficient utilization of materials, besides, it can reduce the volume of waste or accelerate the decomposition process. This way, the residual substrate obtained from the cultivation of edible mushrooms can also be used as soil conditioner, natural fertilizer, or food for animals, closing the exploitation cycle of raw materials (Miles and Chang, 1997), which today is called "zeri" technology, trying to get the maximum use of such material, eliminating the residue of the residue (Chang, 2003).

2.3.1 Nutritional importance

Man has constantly realized the nutritional value of mushrooms, as well as their healthy properties compared to other foods, such as red meat, where mushrooms are more advantageous and important as they are great sources of carbohydrates, proteins, mineral salts, vitamins and essential amino acids, which can help to maintain a good nutritional balance (Crisan and Sands, 1978; Garcia et al., 1993; Miles and Chang, 1997).

Nutritional Analyses of mushrooms have shown their importance. They contain more protein than vegetables. Sources of protein such as meat, chicken, have a high level of cholesterol and fat, which are known to cause increase in weight and cardiovascular diseases. For this reason, the proteins from other sources became more popular in recent years, such as proteins from fungi, algae, bacteria and yeast (Lajolo, 1970; Chang and Haynes, 1978; Urben et al., 2003).

Studies carried out by Lintzel (1941; 1943), according to Crisan and Sands (1978), indicated that approximately 200 g of mushrooms (dry weight) are sufficient to feed a normal human being weighing approximately 70 Kg, providing a good nutritional balance. Nutritionally, these macrofungi are a good food source. The composition of fats, carbohydrates, vitamins, etc., varies according to species, the cultivation method and also with the substrate used in cultivation (Crisan and Sands, 1978; Przybylowicz and Donoguue, 1990; Bononi et al. 1999; Miles and Chang, 1997; Andrade, 2007).

Mushrooms are excellent foods for the diets, because they nourish and do not accumulate fat in the organism. They are sources of all essential and some nonessential amino acids. They contain minerals like calcium, potassium, iodine, phosphorus and vitamins including thiamine, riboflavin, niacin, and ascorbic acid, and others related to the B complex (Molena, 1986; Miles and Chang, 1997 Bononi et al., 1999). They also have a high unsaturated fat content (Miles and Chang, 1997).

Mushrooms with larger nutrition index (based on essential amino acid index) have nutritional value similar to meat and milk, while those with a smaller nutrition index compare to some vegetables such as carrots and tomatoes. The nutritional index of these fungi outperforms those of plants and vegetables, except soy (Crisan and Sands, 1978). In general, the protein content of fresh mushrooms is twice higher than cabbage, four times greater than the content of protein of the orange and twelve times that of the Apple (Chang, 1980).

Research carried out in India by Garcia, et al. (1993), where the authors compared the nutritional levels of *Agaricus* and *Pleutotus*, revealed the importance of the amino acids of these mushrooms for people that are lacking animal protein, for religious reasons, and whose main food source comes from vegetables and grains usually poor in essential amino acids. Food supplementation with mushrooms is of fundamental importance in the diet of this kind of people.

In addition, there is also a great interest in the cultivation of the mycelium in a submerged condition to obtain flavoring and fragrant compounds of great value to the food industry. For this purpose, the mycelium is grown submerged, using a variety of substrates, according to the type of the desired compound. This flavoring property is characteristic of some lignolitic mushrooms, such as the *Pleurotus* genus (Gurtiérrez et al., 1994)

2.4 Factors inherent to the nutritional needs of the mushroom
2.4.1 Carbon

The main source of carbon and energy of a plant tissue, used by fungi for their development, are the polysaccharides and lignin in the cell wall, although other polymeric compounds such as lipids and proteins can also be used. Approximately 50-60% of the dry weight of wood is made of cellulose; 10-30% of hemicellulose and 20-30% of lignin. Cellulose, which is attacked by both brown-rot fungi as well as white-rot fungi, is made up of glucose molecules. On the other hand, the hemicellulose consists of molecules of arabinnose, galactose, mannose, xylose and uronic acids. The lignin has a more complex structure and has not yet been fully described, being basically units of phenyl-propane with a benzene ring bonded to a hydroxyl group and one or two metoxilic groups. The links in this molecule are highly resistant to chemical degradation. Therefore, there are few microorganisms that can use this substance for their nutrition (Mason, 1980).

In relation to the degradation of wood and other lignocellulosic materials, it is generally known that the most efficient natural decomposers of lignin are the white rot-fungi, which are mostly the basidiomycets. This name comes from the white color that wood acquires in advanced stages of degradation (Capelari, 1996). Such organisms degrade cellulose, hemicellulose and lignin, but the lignin is preferentially attacked and these are the only organisms able to metabolize the molecule of lignin in CO_2 and water (Zadrazil, 1978). The degradation is derived from the excretion of enzymes metabolized through the hyphae of fungi (Miles and Chang, 1997).

As a typical white-rot fungi, with decomposing activities of wood, the fungus studied in this work: *Lentinus strigosus*, grows in nature, in favorable conditions, and produces mushrooms through the degradation of the wood substrate or any substrate containing cellulose. From this degradation, the fungi can absorb the nutrients needed for their development and reproduction. The success of mushroom production depends on the understanding of the biology of the fungus and how the environment can influence its growth and development. The domestication of a strain is not a very easy task, when trying to reproduce in the laboratory the ideal conditions for its development, which requires preliminary tests to try to understand its physiology.

2.4.2 Nitrogen

Although wood is the natural substrate for fungi, this substrate does not have a high nitrogen content, and this is necessary for the synthesis of all nitrogen compounds (proteins, purines, pyrimidines and the cell wall chitin of the fungus). The main sources are: salts of ammonia, nitrate, urea nitrogen, and organic compounds like amino acids (Miles and Chang, 1997). However, the need for nitrogen by wood-rot fungi is not very great.

It should be taken into account that when using a salt as a source of nitrogen, there is the release of the ion that integrates the substrate molecules, and this can change the pH of the medium if it is not metabolized at the same rate as nitrogen, since an accumulation of this ion will take place. The same phenomenon occurs when other salts are used as a supplement. Therefore, the various species and strains may respond differently to the addition of these supplements. Urea, ammonia phosphate, tartarate of ammonia and potassium nitrate, apparently are those with best results according to a research carried out by Maziero (1990). Peptone provides better growth of the fungus when compared with other sources of organic nitrogen.

Some authors (Rangaswami et al., 1975; Ginterová and Lazarová, 1987) cited by Maziero (1990), argued that *Pleurotus* has the ability to fix atmospheric nitrogen into organic compounds, because some experiments conducted with pasteurized substrates showed that the total nitrogen content has increased. Kurtzman (1979), cited by Maziero (1990) however, discussed the improbable ability of an eukaryote organism to fix nitrogen. The author suggested the hypothesis that the spores of nitrogen fixing bacteria are stimulated to develop during the process of pasteurization of the substrate, generating bacteria responsible for the nitrogen fixation.

Care should be taken to avoid excessive nitrogen supplements, which can inhibit the development of the fungus. Montini (2001) reports that tested substrates with high concentrations of cereal bran inhibited the formation of the mushroom and consequently, the number of cultivated mushrooms *Lentinula edodes* in axenic conditions (cultivation with substrate sterilized and under controlled environmental conditions). In Taiwan, the substrate for cultivation of *Pleurotus* mushroom is prepared with 84% of sawdust, 5% of rice bran, 5% wheat straw, 3% soya bran and 3% calcium oxide (Przybylowicz and Donoghue, 1990).

2.4.3 Mineral salts

In general, the mineral elements necessary for the fructification of the mushroom are the same as those required by any cultivated plant, which are major elements and microelements (Molena, 1986). Phosphorus, potassium, magnesium and sulphur are major nutrients needed for the growth of various fungi (Miles and Chang, 1997). Molena, (1986), cites the calcium as one of these elements. In addition to increased growth of mycelium, some minerals such as sodium chloride, magnesium, and calcium also stimulate the early formation of fruiting bodies (Kurtman and Zadrazil, 1989).

Among the more studied microelements (trace elements) and essential for the growth of many species of fungus are: iron, zinc, aluminium, manganese, copper, chrome and molybdenum (Molena, 1986; Miles and Chang, 1997). Experimentally, it is not easy to determine the required quantity of these elements because the element under test may be present in sufficient quantities in an impure form in any ingredient of the cultivation medium or may have been introduced through the inoculum. These elements are constituents or enzyme activators (Miles and Chang, 1997).

2.4.4 Vitamins

Vitamins play an important role in the metabolism of fungi, acting as coenzymes. Fungi are capable of producing sufficient quantities of most of the vitamins they need (Miles and Chang, 1997).

Maziero (1990), in some studies testing several vitamins (Vitamin C, folic acid, calcium pantothenate, niacin, pyridoxine, riboflavin and thiamine) in relation to the mycelial growth of *Pleurotus*, observed a better growth of the mycelium on all vitamins tested, but the best result was to thiamine. Kurtzman and Zadrazil (1989) say that there is no need for the addition of thiamine or other vitamins in "not sterile" substrates, because the other present organisms will normally synthesize them. Molena, (1986) experimented various combinations of vitamins, but their high cost did not compensate for the increased production of mushrooms. (Eira and Minhoni, 1991), report that the vitamins and other growth factors are normally excreted by many microorganisms that live in synthrofy during composting, pasteurization and incubation of the substrate, therefore there was no need of vitamin supplements.

2.5 Physical factors

The growth and development of the fungus are not affected only by nutritional factors, but also by physical factors such as temperature, humidity, light, aeration and gravity. There is a range that varies from minimum, maximum, and optimum growth in relation to these physical factors. Certainly these factors are influenced by other factors such as nutrition, medium conditions, genetic characteristics of the strain and mycelial growth stage (Miles and Chang, 1997).

2.5.1 Temperature

The influence of temperature on mycelial growth and production of fruiting bodies is dependent on the species and strains in question, i.e. there is an ideal temperature for the proper development of the metabolism of the fungus, which is a characteristic of each strain. Nonetheless, there is an interval that varies between 10-40° C, which must be respected, because exceeding these limits, it is going to cause the death of the mycelium (Maziero, 1990). The optimum temperature for growth also varies with the purpose of cultivation. So, the ideal temperature to produce the fruiting body (Miles and Chang, 1997) is different from that intended for the production of metabolic products such as those intended for medicinal compounds as polysaccharides/polypeptides immune-regulatory compounds (PSPC). Temperature extremes are important in determining the survival and dispersion of species in nature (Miles and Chang, 1997).

Kaufer (1935), cited by Maziero (1990) cultivated *Pleurotus corticatus* in laboratory and according to their results, the ideal temperature for the growth of mycelium was 27°C. Clock et al., (1959) according to Maziero (1990), obtained a good growth of mycelium of *P. ostreatus* in the range of 22-31° C. At 37° C the mycelium still was able to grow, but abnormally, while at 17° C no growth was observed. The lethal temperature for *P. "florida"*, *P. ostreatus* and *P. eringii* is 40° C when exposed to more than 24 hours (Zadrazil, 1978). Maziero (1990) studying different strains of *Pleurorus* observed that the better mycelial growth happened between 25 and 30° C.

In relation to the emergence of primordia, Block et al. (1959) cited by Maziero (1990) report that the strain of *P. ostreatus* fructified at a 26° C, however at 31° C, although the fruiting body continues to   develop, there was no emergence of primordia. For Kurtzman and Zadrazil (1989), the authors must have used in their work, a strain of *P. "florida"* since the fruiting temperature at 26° C is very high, being more appropriate for *P. "florida"*.

Eira and Minhoni (1997) report that the control of temperature in a cultivation chamber is decisive for a good harvest. For a good growth of the mycelial mass on substrate cultivation, the ideal temperature for *Pleurotus* spp should be between 24 and 26° C. After that the primordia initiation and growth phases start, when the temperature inside the cultivation chamber must be between 15 and 24ºC, considering that, the lowest are ideal for cultivation of shimeji or *Pleurotus* spp strains that are more demanding and also minimizes the incidence of pests and diseases. According to the same authors, some strains of hiratake usually fructify in hot weather (up to 30° C). For the most demanding strains, temperature and relative humidity control in the chamber of cultivation can be achieved with an automated central air-conditioning associated with a ventilation system, to ensure the ideal climatic conditions for the development of the mushroom.

2.5.2 Moisture

Most fungi require high moisture content. Guzmán et al. (1993) report that fungi have an optimum growth on substrates with 70 to 80% humidity. Urben et al. (2003) cite a good humidity range for *Lentinula edodes* cultivated with Jun-Cao technique between 55-70%. It has to be taken into consideration, not only the moisture content of the substrate, but also the relative humidity of the air. It should also be taken into account that the mushrooms are composed of approximately 90% water, therefore, water is very important to its development, besides the fact that they do not have special structures to protect themselves against water loss, since they lose water easily to the environment, mainly the vegetative mycelium (Maziero, 1990).

There is an optimum water content, both in the compost and in the air. Low relative air humidity causes the mushroom to lose water to the environment, which can even prevent it from growing properly. The outer layers of the mushroom begin to dry and yellow. This way, there is a loss of quality or a loss of production. Low air humidity also causes the compost to lose moisture to the environment, reducing the availability of water for the formation of the mushroom. In the case of *Pleurotus*, if the superficial mycelium of the compost suffers a very intense dryness, it dies and the primordia are aborted (Eira and Minhoni, 1997; Bononi et al., 1999). The relative humidity of the production room is around 80-90% and can be maintained that way by waterproof walls and by sprinkling water (Eira and Minhoni, 1997). There are highly sophisticated systems of cultivation on a commercial scale in Europe, Canada, United States and Japan, where patterns of moisture, temperature, O_2 and CO_2 are monitored by computers. Currently there are automated systems in South and Southeast of Brazil, but not as much as in those countries. In rustic cultivations in Brazil, it is customary to keep the floor and sides of the cultivation shed damp, so that normal evaporation maintains the relative humidity the air. We consider that, in addition to other factors already mentioned, the humidity is the key factor in the cultivation process of edible mushrooms.

2.5.3 Lighting

Even though it is not a photosynthesizer organism, luminosity is essential to many species of fungi. It can retard the primordia formation in some species while in others, it is essential for fruiting. For *Pleurotus* and *Lentinus* cultivation, as well as for many other edible fungi, there must be some light to induce the formation of primordia and also for the normal development of fruiting bodies. The recommended luminosity for *Pleurotus*, after the incubation period and the opening of the cultivation bags is 2000 lux/hour, 12 hours a day (Bononi et al., 1999). Nevertheless, it can vary according to the mushroom species.

Miles and Chang (1997) mention that ultraviolet light in the range from 200 to 300nm affects the growth of the fungus, it can be lethal or induce mutation, since this wavelength is absorbed by the DNA. The authors report that the effects of ultraviolet light can be reversible by the photo reactivation process, provided that these mycelia are exposed to visible light at a wavelength between 360 and 420nm.

For Przybylowicz and Donoghue (1990), shiitake mushroom needs light in both stages: vegetative growth and fruiting. Light exposure during vegetative growth, according to Ishikawa (1967) cited by Przybylowicz and Donoghue (1990), is a prerequisite for the fruiting stage. The duration is not well defined. However, Przybylowicz and Donoghue (1990) suggest that a brief exposure of 20 minutes per day can be enough. For these authors, the growth of shiitake responds well to a range between 180-940 lux, with an optimum value of 500 lux. Rajarathnan and Bano (1987), cited by Eira and Minhoni (1997), stated that the presence of light is required for the formation of fruiting bodies. However, there may be changes in the color of the pileus, where *Pleurotus* species can change from white to opaque and dark color in the presence of light, due to the release of fenoloxidases that oxidize phenol and form melanoidins.

Urben et al. (2003) report that the light affects the growth of mycelium and spores of *Lentinula edodes*, and therefore, it needs a dark environment for its development. Under a light intensity of 50 to 270 lux and a suitable temperature, the mycelium forms a membranous brown layer for substrates made with Jun-Cao and sawdust. According to the same authors, for the formation of the fruiting body (mushroom), little diffused light is necessary. On the contrary, in a very bright environment the fruiting body becomes pale with a long stipe and a deformed pileus. In very bright environments the authors advise to use plastic bags in the green house to cover the mushrooms during the day.

2.6 Chemical factors
2.6.1 Gaseous exchanges
Requirements during the growth phases of the vegetative mycelium of a fungus are different from those during the fruiting stage. The rate of CO_2 that occurs naturally inside a trunk colonized by *Pleurotus* in the forest will surely be higher than the rate of fruiting. However, it does not cause damage to growth. It is a self-regulated system.

Zadrazil (1975), studied various *pleurotus* species, relating to the effect CO_2, and noted that all studied species grew faster in higher concentrations of CO_2, limited to approximately 22%. The good performance of these strains in high rates of CO_2 demonstrates their significant competitive advantage against other microorganisms which do not grow or do not survive in such conditions, especially if the substrate is colonized in not axenic conditions. On the other hand, high concentrations cause a deformation of the fruiting body, being similar to that which occurs when there is light deficiency in the development of the fruiting body. The stipe grows sharply and the pileus stays reduced, similar to the process of etiolation in plants (Zadrazil, 1978). Oxygen also influences the growth of mycelium. Despite the fact that *Pleurotus* mycelium develops in semi-anaerobic conditions, a certain rate of O_2 is required, otherwise, the growth will be nil (Zadralil, 1978). For the development of fruiting body oxygen is essential.

Adequate ventilation is essential to reduce the carbon dioxide content (generated during the development stages of the fungus) to a desirable level in the mushroom production phase. Concentrations above 2% may cause delays in the mycelial growth and, consequently, decrease productivity (Eira and minhoni, 1997). Concentrations of CO_2 below 0.2% are considered optimum for development. During a peak of growth, the ventilation must be intense and constant, since large quantities of mushrooms in rapid growth give off large amounts of CO_2 (Eira and minhoni, 1997). The same authors reported that in cultivations carried out in The Mushrooms Module of the Faculty of Agricultural Sciences of the "Universidade Estadual Paulista" (FCA/UNESP) it is possible to cultivate strains that usually demand cold weather, provided that climatic chambers for thermal shock are used.

2.6.2 pH
Its importance is primarily related to the metabolism of nutrients. Most mushrooms have a good development with pH levels between 6.5 and 7, but there are variations according to the species and strains (Miles and Chang, 1997). The microbiota present in the substrate, according to Zadrazil and Grabbe, (1983), is distinctly influenced by the initial pH level: values below 7.0 usually are good for the development of the mushroom mycelium, but most fungi can develop at pH levels above 7.0. Urben et al. (2003), reporting about the cultivation of *Lentinula edodes* by Jun-Cao technique, stated that the mycelium can grow in pH levels between 3.0 to 6.5, while the ideal range is between 4.0 and 5.5. However, the pH

value between 3.5 and 5.0 is the best for the formation of primordia and development of the fruiting body. For this reason, the pH value should always be monitored when choosing the materials that compose the substrate, the cultivation and the source of water supply (Urben et al., 2003).

The pH is directly linked to the enzymatic reaction of fungus and wood. Each enzyme has its optimum pH value. The pH affects the solubility of the compost which in turn determines its availability to the fungus (Przybylowicz and Donoghue, 1990). The optimum pH value for the wood-rotting fungi *Lentinus* and *Pleurotus* is between 4.5 and 5.5. The pH of the wood is usually 4.5 to 5.0, increasing the acidity with its decomposition. The optimum pH value for fruiting lies between 3.5 to 4.5 (for laboratory culture or artificial medium) and 5.0 for compost with sawdust for *Lentinula edodes* (Przybylowicz and Donoghue, 1990)

2.7 Steps to be followed in the cultivation of mushrooms

For the cultivation of edible fungi the following steps are generally adopted: obtaining primary matrix, the production of seed or Spawn (matrix that will serve as inoculum for the substrate), preparation of substrate or compost, sterilization or pasteurization (when cultivation is done in natural conditions), inoculation and colonization of substrate, inducing primordia (with thermal or water shock when necessary), fruiting and harvest. The production aspects of primary decomposition fungi like *Pleurotus*, *Lentinula edodes*) will be covered here.

2.7.1 Obtaining primary matrix and spawn production

For most mushrooms, the production matrix or mycelium follows the same techniques and recommendations for the cultivation of champignon (*Agaricus*), oyster mushroom (*Pleurotus*), shiitake (*Lentinula edodes*) and jewish ear (Auricularia), with some exceptions (Urben et al., 2003). Two distinct steps are fundamental for the preparation of the matrix: obtaining pure inoculum of the fungus and the preparation of the "spawn" or matrix itself.

Obtaining the primary matrix of mushroom can be performed both by sexual or by asexual process. In this work it will be related as an asexual process. It is relative to the mycelial or vegetative phase of the fungus colonizing a previously sterilized nutritional substrate (growth medium). Its production starts by the isolation of a fungus using tiny fragments of a mushroom, placed in sterile culture medium under aseptic conditions. After mycelial growth in the dark, with a temperature of 24 ± 1°C (depending on the strain), fragments of this culture (primary matrix) are transferred to the cereal grain or bran or sawdust enriched with bran and incubated for 30 days in the dark at 24 ± 1 °C. This step corresponds to the production of the "seed" or spawn (Molena, 1986; Eira and Minhone, 1997; Eira and Montini, 1997). The main function of the grain is to serve as means of dispersion of mycelium, since it is impossible to handle the mycelium without damaging the fragile structure of the hyphae walls (Maziero, 1990).

Although the most used media to obtain the primary matrix are potato-dextrose-agar and malt extract (Bononi et al., 1999), the sawdust-dextrose-agar (SDA) medium is the most indicated by avoiding the physiological adaptation that can occur when the used culture medium has very different characteristics of production substrate (Eira and Minhoni, 1997, Eira and Montini, 1997).

The current trend is to produce inoculum from the cultivation substrate. When working with sawdust it is possible to produce the inoculum ("seed") with grain mixed with sawdust or with sawdust only. In this case "spawn" or "seed" is the substrate colonized by mushroom mycelium, with the goal of facilitating the distribution of the inoculum in different points of cultivation, thereby contributing to a more uniform and rapid colonization of the substrate, reducing the possibility of contamination.

2.7.2 Substrate cultivation

Currently there is a growing tendency to use agro-industrial residues for the cultivation of edible and medicinal fungi. However, traditional methods are still being used like the cultivation of *Lentinula edodes* (shiitake), by some Japanese and Chinese farmers, using oak and hazel logs, although the cultivation in cylindrical tubes (in polypropylene or high density polyethylene-HDPE bags) with enriched sawdust is the most widely used technique.

The technique for the production in sawdust was developed mainly in Japan. Other countries like the Netherlands and the United States are also using this method for the production of *Lentinula edodes* (shiitake), on a large-scale (Bononi at al., 1999). In Brazil the traditional cultivation is done normally on eucalyptus logs and it may also be grown on logs of avocado, mango, walnut, hazel and oak (the last two being widely used for cultivation in Japan) (Eira and Minhoni, 1997). Eucalyptus sawdust is already used in Brazil for the production of this mushroom.

The material used for production of mushroom has to be preferably a residue, easily available, and produced not far away from the cultivation place to lower the production costs. Care should be taken to observe that the waste should be free of chemicals that could affect the growth of the mycelium and not offering toxicity. If a low productivity residue is used, supplementation has to be made with cereal grains or cereal bran (Eira and Minhoni, 1997; Bononi et al., 1999; Przybylowicz and Donoghue, 1990; Stames and Chilton, 1983; Stames, 2000).

The supplements contain a mixture of protein, carbohydrate and fat, where the protein is the main source of nitrogen. They contain minerals and vitamins that also influence the growth of the fungus. The addition of these supplements aims mainly to increase the levels of nitrogen and carbohydrates available. Sugars and starch which are readily available carbohydrates, speed up colonization and the consequent degradation of the substrate, reducing the time of fruiting since the mycelium easily converts these carbohydrates in reserve for the fructification, increasing productivity (Przybylowicz and Donoghue, 1990). Other supplements like limestone ($CaCO_3$) must be added to the cultivation medium, to get the correct pH favorable for the growth of the fungus during the last stages of decomposition since there is an increase in acidity caused by the fungus metabolism. Gypsum is widely used in the mushroom industry to improve the physical structure of the compost and to change the pH value, also acting as a source of calcium (Przybylowicz and Donoghue, 1990). The concentration of 5% (in relation to the dry weight of the substrate) is ideal for the cultivation of shiitake in sawdust, improving structure and porosity of the substrate (Stames and Chilton, 1993)

When working with primary-decomposer fungi as *Pleurotus* and *Lentinus*, i.e. fungi that degrade the structural elements of the residue, it is important to ensure that the material to be used in cultivation has not undergone decomposition by microorganisms during storage. If it is already degraded, colonization by these fungi will be hampered and the attack of other organisms will be facilitated, causing a reduction in productivity (Maziero, 1990).

The sawdust used to prepare the substrate is usually from hardwoods. Sawdusts of conifers are used for *Lentinula* cultivation (shiitake) in areas where there is shortage of hardwood sawdust and it is therefore necessary to make a mixture of the two kinds of sawdust (Przybylowicz and Donoghue, 1990). Many conifers contain resin and phenolic compounds which inhibit the growth of the fungus. These compounds must be degraded or removed before using this kind of sawdust, or it can be changed with the addition of sodium carbonate to remove these compounds (Przybylowicz and Donoghue, 1990).

Various types of substrates have been used for the production of edible fungi (Guzmán and Martinez; 1986; Guzmán et al., 1993; Maziero, 1990; Bononi et al., 1999; Eira and minhoni, 1997; Miles and Chang, 1997; Stames and Chilton, 1983; Stames, 1993; Urben, 2001; Urben et al., 2003; Zhanhua and Zhanxi, 2001). The most used are: sawdust, wheat straw, corn, rice; corn cobs, sugar cane bagasse, various grasses, supplemented with cereal grain or bran. The choice of one or more residues as supplement to sawdust will depend, among other factors, on cost and availability of these materials (Maziero, 1990; Eira and Minhoni, 1997; Eira and Montini, 1997; Guzmán et al., 1993; Urben et al., 2003; Stames and Chilton, 1983; Stames, 1993).

When using the bagasse of sugar cane it is important to make sure that this residue is not very old, which can reduce productivity. However, the fresh ground bagasse is rich in carbohydrates, allowing other competitors or pathogens to colonize the substrate more quickly. To avoid this problem, the residue should be pre-treated through a process of fermentation or washing (Kurtzman and Zadrazil (1989)

Japanese producers of *Flammulina Velutipes*, *Auricularia* and *Pleurotus ostreatus* use a standard formula with a ratio 4:1 of sawdust and bran respectively, where the sawdust is aged for one year, with the purpose to improve the water retention capacity. An immersion of sawdust in water before mixing with the bran is an effective way used by these producers to achieve an optimum of 60% humidity (Samets and Chilton, 1983). This method, according to Lizuka and Takeuchi (1978) cited by Przybylowicz and Donoghue (1990) is widely used in Asia. In the United States, 80% of sawdust, 10% bran and 10% grain (usually wheat or millet). In Taiwan, the substrate for the cultivation of shiitake is done with 84% of sawdust, 5% of rice bran, 5% wheat straw, 3% soy bran and 3% calcium oxide (Przybylowicz and Donoghue, 1990). The substrate formulations have become unlimited in terms of raw material and agro-industrial residues (Stames, 1993).

Currently, studies are trying to develop a technology that allows the cultivation of edible mushrooms in substrates of low cost and easily available. Perhaps this explains why in Brazil the largest edible fungi producing region is located in São Paulo, where there is a big sugar and alcohol production. This bagasse comes from sugar cane mills, it is homogeneous, it has a fibrous characteristic, and when it is pressed allows aeration for mycelial growth (Rossi, 1999). The sawdust is also a material in abundance in the Amazon region, because of its timber industry.

2.7.3 Pasteurization/sterilization

The pasteurization process is a heat treatment given to the compost for the removal of possible organisms that could compete with the fungus to be cultivated (Maziero, 1990). It can be done in a natural way, in a pasteurizing tunnel or room without heated steam, using only the thermogenesis, with the control of the air that gets in and out in order to control the temperature inside the room, or it can be made with heated steam produced by boilers heated with firewood, diesel or gas.

Pasteurization for the cultivation of lignicol fungi as *Pleurotus*, occurs when after the revolving process of the compost, the temperature of thermogensis, produced by the action of microorganisms, falls below 45 to 50⁰ C (Eira and Minhoni, 1997). The compost is then introduced in the pasteurization chamber. The pasteurization temperature for *Pleurotus* is more severe than for *Agaricus*, being raised to 75⁰C during the first 6 hours. After cutting the steam, the temperature falls to 40 to 45⁰ C, maintaining a constant ventilation to cool the compost and then proceeding to the inoculation process (Eira and Minhoni, 1997). Depending on the type of cultivation, the substrate to be inoculated can be packed in plastic bags, put in wooden or plastic boxes, shelves or "bed", pressed blocks covered with plastic sheets or in special containers (Maziero, 1990).

2.7.4 Inoculation of the substrate
The sterilization process (used in axenic cultivation) or pasteurization (when working with composted natural substrate) is followed by the inoculation of the substrate, which is made immediately after the cooling of the substrate in aseptic conditions (laminar flow chamber) in case of cultivation in totally axenic conditions. Under these conditions, the substrate to be inoculated is autoclaved at 121⁰ C for 2 to 4 hours (Eira and Minhoni, 1997). For cultivation under natural conditions (not axenic), the substrate is inoculated after pasteurization and cooling of substrate that is around 30⁰ C, in aseptic place.
There are several types of inoculum ("seed"): with grains, grains with sawdust, and less used liquid inoculum. There are still those that are made of small wooden dowels (wooden rods inoculated with fungus, used for the cultivation of shiitake in logs). Inoculum of sawdust/grain is used for *Pleurotus* and *Lentinus* cultivation, when using sawdust for cultivation. It is important that the inoculum is the same sawdust from the cultivation substrate. The quantity used for the cultivation substrate is also variable. It is usually 0.5 to 5% (v/v) (Chang and Miles, 1997). (Zadrazil and Grabbe, 1983), recommend 0.5 to 5% of wet substrate. Urben et al. (2003), recommend 0.5 to 5% of the wet weight of the substrate for the *Pleurotus* cultivation when using the Jun-Cao technique. Gonçalves (2002) studying the effect of mycelial fragmentation in order to obtain inoculants in suspension (liquid fermentation) for cultivation of shiitake in axenic cultivation, found that inoculants fragmented up to 10 seconds provided greater biological productivity and efficiency in comparison with usual solid inoculants. The author reports that cultivation using liquid inoculants has the advantage of reducing the time for fruiting. However, it has the disadvantage of having predisposition for degeneration and mutation after successive crops (Itaavara, 1993), cited by Gonçalves (2002).

2.7.5 Substrate incubation
Incubation period, also known as the "mycelial race", is the development of the vegetative mycelium on the substrate (Przybylowicz and Donoghue, 1990). It is the mechanism in which the mycelium of the fungus, through an enzymatic process, digests the substrate and stores reserves for fruiting. During this process the mycelium develops and colonizes the whole compost, forming a compact white mass. It is a complex process, characterized by intense biological activity in which molecules of cellulose, hemicellulose and lignin of the compost are attacked by fungal enzymes such as cellulase and lacase that reduce these molecules to phenols and simple sugars which are more easily assimilated. This enzymatic activity lasts from the beginning of colonization until the production of mushrooms, but during the period of growth of the mycelium production it is greater (Bononi et al., 1999). Incubation usually occurs in a room that can be dark or not, depending on the light requirement of the fungus, at a temperature between 22 to 25⁰ C for *Pleurotus* (Maziero, 1990).

Guzmán et al. (1993) uses the range 25 to 30^0 C for the cultivation of several *Pleurotus* species in Mexico. Bononi et al. (1999) report that the ideal temperature for the incubation of these fungi varies between species, but in general it should be kept between 25 and 28^0 C (the temperature of the compost). The range 25-30^0 C is also used for the *Pleurotus* cultivation, and 22-25 ^{0} C for *L. edodes* cultivation in Jun-Cao (Urben et al., 2003). Przybylowicz and Donoghue (1990), reported temperatures of 25^0 C that are ideal for *L. edodes*.

It is important to monitor the temperature during the mycelial race to maintain the optimum temperature for the growth of the mycelium. If there is an excessive rise in temperature (a phenomenon that occurs during metabolic activity of *Pleurotus* and micro-organisms present in the substrate) mycelial growth retardation or even its death may occur. Containers with large amounts of substrate mass are avoided since of heat loss is hampered, and generates an increase in temperature. (Maziero, 1990).

The incubation period is approximately three weeks for *Pleurotus* (Maziero, 1990; Bononi et al., 1999). At low temperatures, such as 4-5^0 C, the mycelium of most species ceases its activity, entering "latency", and at temperatures over 35-40^0 C can be lethal to certain species (Bononi et al., 1999). To avoid excessive internal temperature of the substrate during the incubation period, Bononi et al. (1999) recommend keeping the room temperature between 20-22^0 C, and avoid to clutter the bags of the substrates.

The incubation period is variable, because the development of mycelium occurs within variable time, according to the type of the inoculum, the quality of the compost and conditions of the cultivation chamber, but it generally oscillates between 20 and 30 days for *Pleurotus* (Eira and Minhoni, 1997). Urben et al. (2003) report 20-45 days for the total development of the mycelium with Jun-Cao technique. For the cultivation of shiitake in logs, Eira and Minhoni (1997) report that, after two to three months from log incubation, there is already a significant mycelial growth, which can be indicated by a yellow color in the region of the inoculated holes and the region around those holes become soft. In natural conditions of cultivation on logs, this period of maturity of the mycelium for mycelia production, ranges from six months to a year (Przybylowicz and Donoghue, 1990).

During the colonization of the substrate in the cultivation of *Lentinula edodes* using sawdust enriched with rice bran, packed in plastic bags, Bononi et al. (1999) recommend cycles of alternating light and dark, with at least 8 hours of light per day during a period of four to six weeks. The wavelengths between 370 to 420nm and light intensity between 180 to 500 lux are more efficient during the process of colonization, and it can be achieved with cold fluorescent lamps (Przybylowicz and Donoghue, 1990). According to Bononi et al. (1999), after the total colonization of the substrate, the plastic bags are cut and the surface of mycelium begins to turn into a brown skin. The air humidity must be maintained around 80 to 90% and between 40 to 50 days after the opening of the bags the production starts, after the induction of primordia through thermal shock for 24 to 48 hours at 10^0 C.

At the end of the mycelial race for shitake there is a period of mycelial stability or mycelium maturation, which lasts until the hardening and darkening of the mycelial skin that becomes brownish grey (Chang and Miles, 1989). The formation of mycelial cover is very important because it acts as a barrier to moisture loss, being also a defense against contaminants, resulting from the oxidation of poliphenol oxidase, a reaction to light and oxygen (Przybylowicz and Donoghue, 1990).

2.7.6 Induction of primordia, fruiting and harvesting

The induction of the primordia occurs naturally in nature. The sudden change of external physical conditions stimulates primordia formation, which will develop, forming the fruiting body (Bononi et al., 1999). On the cultivation of mushrooms, it is used to stimulate or speed their formation. During the induction phase and the production of mushrooms, physical factors such as temperature, lighting, gas exchange, water availability in the compost, relative humidity and the methods of induction are aspects that influence the production and the quality of mushrooms (Zadrazil and Grabbe, 1983).

Sudden changes in temperature usually cause induction of primordia. However, there are differences according to the strains (Przybylowicz and Donoghue, 1990). Low temperatures may indirectly induce fruiting in strains of shiitake, because of the reduction of metabolic activity, reducing therefore the available nutrients, leading to a condition of "stress". On the other hand, in other fungi, temperature can have a direct effect, favoring specific metabolic processes that trigger the induction (Przybylowicz and Donoghue, 1990).

Hawker (1966), cited by (Przybylowicz and Donoghue, 1990) reports that studies with various fungi showed that reducing sugars readily available on the substrate (end of vegetative growth) favors the fruiting. During the entire cycle of fruiting, the primordia phase is the most sensitive to environmental changes. The moisture content of the substrate, temperature and relative humidity are important in this process. On the cultivation of shiitake in logs, moisture for primordia induction should be around 55-65% and the temperature depends on the strain (Przybylowicz and Donoghue, 1990).

There are several artificial induction mechanisms of primordia. It can be done by changing the temperature of incubation ($\pm$ 25^0 C) to lower temperatures ($\pm16^0$ C) in *Pleurotus* cultivation "shimeji" (Eira and Minhoni, 1997). Some strains respond well to this temperature variation, others produce more when subjected to thermal shock.

Marino (2002) in a study about genetic improvement with *Pleurotus ostreatus* aiming the axenic cultivation of strains resistant to heat obtained strains that stood out by their early fruiting and productivity, with two production cycles and without the need for thermal shock, using water immersion only.

For the cultivation of shiitake in sawdust, according to Leatham (1985), cited by Przybylowicz and Donoghue (1990), the thermal shock can be done by cooling the cultivation blocks (packed in bags of polypropylene) at a temperature of 5^0-8^0 C for five to twelve days or by putting them into cold water (5-16^0 C) for 12 to 24 hours, packing them later in the fruiting room (16^{0}C). After some time primordia will appear at the top of the bags. After the development (3-4 days), according to Eira and Minhoni (1997), the mushrooms are ready to be collected.

Additional flushes of fruiting will emerge without the need for new inductions, provided they are kept in conditions of fruiting. Producers can control the flush making synchronized induction by heating the blocks, followed by reduction of temperature or thermal shock. Sprinkling or immersion can also induce the flush (Przybylowicz and Donoghue, 1990).

Treatment for production of mushrooms (Eira and Minhoni, 1997) is done by reducing temperature and/or water logging (covering with clean cold water for 2 to 4 hours) and removing the bag after water drainage.

The thermal shock for *Lentinula edodes* in modified Jun-Cao technology is made by dipping the miceliated substrates in cold or icy water during 7-8 hours. Then, the bags are packed in a shed or green house. When the buttons (primordia) begin to emerge, the plastic bags (high-density polyethylene) are removed, and the substrates are watered twice a day. After spraying, the bags are covered with a plastic for two hours or until the environment is agreeable (Urben et al., 2003). Regional climatic variations need to be considered. The relative humidity varies with the location of cultivation.

In Brazil, the necessary time for the complete development of the shiitake mushroom is not well defined due to climatic variations. The fruiting occurs over a period between three and twelve months after the inoculation, depending on the temperature of the region and the maintenance of moisture in the log (Eira and Minhoni, 1997; Eira and Montini, 1997). To accelerate this process in the cultivation on eucalyptus logs, the authors recommend soaking the miceliated logs for induction, after the incubation period when the first signs of primordia emission (callus or popcorns) which usually appear after 2 to 3 months. Mineral supplementation in water immersion increased the productivity of this mushroom. However, the increase of productivity and the efficiency of energy conversion were only possible in logs well colonized by the fungus (Queiroz, 2002; Eira and Minhoni, 1997; Eira and Montini, 1997).

In relation to water temperature for immersion, there is a controversy, probably because of environmental differences, and observations often without experimental parameters (Eira and Minhoni, 1997; Eira and Montini, 1997). Some Brazilian producers who own cooling bath system report positive results since this system causes a steady temperature differential of 5-10⁰ C. However, experiments performed in the Module of Mushrooms of the Faculty of Agricultural Sciences "Universidade Estadual Paulista" (FCA/UNESP), in Botucatu, São Paulo State, these same authors report that, in regions with mild climate and thermal amplitude greater than 10⁰ C, the use of ice for cooling did not show significant difference in relation to normal bath immersion.

Induction time depends on environmental conditions and age of the logs and the fruiting temperature varies from 5 to 30⁰ C depending on the strain and the spawn used for cultivation. The relative humidity of the location of the logs should be between 80 and 90%. The emergence of primordia will be within two to three days and harvesting can be made after seven to ten days, and in cool seasons the metabolism of the fungus is reduced, increasing the time before harvest (Eira and Minhoni, 1997). The induction bath can be done in stages, depending on the needs of the producer, thus inducing bath of logs can be programmed as a function of demand (Eira and Minhoni, 1997; Eira and Montini, 1997).

3. Material and methods

The study was carried out at the Edible Mushroom Cultivation Laboratory from the Department of Forest Products in the "Instituto Nacional de Pesquisas da Amazônia" - INPA, in the following steps:

3.1 Collection, drying and preparation of material

Wood residue (sawdust) was chosen based on the generation of the wood waste produced by local lumber industry. Collection, drying and preparation of materials were done at CPPF/INPA, using sawdust of *Anacardium giganteum* Hanck ex Engl (cajuí). After the collection of the residues, they were dried (12% of humididy) in a solar dryer at CPPF/INPA, and packaged into plastic bags of 100 L until the preparation of the substrates.

3.2 Production of a primary and secondary matrix and the "spawn"

The strain of *Lentinus strigosus (Schwinitz) Fries* was taken from the collection of fungi at INPA Institute. Mycelial fragments of fungus (stored in test tubes) were transferred to a Petri dish containing malt medium and incubated at 27 °C until colonization by the fungus (primary matrix) that was used as a source of inoculum for the secondary matrix . Mycelial disks, 9 mm in diameter, were removed from the primary matrix and transferred to Petri dishes containing SDA medium (sawdust-dextrose-agar), prepared according to Sales-Campos (2008), named secondary matrix. "Spawn" is the source for inoculation of the cultivation substrate, considered here as a tertiary matrix. This matrix was produced from cajuí sawdust, with humidification of 75%. The pH was corrected to approximately 6.5, by adding $CaCO_3$. Then that substrate was deposited on glass bottles of 500 mL, in 200 g portions, which were autoclaved at 121 °C for 45 minutes. After cooling, the substrate was inoculated with the secondary matrix. The bottles were partially closed, and kept in special chamber with biochemical oxygen demand (BOD) at 25 ± 2 °C until the complete colonization of the substrate by fungus. This matrix served as a source of inoculation for the cultivation substrates for the production of L. strigosus mushrooms

3.3 Preparation of cultivation substrate and processing
The cultivation substrate was prepared from the same residue (cajuí sawdust) as the spawn inoculums. It consisted of 88% of sawdust + 10% of the soy bran as a protein source + 2% of $CaCO_3$, for pH adjustment (6.5). The material was homogenised and humidified to 75%, and packed into bags of high density polyethylene-HDPE (1 kg capacity). Only 500 g of the substrate (wet basis) were put into each bag, with ten repetitions. The substrates were autoclaved at 121 °C for one hour. After that, they were cooled and inoculated with a tertiary matrix under axenical conditions. Each experimental unit (the bag containing the substrate) received 3% of the inoculum in relation to the wet weight of the substrate. They were taken to an incubating chamber until the colonization of the substrate by the fungus. Afterwards, they were transferred to a production chamber. The control samples were also prepared as above, but without inoculation by the fungus. The bags were taken to an oven with air circulation at 55 ± 5 °C and dried to a constant weight, in order to obtain the dry mass of the initial substrate (DMIS) so that they were used to calculate the productivity, based on the biological efficiency index of substrate (BE) and the loss of organic matter (LOM).

3.4 Experimental conditions
The experiment was conducted indoors. The bags contained the substrates were incubated in a climatic chamber at the temperature of 25 ± 3 °C, in the absence of light and at around 80-85% humidity, in order to allow substrate colonization until the production of primordia. Then, they were transferred to the production chamber. The temperature was reduced from 25 °C to 22 °C to induce primordial emission and to allow the production of basidioma (fruit body of the mushrooms) in a way that it would be as uniform as possible. Light intensity was maintained at 2000 Lux, with a photoperiod of 12 hours per day. The relative humidity was scheduled to 95% during the "fructification. The total period of cultivation was 100 days. After "fructification", the mature mushrooms were collected and weighed, and then oven dried for the determination of moisture, dry mass and chemical analyzes. During cultivation, the variables analyzed were: biological efficiency (BE), and loss of organic matter (LOM). Biological efficiency (used to express the productivity of fungus), was calculated according to Tisdale et al. (2006) and Das and Mukherjee (2007):

$$BE = \frac{FMM}{DMS} * 100$$

Where:
BE= Biological efficiency, %
FMM= Fresh mass of mushrooms, g
DMIS= Dry mass of the initial substrate, g

The loss of organic matter (LOM) is the index that evaluates the substrate decomposition by the fungus. It was evaluated according to Sturion (1994), expressed by the following formula:

$$LOM = \frac{DMIS - DMSS}{DMSS} * 100$$

Where:
LOM = loss of organic matter,%
DMSS = Dry mass of the spent substrate, g
DMIS = Dry mass of the initial substrate, g

4. Results and discussion

Table 2 presents the development of the *L. strigosus*. The fructification happened three to five days after the primordia initiation (Table 2), with the development of vigorous mushrooms.

Period (days)						
Mycelial growth	Primordium emission	Fruiting	Cultivation time	Number of flushes	Pileus cm	Stipe cm
11 to 15	33 to 34	35 to 38	100	6	2 to 7	1

Table 2. Profile of the *Lentinus strigosus* mushroom cultivated in cajuí sawdust with soy bran as protein source during 100 days.

The high biological efficiency (BE) of the substrate formulated with cajuí sawdust, supplemented with soy bran demonstrates good productivity of the substrate (Table 3). The result (80%) is superior to other substrates used in cultivation of *P. ostreatus* mushroom formulated with sawdust of *Fagus orientalis* (Yildz et al., 2002), and with *Eucalyptus* sp. according to Marino et al. (2002) which presented BE equal to 8.6 to 64.3% and 11.4 to 43% in their respective studies. The good productivity of this substrate is the result of the quantity of material readily available and absorbed by fungus during the mycelial development process and the soy bran was a good source of protein.

Philippoussis *et al.* (2003) showed that the mycelial growth rate is related to the bio-availability of nitrogen and that the formulation of the substrate influences nutritional levels and porosity (availability of O_2) and Gbolagade *et al.* (2006) stated that each fungus utilizes a specific C/N ratio. The soy bran provided a good source of protein for the fungus as we can see at the table 3 (20%). The results are very important for this edible mushroom, since they present low lipid content (2.5%) and a high fiber level (18%).

The use of alternative substrates, easily obtained at low cost for the cultivation of edible mushrooms have been investigated in many publications (Özçelik and Pekşen, 2007; Philippoussis *et al.*, 2007; Royse and Sanchez, 2007). The supplementation of the substrates with a nitrogen source, mainly with cereal bran, has been adopted to achieve a C/N ratio good for the production of mushrooms.

Özçelik and Pekşen (2007), analyzing the application of hazelnut shells in the formulation of substrate for mushroom cultivation *Lentinula edodes*, reported that the biological efficiency of the substrate made with hazelnut shells only, was considered to be low (43.73%). However, when the proportion of hazelnut shells was reduced and combined with wheat straw (25:75) the biological efficiency was considered good (62.24%). The result however, is less than that reported in this study (Table 3).

Philippoussis et al. (2007) tested the productivity of agricultural residues (sawdust of oak, wheat straw and corncobs) in the cultivation of *Lentinula edodes* and found that corncobs and wheat straw presented higher rates of biological efficiency: 80.64% and 75.23% respectively, which were similar to those presented in this research. Alberto and Lechner (2007) however, obtained lower BE (61.93%), cultivating *Lentinus tigrinus* with Salix sawdust.

Royse and Sanchez (2007) tested three formulations for the cultivation of *L. edodes*, combining wheat straw and oak residues. They found that the substrate with higher proportions of wheat straw (in relation to oak residue), provided the best biological efficiency (98.9%) at the end of 4 harvests. These results are superior to the ones obtained in this work. However, 80% of Biological Efficiency presented by *L. strigosus* cultivated in cajuí sawdust supplemented with soy bran in the present study is considered high.

Substrate Cajuí Sawdust	Productivity Results		Nutritional Composition			
	Biological Efficiency (BE) (%)	Loss of Organic Matter (LOM) (%)	Total Protein (%)	Total Fiber (%)	Lipid (%)	Ash (%)
Standard deviation	6.94	4.85	1.00	2,00	0.30	1.00
Average	80.0	47.51	20	18	2.5	5

Table 3. Results of Productivity and Nutritional Composition of the edible mushroom *Lentinus strigosus* cultivated on cajuí wood waste, supplemented with soy bran.

5. Conclusions

The high biological efficiency of the mushroom in this substrate, formulated with the cajuí sawdust supplemented with soy bran, makes its use feasible for the cultivation of *Lentinus strigosus* mushroom from the Amazon Region. The soy bran provided a good source of protein for the fungus.

The findings presented herein point out the utilization of the Amazon wood waste as substrate for the mushroom cultivation, which will certainly promote the improvement of the social and economical conditions of its people and the sustainability of the biodiversity resources, enabling the establishment of a new economical niche in the region.

L. strigosus can be considered an important food in terms of their characteristics: rich in protein and low in fat, important for nutrition and human health.

6. References

Bononi, V. L.; Capelari, M.; Maziero, R. & Trufem, S. F. B. S. (1999). *Cultivo de cogumelos comestíveis*. 2nd Ed., Ícone , ISBN: 85-274-0339-0, S. Paulo.

Crisan, E. V.; Sands, A. (1978). A nutritional value, In: *The biology and cultivation of edible mushroom*, Chang, S. T.; Hayes, W.A. (Eds), pp.137-168, Academic Press, ISBN: 0-12-168050-9, New York.

Chang, S. T.; Hayes, W.A. (1978). *The biology and cultivation of edible mushroom*, Academic Press, ISBN: 0-12-168050-9, New York.

Chang, S. T. (1980). Mushroom as human food. *Bioscience*, vol. 30, No. 6, jun, pp. 399-401, ISSN: 0006-3568

Chang, S. T.; Miles, P. G. (1984). A new look at cultivated mushroom *Bioscience*, vol. 34, No. 6, pp. 358-362, 1984, ISSN: 0006-3568

Chang, S. T. O (2003). Mushroom cultivation using the "Zeri" principle: potential for application in Brazil, Proceeding of First International Symposium on Mushroom in food, health, technology and the environment in Brazil, ISBN: 0102-0110 Brasília-DF, dec, 2002.

Chang, S. T.; *Miles*, P. G. (1987). Historical record of the early cultivation of *Lentinus* in China. *Mushroom Journal for the Tropics*, vol. 7, pp. 31-37.

Delmas, J. (1978). Cultivation in Westerns Countries: Growing in Caves. In: *The biology and cultivation of edible mushroom*, Chang, S. T.; Hayes, W.A. (Eds), pp. 252-299, Academic Press, ISBN: 0-12-168050-9, New York.

Chang, S. T. (1993). Mushroom biology: The impact on mushroom production and mushroom products, In: *Mushroom biology and mushroom products*, Proceeding of the first international conference on mushroom biology and mushroom products, Chang, S. T.; Buswell, J. A.; Chiu, S. W. (Eds), pp. 3-20, The Chinese University Press, ISBN: 962-201-610-3, Hong Kong.

Eira, A. F.; Minhoni, M. T. A. (1991). *Manual teórico/ prático de biotecnologia e microbiologia agrícola: cultivo de cogumelos comestíveis*. Fundação de Estudos e Pesquisas Agrícolas e Florestais, Universidade Estadual Paulista FEPAF-UNESP, Botucatu, São Paulo.

Eira, A. F.; Minhoni, M. T A. (1997). *Manual do cultivo do "Hiratake" e "Shimeji" (Pleurotus spp)*. Fundação de Estudos e Pesquisas Agrícolas e Florestais, Universidade Estadual Paulista FEPAF-UNESP, Botucatu, São Paulo.

Eira, A. F.; Montini, R. M. C. (1997). *Manual do cultivo do shiitake (Lentinula edodes (Berk) Pegler)*. Fundação de Estudos e Pesquisas Agrícolas e Florestais, Universidade Estadual Paulista FEPAF-UNESP, Botucatu, São Paulo.

Eira, A. F.; Minhoni, M. T. A.; Braga, G. C.; MontinI, R. M. C.; Ichida, M. S.;Marino, R. H.; Colauto, N. B.; Silva, J.; & Neto, F. J.(1997). *Manual teórico/ prático do cultivo de cogumelos comestívei*s, Fundação de Estudos e Pesquisas Agrícolas e Florestais, Universidade Estadual Paulista FEPAF-UNESP, Botucatu, São Paulo.

Eira, A. F. (2000). Cultivo de cogumelos (compostagem condução e ambiente). In: *Anais da III Reunião Itinerante de Fitossanidade do Instituto Biológico*, pp. p. 83-95. Mogi das Cruzes, São Paulo.

Das, N. and Mukherjee, N. (2007), Cultivation of *Pleurotus ostreatus* on weed plant. *Bioresource Technology*, vol. 98, No. 14, pp. 2723-2726, ISBN: 0960-8524

Fidalgo, O., Guimarães, S. M. P. B. (1985). A situação do cogumelo comestível no Brasil e no exterior, *Anais do I Encontro Nacional sobre Cogumelos Comestíveis*, pp. 7-23, Secretaria de agricultura e abastecimento- Instituto de Botânica, Mogi das Cruzes, São Paulo, 15-17 out, 1980.

Garcia, H. S., Khanna, P. K. & Soni, G. L.(1993). Nutritional importance of the mushrooms, In: *Mushroom biology and mushroom products*, Proceeding of the first international conference on mushroom biology and mushroom products, Chang, S. T.; Buswell, J.

A. & Chiu, S. W. (Eds), pp. 227-236, The Chinese University Press, ISBN: 962-201-610-3, Hong Kong.

Gbolagade, J., Ajayi, A., Oku, I., Wankasi, D. (2006), Nutritive value of common wild edible mushrooms from northern Nigeria. *Global Journal of Biotechnology & Biochemistry*, vol. 1, No. 1, pp. 16-21, ISSN: 2078-466X.

Gonçalves, C. R. (2002). *Efeito da fragmentação do micélio visando a obtenção de inoculantes em suspensão para cultivo de shiitake.* Thesis (Doctorate in Biotchnology) Instituto de Química. Universidade Estadual Paulista, UNESP, Araraquara, São Paulo.

Guzmán, G., Mata, G., Salmones, D., Soto-Velasco,C. & Guzmán-Dávalos, L (1993). *El cultivo de los hongos comestibles.* Instituto Politécnico Nacional, ISBN: 968-29-4492-9. México, D. F.

Gutiérrez, A. Caramelo, L., Prieto, A., Martinez, M. J., Martinez, A. T. (1994). Anisaldehyde production and aryl-alcohol oxidase and dehydrogenase activities in lignolytic fungi of the genus Pleurotus. *Applied and Environmental Microbiology*, Vol. 60, n. 6, pp. 1783-1788, 1994, ISSN: 0099-2240.

Kurtzman JR. R. H., ZadraziL, F. (1989). Physiological and taxonomic consideration for cultivation of *Pleurotus* mushrooms, In: *Tropical Mushrooms*, Chang, S. T., Quimio, T. H., (Eds.), pp. 299-348, The Chinese University Press, ISBN: 962-201-264-7, Hong Kong.

Lajolo, R. M (1970). Fungos como alimentos, In: *O grande mundo dos cogumelos*, Lacaz, C. S., Minami, P. S., Purchio, A.(Eds.), p. 113-124, Nobel, São Palo.

Lechner, B. E. and Albertó, E. (2007). Optimal conditions for the fruit body production of natural occurring strains of *Lentinus tigrinus. Bioresource Technology*, vol. 98, No. 9, pp. 1866-1869, ISSN: 0960-8524.

Maziero, R. (1990). *Substratos alternativos para o cultivo de Pleurotus* spp. 1990. 136p., Dissertação (Mestrado em Ciências Biológicas/Botânica), Instituto de Biociências USP, Universidade de São Paulo, São Paulo.

Matheus, D. R., Okino, L. K. (1999). Utilização de Basidiomicetos em processos biotecnológicos, In: *Zigomiceto, Basidiomicetos e Deuteromicetos: noções básicas de taxonomia e aplicações biotecnológicas*, Bononi, V. L. R.; Grande, R. A. P. (Eds.), pp.107-139, Instituto de Botânica/Secretaria do Meio Ambiente, ISBN: 85-85662-08-5, São Paulo.

Mason, C. F. (1980). *Decomposição*. 1rst Ed, EPU/EDUSP, ISBN: 10 8512921803, São Paulo.

Montini, R. M. C. (2001). *Efeito de linhagens e substratos no crescimento micelial e na produtividade em cultivo axênico de shiitake (Lentinula edodes* (Berk.) Pegler). Thesis (Doctorate) Faculdade de Ciências Agronômicas. Universidade Estadual Paulista, UNESP, Botucatu, São Paulo.

Marino, R. H (2002). *Melhoramento Genético de Pleurotus ostreatus visando o cultivo axênico de linhagens resistentes ao calor.* Thesis (Doctorate) – Instituto de Química, Universidade Estadual Paulista, UNESP, Araraquara – São Paulo.

Molena, O. (1986). *O moderno cultivo de cogumelos.* Nobel, ISBN: 85-213-0377-7, S. Paulo,

Özçelik, E. and Pekşen, A. (2007), Hazelnut husk as a substrate for the cultivation of shiitake mushroom (*Lentinula edodes*), *Bioresource Technology*, vol. 98, No.14, pp. 2652–2658, ISSN: 0960-8524.

Philippoussis ,A., Diamantopoulou, P. & Zervakis, G. (2003). Correlation of the properties of several lignocellulosic substrates to the crop performance of the shiitake mushroom

Lentinus edodes", *World Journal of Microbiology and Biotechnology*, vol. 19, No. 6, pp. 551–557, ISSN: 0959-3993.

Philippoussis, A., Diamantopoulou, P. & Israilides, C. (2007). Productivity of agricultural residues used for the cultivation of the medicinal fungus *Lentinula edodes*, *International Biodeterioration & Biodegradation*, vol. 59, No.3, pp. 216–219, ISSN: 0964-8305

Przybylowicz, P., Donoghue, J. (1990). *Shiitake growers handbook: The art and science of mushroom cultivation*, Kendall/Hunt Publishing Company, ISBN: 0-8403-4962-9 Dubuque, Iowa.

Royse, D. J., Sanchez, J. E. (2007). Ground wheat straw as a substitute for portions of oak wood chips used in shiitake (*Lentinula edodes*) substrate formulae. *Bioresource Technology*, vol. 98, No. 11, pp. 2137–2141, ISSN: 0960-8524

Souza, E. (2011). Cultivo de cogumelos no Brasil: situação atual e perspectivas futuras, In: *Bioconversão de resíduos lignocelulolíticos da Amazônia para cultivo de cogumelos comestíveis*, Sales-Campos, C. (Ed.), PP. 77-94, Editora INPA, ISBN: 978-85-211-0068-3, Manaus-AM, Brasil.

Stamets, P. Chiiton, J. S. (1983) *The Mushroom cultivator: a practical guide to growing mushroom at home*, Agarikon Press, ISBN: 0-9610798-0-0, Olympia, Washington.

Stamets, P. (2000). *Growing gourmet and medicinal mushrooms*, 3 ed, Ten Speed Press, ISBN: 1-58008-175-4, Barkeley, Ca.

Sturion, G. L. (1994). *Utilização da folha da Bananeira como substrato para o cultivo cogumelo (Pleurotus spp.)*, Dissertação (MSc.), ESALQ/ USP, Piracicaba, São Paulo. 147pp.

Tisdale, T. E.; Miyasaka, S. C. & Hemmes, D.E. (2006), Cultivation of oyster mushroom (*Pleurotus ostreatus*) on wood substrates in Havai, *Word Journal of Microbiology & Biotechnology*, vol. 22, No. 3, pp. 201-206, ISSN: 0959-3993.

Trufem, S. F. B. (1999). Utilização de zigomicetos em processos biotecnológicos. In: *Zigomiceto, Basidiomicetos e Deuteromicetos: noções básicas de taxonomia e aplicações biotecnológicas*, Bononi, V. L. R.; Grande, R. A. P. (Eds.), pp.51-67, Instituto de Botânica/Secretaria do Meio Ambiente, ISBN: 85-85662-08-5, São Paulo.

Urben, A. F.; Oliveira, H. C. B.; Vieira, W., Correa, M. J. & Uriartt, A. H (2001). *Produção de cogumelos por meio de tecnologia chinesa modificada*, Urben, A. F. (Ed.), Editora EMBRAPA - Recursos Genéticos e Biotecnologia, ISBN: 85-87697-08-0, Brasília-DF.

Urben A. F.; URIARTT, A. H.; AMAZONAS, M. A. L.; OLIVEIRA, H. C. B.; CORREA,M. J.; VIEIRA, V. (2003). *Cultivo de cogumelos comestíveis e medicinais* (apostila do curso) EMBRAPA. 169p Brasília, DF.

Vianez, B. F., Barbosa, A. P. (2003). *Estudo de alternativas de uso dos resíduos gerados pela indústria madeireira em Manaus e Itacoatiara, Estado do Amazonas*. Report from Department of Forest Products /INPA, Manaus, 50 p.

Vedder, P.J.C. (1996). *Cultivo moderno del champiñon*. Tradução: Martinez, J. M. G. Ediciones Mundi-Prensa, ISBN: 84-7114-074-8, Madrid.

Zhanxi, L., Zhanhua, L. (2001) *Juncao Technology*, Xiangzhou, L. (Ed.), translated by Dongmei, L; Rui, T, China Agricultural Scientech Press, ISBN: 7-80167-210-0, the People's Republic of China.

Zadrazil, F. (1978). Cultivation of *Pleurotus*, In: *The biology and cultivation of edible mushrooms*, Ghang, S. T.; Hayes W. A. (Eds.), pp. 521-557, Academic Press, ISBN: 0- 12- 168050-9, New York.

ZadraziL, F.; Grabbe, K. (1983) Edible mushrooms. *Biotechnology*, Weinheim, Vol.3, N. 1, pp.145-187, ISSN: 0733-222X.

Permissions

The contributors of this book come from diverse backgrounds, making this book a truly international effort. This book will bring forth new frontiers with its revolutionizing research information and detailed analysis of the nascent developments around the world.

We would like to thank Prof. Dora Krezhova, for lending her expertise to make the book truly unique. She has played a crucial role in the development of this book. Without her invaluable contribution this book wouldn't have been possible. She has made vital efforts to compile up to date information on the varied aspects of this subject to make this book a valuable addition to the collection of many professionals and students.

This book was conceptualized with the vision of imparting up-to-date information and advanced data in this field. To ensure the same, a matchless editorial board was set up. Every individual on the board went through rigorous rounds of assessment to prove their worth. After which they invested a large part of their time researching and compiling the most relevant data for our readers. Conferences and sessions were held from time to time between the editorial board and the contributing authors to present the data in the most comprehensible form. The editorial team has worked tirelessly to provide valuable and valid information to help people across the globe.

Every chapter published in this book has been scrutinized by our experts. Their significance has been extensively debated. The topics covered herein carry significant findings which will fuel the growth of the discipline. They may even be implemented as practical applications or may be referred to as a beginning point for another development. Chapters in this book were first published by InTech; hereby published with permission under the Creative Commons Attribution License or equivalent.

The editorial board has been involved in producing this book since its inception. They have spent rigorous hours researching and exploring the diverse topics which have resulted in the successful publishing of this book. They have passed on their knowledge of decades through this book. To expedite this challenging task, the publisher supported the team at every step. A small team of assistant editors was also appointed to further simplify the editing procedure and attain best results for the readers.

Our editorial team has been hand-picked from every corner of the world. Their multi-ethnicity adds dynamic inputs to the discussions which result in innovative outcomes. These outcomes are then further discussed with the researchers and contributors who give their valuable feedback and opinion regarding the same. The feedback is then collaborated with the researches and they are edited in a comprehensive manner to aid the understanding of the subject.

Apart from the editorial board, the designing team has also invested a significant amount of their time in understanding the subject and creating the most relevant covers. They scrutinized every image to scout for the most suitable representation of the subject and create an appropriate cover for the book.

The publishing team has been involved in this book since its early stages. They were actively engaged in every process, be it collecting the data, connecting with the contributors or procuring relevant information. The team has been an ardent support to the editorial, designing and production team. Their endless efforts to recruit the best for this project, has resulted in the accomplishment of this book. They are a veteran in the field of academics and their pool of knowledge is as vast as their experience in printing. Their expertise and guidance has proved useful at every step. Their uncompromising quality standards have made this book an exceptional effort. Their encouragement from time to time has been an inspiration for everyone.

The publisher and the editorial board hope that this book will prove to be a valuable piece of knowledge for researchers, students, practitioners and scholars across the globe.

List of Contributors

Ernandes Rodrigues de Alencar
Faculdade de Agronomia e Medicina Veterinária, Universidade de Brasília, Brasília, Distrito Federal, Brasil

Lêda Rita D'Antonino Faroni
Departamento de Engenharia Agrícola, Universidade Federal de Viçosa, Viçosa, Minas Gerais, Brasil

Yves Bertheau
Inra SPE, route de Saint Cyr, 78 026 Versailles cedex, France

John Davison
Inra, route de Saint Cyr, 78 026 Versailles cedex(retired), France

Kashif Ghafoor and Fahad Y.I. AL-Juhaimi
Department of Food and Nutrition Sciences, King Saud University, Riyadh, Saudi Arabia

Jiyong Park
Department of Biotechnology, Yonsei University, Seoul 120-749, South Korea

Juliana M.C. Borba, Maria Surama P. da Silva and Ana Paula Rocha de Melo
Federal University of Pernambuco, Brazil

Veronica Sanda Chedea
Laboratory of Animal Biology, National Research Development Institute for Animal Biology and Nutrition (IBNA), Romania

Mitsuo Jisaka
Faculty of Life Sciences and Biotechnology, Shimane University, Japan

Branka Šošić-Jurjević, Branko Filipović and Milka Sekulić
University of Belgrade, Institute for Biological Research Siniša Stanković, Serbia

Nada Nikolić and Miodrag Lazić
University of Niš, Faculty of Technology, Leskovac, Serbia

Monica I. Cutrignelli, Serena Calabrò, Raffaella Tudisco, Federico Infascelli and Vincenzo Piccolo
Department of Animal Science and Food Control, University of Naples Federico II, Naples, Italy

H.K. Dei
Department of Animal Science, Faculty of Agriculture, University for Development Studies, Tamale, Ghana

Leilane Rocha Barros Dourado, Leonardo Augusto Fonseca Pascoal, Nilva Kazue Sakomura, Fernando Guilherme Perazzo Costa and Daniel Biagiotti
Department of Animal Science, Brazil

Leilane Rocha Barros Dourado and Daniel Biagiotti
Federal University of Piauí, Brazil

Leonardo Augusto Fonseca Pascoal and Fernando Guilherme Perazzo Costa
Federal University of Paraíba, Brazil

Nilva Kazue Sakomura
São Paulo State University, Brazil

Morio Matsuzaki
National Agricultural Reseach Center (NARC), National Agriculture and Food Research Organization (NARO), Japan

Ceci Sales-Campos, Bazilio Frasco Vianez and Raimunda Liége Souza de Abreu
National Institute for Amazonian Research –INPA/ Department of Forest, Products – CPPF, Aleixo CEP-Manaus, AM Brazil

Printed in the USA
CPSIA information can be obtained
at www.ICGtesting.com
JSHW011815301024
72690JS00002B/96